Leibniz Algebras

Structure and Classification

Leibniz Algebras

Structure and Classification

Shavkat Ayupov
Institute of Mathematics Uzbekistan Academy of Sciences

Bakhrom Omirov
National University of Uzbekistan

Isamiddin Rakhimov
University Technology MARA (UiTM)

CRC Press
Taylor & Francis Group
Boca Raton London New York

CRC Press is an imprint of the
Taylor & Francis Group, an **informa** business
A CHAPMAN & HALL BOOK

CRC Press
Taylor & Francis Group
6000 Broken Sound Parkway NW, Suite 300
Boca Raton, FL 33487-2742

First issued in paperback 2022

ISBN-13: 978-0-367-35481-7 (hbk)
ISBN-13: 978-1-03-233772-2 (pbk)
DOI: 10.1201/9780429344336

Publisher's Note

The publisher has gone to great lengths to ensure the quality of this reprint but points out that some imperfections in the original copies may be apparent.

Library of Congress Control Number: 2019953041

Visit the Taylor & Francis Web site at
http://www.taylorandfrancis.com

and the CRC Press Web site at
http://www.crcpress.com

Dedicated to the memory of Professor J.-L. Loday (1946–2012)

Contents

Preface

The book is designed to introduce the reader to the theory of Leibniz algebras. Knowledge of a standard course of linear algebra is presupposed, as well as some acquaintance with the methods of Lie algebras is required. Due to this we omitted the proofs of the facts from the theory of Lie algebras. Leibniz algebra is a generalization of the Lie algebras and these algebras preserve a unique property of Lie algebras that the right multiplication operators are derivations. They were first called D-algebras in papers by A.M. Blokh published in the 1960s to indicate their close relations with derivations. The theory of D-algebras did not receive great attention immediately after their introduction. Later the same algebras were introduced by J.-L. Loday in 1993, who called them Leibniz algebras due to the identity they satisfy. The main motivation to the introduction of Leibniz algebras was to study periodicity phenomena in algebraic K-theory. Nowadays the theory of Leibniz algebras is one of actively developing areas of modern algebra.

Along with cohomological, structural and classification results on Leibniz algebras some papers with their various applications also appear. But the main goal of the book is specific. We focus mainly on description problems of Leibniz algebras. Particularly, in the book we propose a method of classification of a subclass of Leibniz algebras based on algebraic invariants. The method is applicable in the Lie algebras case as well.

Our motivation to write this book came from a 2001 book titled *Dialgebras and Related Operands* by Jean-Louis Loday et al. (volume 1763 in Springer's Lecture Notes in Mathematics Series). Many readers are unaware of recent results obtained in the theory of Leibniz algebras. We hope that the review and citations given in Chapter 1 will provide such awareness.

Nevertheless, we had to omit many topics which are out of the description problem of algebras, i.e., (co)homology, analogues of some results from Lie algebras, classification of graded Leibniz algebras (p-filiform, quasi-filiform, algebras of maximum length, etc.) and over non-algebraically closed fields, Leibniz algebras in prime characteristic, descriptions of the derivations and automorphisms of algebras, recent results on geometric classifications and much more. We also hope the reader will be stimulated to pursue these topics in the references listed in the Bibliography at the end of this book.

We start Chapter 2 with some basic properties of Leibniz algebras followed by introducing the notion of semisimplicity of Leibniz algebras and presenting the structure of semisimple Leibniz algebras. Here we discuss the properties of Lie algebras that can be extended to the Leibniz algebras case and those which fail to extend, and we provide examples for each case. Chapter 3 deals with the problem of algebraic and geometric classification of low-dimensional complex Leibniz algebras and their subclasses. Chapter 4 of the book is devoted to the classification problem of some classes of complex Leibniz algebras in arbitrary finite dimension. Starting with Chapter 5, we focus on a subclass of nilpotent Leibniz algebras called *filiform* Leibniz algebras. We propose a classification technique based on algebraic invariants under action of base change ("transport of structure"). In Chapter 6 we apply the classification method proposed to low-dimensional cases. In the appendices, we provide summaries of basic linear algebra, representation theory and the Zariski topology.

Acknowledgments

Several people have played important roles at various stages of the work that culminated in this book. They are S. Albeverio, U.D. Bekbaev, L.M. Camacho, J.M. Casas, J.R. Gómez, A.Kh. Khudoyberdiyev, K.K. Kudaybergenov, M. Ladra, K.A. Mohd Atan, Sh.K. Said Husain, I. Rikhsiboev, and others. The authors are grateful to them for fruitful discussions and collaborations. Some of the facts given in Section 2.1 are results of private discussions of one of the authors (B.O.) with V. Gorbatsevich to whom we are grateful. We also are indebted to Mrs. Elena E. Rakhimova for typing of an earlier version of the book. The authors' special thanks go to Professor M. Ladra for the expert design, reading and in bringing the manuscript to its final form.

Should errors appear in this book, the authors accept complete responsibility for their occurrence, and they will be rectified subsequently once they are brought to the authors' attention. Any feedback from the readers in this context will be gratefully acknowledged.

Shavkat Ayupov
Bakhrom Omirov
Isamiddin Rakhimov

Authors

Shavkat Ayupov is the Director of the Institute of Mathematics at Uzbekistan Academy of Sciences, Professor at the National University of Uzbekistan, and Doctor of Sciences in Physics and Mathematics and Fellow of TWAS (The World Academy of Sciences). His research interests include functional analysis, algebra and topology. Professor Ayupov is the recipient of several international titles and awards, and his main research deals with the study of operator algebras, Jordan and Lie structures on von Neumann algebras, derivations and automorphisms on operator algebras, structure theory of Leibniz algebras and superalgebras and other non-associative algebras. Professor Ayupov is an organizer of CIMPA research school workshops, international conferences on nonassociative algebras and applications and on operator algebras and quantum probability. He has spoken at numerous plenary sessions and has been invited to numerous talks in various international conferences and workshops. Professor Ayupov is also the chief editor of the *Uzbek Mathematical Journal* and has authored 4 textbooks, 5 monographs, and more than 150 research papers which have been published in several international journals.

Bakhrom Omirov is Professor at the National University of Uzbekistan, Doctor of Sciences in Physics and Mathematics, and Research Fellow at the Institute of Mathematics of the Uzbekistan Academy of Sciences. His research interests include non-associative algebras, Lie (super)algebras, Leibniz (super)algebras, n-Leibniz algebras, structure theory of algebras, p-adic analysis, evolution algebras and their applications. Professor Omirov has received several local and international awards. He is currently leading several international research projects and collaborations and has been invited as a speaker to many workshops and universities abroad. He has also authored more than 100 research papers in high impact international journals.

Isamiddin Rakhimov is Professor at the University Technology MARA (UiTM), Malaysia, Doctor of Sciences in Physics and Mathematics, and Research Fellow at the Institute for Mathematical Research (INSPEM), Universiti Putra Malaysia. His research interests focus on the theory of finite-dimensional algebras and its applications. He has been invited to speak at a number of international conferences and workshops. He obtained a PhD in algebra from the Sankt Petersburg University of Russia. He has organized several international mathematical events and is on the editorial boards of a few international journals. He has also authored 1 textbook, 2 monographs and more than 70 scientific papers published in international cited journals.

Notations and Conventions

- $\mathbb{N}, \mathbb{Z}, \mathbb{Q}, \mathbb{R}, \mathbb{C}$ - natural numbers, integers, the fields of rational, real and complex numbers, respectively.
- $LB_n(\mathbb{F})$ - class of Leibniz algebras over $\mathbb{F}$ in dimension n.
- $LN_n(\mathbb{F})$ - class of nilpotent Leibniz algebras over $\mathbb{F}$ in dimension n.
- $Lb_n(\mathbb{F})$ - class of filiform Leibniz algebras over $\mathbb{F}$ in dimension n.
- $\mathrm{FLb}_n, \mathrm{SLb}_n$ and TLb_n - subclasses of Lb_n.
- NGF_i, $i = 1, 2, 3$ - classes of naturally graded filiform Leibniz algebras.
- $\mathrm{Cent}(L)$ - center of the Leibniz algebra L.
- $\mathrm{Ann}_l(L)$ - left annihilator of L.
- $\mathrm{Ann}_r(L)$ - right annihilator of L.
- $\mathrm{Rad}(L)$ - radical of L.
- $\mathcal{I}$ - liezator of a Liebniz algebra L.
- $\mathrm{Aut}(L)$ - group automorphisms of L.
- $\mathrm{Der}(L)$ - algebra of derivations of L.
- $\mathrm{Nor}_l(H)$ and $\mathrm{Nor}_l(H)$ - left and right normalizer of a subalgebra H in a Leibniz algebra L, respectively.
- $K(x, y)$ - Killing form.
- $\mathrm{Orb}(\mu)$ - orbit of an algebra μ.

- $Z^n(L, M)$ - the space of n-cocyles.
- $B^n(L, M)$ - the space of n-coboundaries.
- $H^n(L, M)$ - n^{th}-(co)homology group of an algebra L with coefficients in a L-module M.
- $ZL^n(L, M)$, $BL^n(L, M)$ and $HL^n(L, M)$ - the space of n-cocyles, the space of n-coboundaries and n^{th}-(co)homology group of a Leibniz algebra L with coefficients in a L-module M, respectively.
- $C_k = \frac{(2k)!}{(k+1)!k!}$, $k = 0, 1, 2, \ldots.$ - Catalan numbers.

CHAPTER 1

INTRODUCTION

1.1 ALGEBRAS

An *algebra* $\mathfrak{A}$ is a vector space V over a field $\mathbb{F}$ equipped by a bilinear binary operation $\lambda : V \times V \longrightarrow V$ on it. The vector space V is called the underlying vector space of $\mathfrak{A}$. In the book V is assumed to be finite-dimensional over the field of complex numbers, although many of the statements can be proved over the fields of the characteristic different from two.

Example 1.1.

- *Any field is an algebra over itself and over its subfield. Particularly, the number fields* $\mathbb{Q}$, $\mathbb{R}$, $\mathbb{C}$ *are examples of algebras.*
- *The set of polynomials* $\mathbb{F}[x_1, x_2, \ldots, x_n]$ *at variables* $x_1, x_2, \ldots, x_n$ *with coefficients from a field* $\mathbb{F}$ *is an algebra over* $\mathbb{F}$.
- *The free algebra* $\mathfrak{A} = \mathbb{F} < x_1, x_2, \ldots, x_n >$ *at non-commuting variables* $x_1, x_2, \ldots, x_n$ *with coefficients from a field* $\mathbb{F}$. *A basis of this algebra consists of words in letters* $x_1, x_2, \ldots, x_n$ *and multiplication in this basis is simply concatenation of words.*
- *The set of* $n \times n$ *matrices with entries in a field* $\mathbb{F}$ *is an algebra over* $\mathbb{F}$.
- *The set of endomorphisms* $\mathrm{End}_{\mathbb{F}}(V)$ *of a vector space* V *over a field* $\mathbb{F}$ *is an algebra over* $\mathbb{F}$.

- *Let us consider*

$$O(\mathbb{F}) = \left\{ \begin{pmatrix} \alpha & \mathbf{v} \\ \mathbf{u} & \beta \end{pmatrix} \,\middle|\, \alpha, \beta \in \mathbb{F} \text{ and } \mathbf{u}, \mathbf{v} \in \mathbb{F}^3 \right\}.$$

Define the binary operation λ on $O(\mathbb{F})$ as follows:

$$\lambda\left(\begin{pmatrix} \alpha & \mathbf{v} \\ \mathbf{u} & \beta \end{pmatrix}, \begin{pmatrix} \gamma & \mathbf{z} \\ \mathbf{w} & \delta \end{pmatrix} \right) = \begin{pmatrix} \alpha\gamma + (\mathbf{u}, \mathbf{v}) & \alpha\mathbf{z} + \delta\mathbf{u} - \mathbf{v} \times \mathbf{w} \\ \gamma\mathbf{v} + \beta\mathbf{w} - \mathbf{u} \times \mathbf{z} & \beta\delta + (\mathbf{v}, \mathbf{z}) \end{pmatrix},$$

where $(\mathbf{x}_1, \mathbf{x}_2)$ and $\mathbf{x}_1 \times \mathbf{x}_2$ are the inner and cross products of the vectors $\mathbf{x}_1, \mathbf{x}_2 \in \mathbb{F}$, respectively. Then $(O(\mathbb{F}), \lambda)$ is an algebra over $\mathbb{F}$ called Cayley-Dickson matrix algebra.

If $\{e_1, e_2, \ldots, e_n\}$ is a basis of the vector space V then the coefficients γ_{ij}^k, where $i, j, k = 1, 2, \ldots, n$ of the linear combinations

$$\lambda(e_i, e_j) = \sum_{k=1}^{n} \gamma_{ij}^k e_k \text{ for } i, j = 1, 2, \ldots, n$$

are said to be *the structure constants* of the algebra $\mathfrak{A}$ on the basis $\{e_1, e_2, \ldots, e_n\}$. Therefore, if we fix a basis of the underlying vector space V over a field $\mathbb{F}$ then all possible algebra structures over V can be identified by points $\{\gamma_{ij}^k\}$ of n^3-dimensional affine space $\mathbb{F}^{n^3}$. Note that the structure constants completely determine the product of any two elements $x = \alpha_1 e_1 + \alpha_2 e_2 + \cdots + \alpha_n e_n$ and $y = \beta_1 e_1 + \beta_2 e_2 + \cdots + \beta_n e_n$ as follows:

$$\lambda(x, y) = \sum_{i,j=1}^{n} \alpha_i \beta_j \lambda(e_i, e_j) = \sum_{i,j,k=1}^{n} \alpha_i \beta_j \gamma_{ij}^k e_k. \tag{1.1}$$

Thus the structure of an algebra can be completely determined by providing its structure constants. Such a definition is called *the determination by the structure constants*. The product $\lambda(x, y)$ of elements x and y of an algebra is denoted by $x \cdot y$, $x * y$, $x \star y$, $[x, y]$ or just xy etc., in each the cases above it is clear from the content of discussion what the specific notation means.

Let $(\mathfrak{A}_1, \lambda_1)$ and $(\mathfrak{A}_2, \lambda_2)$ be two algebras with binary operations λ_1 and λ_2, respectively. A function $f : \mathfrak{A}_1 \longrightarrow \mathfrak{A}_2$ is a homomorphism if

$$f(\lambda_1(x, y)) = \lambda_2(f(x), f(y)) \text{ for all } x, y \in \mathfrak{A}_1.$$

A bijective homomorphism is called *isomorphism*, this relation is denoted by "$\cong$". An automorphism of an algebra $\mathfrak{A}$ is an isomorphism onto itself. The set of all automorphisms of an algebra $\mathfrak{A}$ form a group with respect to the composition, the group is denoted by $\mathrm{Aut}\mathfrak{A}$.

A linear transformation $\iota : \mathfrak{A} \longrightarrow \mathfrak{A}$ is called involution if

$$\iota(\iota(x)) = x \text{ and } \iota(xy) = \iota(x)\iota(y) \text{ for all } x, y \in \mathfrak{A}.$$

Example 1.2. *The algebra* $(\mathrm{O}(\mathbb{F}), \lambda)$ *given in Example 1.1 has an involution defined as follows:*

$$\iota\left(\overline{\begin{pmatrix} \alpha & \mathbf{v} \\ \mathbf{u} & \beta \end{pmatrix}}\right) = \begin{pmatrix} \alpha & -\mathbf{v} \\ -\mathbf{u} & \beta \end{pmatrix}.$$

Definition 1.1. *A subspace S of an algebra $(\mathfrak{A}, \lambda)$ is said to be a subalgebra, if*

$$\lambda(S, S) \subset S, \text{ i.e., } \lambda(x, y) \in S \text{ whenever } x, y \in S.$$

Example 1.3. *Let $f : \mathfrak{A}_1 \longrightarrow \mathfrak{A}_2$ be a homomorphism of algebras. Then the image*

$$\mathrm{Im} f = \{y \in \mathfrak{A}_2 \mid \exists x \in \mathfrak{A}_1 : y = f(x)\}$$

is a subalgebra of $\mathfrak{A}_2$.

Definition 1.2. *A subspace I of an algebra $(\mathfrak{A}, \lambda)$ is called a left (respectively right) ideal of $\mathfrak{A}$ if*

$$\lambda(\mathfrak{A}, I) \subset I \quad (\text{respectively } \lambda(I, \mathfrak{A}) \subset I).$$

Definition 1.3. *A subspace I of an algebra $(\mathfrak{A}, \lambda)$ is said to be a two-sided ideal (or just ideal) of $\mathfrak{A}$ if it is both left and right, i.e.,*

$$\lambda(I, \mathfrak{A}) \subset I \text{ and } \lambda(\mathfrak{A}, I) \subset I.$$

Examples.

1. *Any algebra has two trivial ideals, the subspace consisting only of the zero vector and the algebra itself.*

2. *A subset of $(\mathfrak{A}, \lambda)$ defined by*

$$K(\mathfrak{A}) = \{x \in \mathfrak{A} \mid \lambda(x, y) = \lambda(y, x) = 0 \text{ for all } y \in \mathfrak{A}\}$$

is called the annihilator of $\mathfrak{A}$. It is an ideal of $\mathfrak{A}$.

3. *The kernel*
$$\mathrm{Ker}(f) = \{x \in \mathfrak{A}_1 | \ f(x) = 0\}$$
of a homomorphism $f : \mathfrak{A}_1 \longrightarrow \mathfrak{A}_2$ *is an ideal of* $\mathfrak{A}_1$.

4. *Let* I_1 *and* I_2 *be ideals of* $\mathfrak{A}$ *and*
$$I_1 + I_2 = \{x \in \mathfrak{A} | \text{ there exist } x_i \in I_i, i = 1, 2 \text{ such that } x = x_1 + x_2\}.$$
Then
   a. $I_1 + I_2$ *is an ideal of* $\mathfrak{A}$;
   b. I_1 *and* I_2 *are ideals of* $I_1 + I_2$;
   c. $I_1 \cap I_2$ *is an ideal of* I_1 *and* I_2.

5. *Let* $f : \mathfrak{A}_1 \longrightarrow \mathfrak{A}_2$ *be a homomorphism of algebras.*
   a. *If* J *is an ideal of* $\mathfrak{A}_2$ *then* $f^{-1}(J)$ *is an ideal of* $\mathfrak{A}_1$;
   b. *If* f *is an epimorphism and* I *is an ideal of* $\mathfrak{A}_1$ *then* $f(I)$ *is an ideal of* $\mathfrak{A}_2$.

Let $\mathfrak{A}$ be an algebra and let I be its ideal. The quotient vector space $\mathfrak{A}/I$ is an algebra with respect to the multiplication (product) $\overline{\lambda} : \mathfrak{A}/I \times \mathfrak{A}/I \longrightarrow \mathfrak{A}/I$ defined by $\overline{\lambda}(x + I, y + I) = \lambda(x, y) + I$. The operation $\overline{\lambda}$ is well-defined since if $x+I = x' + I$ and $y+I = y' + I$ then $\overline{\lambda}(x+I, y+I) = \overline{\lambda}(x' + I, y' + I)$.

Let I be an ideal of an algebra $\mathfrak{A}$. The canonical map
$$\pi : \mathfrak{A} \longrightarrow \mathfrak{A}/I \text{ defined by } x \longmapsto x + I$$
is a homomorphism from $\mathfrak{A}$ onto $\mathfrak{A}/I$. The kernel
$$\mathrm{Ker}(\pi) = \{x \in \mathfrak{A} | \ \pi(x) = 0\}$$
of the homomorphism π equals I.

Theorem 1.1. *Let* $\mathfrak{A}$ *be an algebra and* I *be its ideal. Then there is one to one order-preserving correspondence between the ideals* J *of* $\mathfrak{A}$ *which contain* I *and the ideals* $\overline{J}$ *of* $\mathfrak{A}/I$ *given by* $J = \pi^{-1}\overline{J}$.

Theorem 1.2. *Let* $f : \mathfrak{A}_1 \longrightarrow \mathfrak{A}_2$ *be a homomorphism of algebras.*

a) i) *For any ideal J of $\mathfrak{A}_1$ contained in* $\mathrm{Ker}(f)$, *there exists an unique homomorphism $\psi : \mathfrak{A}_1/J \longrightarrow \mathfrak{A}_2$ such that the following diagram*

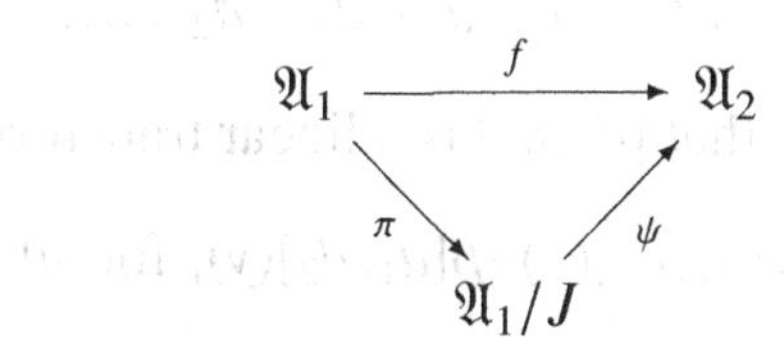

commutes, where π is the canonical homomorphism.

ii) *If* $J = \mathrm{Ker} f$ *then ψ produces an isomorphism between* $\mathfrak{A}_1/\mathrm{Ker} f$ *and* $\mathrm{Im} f$.

b) *If I_1 and I_2 are ideals of an algebra $\mathfrak{A}$ then $(I_1 + I_2)/I_2$ is isomorphic to $I_1/I_1 \cap I_2$.*

c) *Suppose that I_1 and I_2 are ideals of an algebra $\mathfrak{A}$ such that $I_1 \subset I_2$. Then I_2/I_1 is an ideal of $\mathfrak{A}/I_1$ and $(\mathfrak{A}/I_1)/(I_2/I_1) \cong (\mathfrak{A}/I_2)$.*

Definition 1.4. *Let $(\mathfrak{A}_1, \lambda_1)$ and $(\mathfrak{A}_2, \lambda_2)$ be two algebras over a field $\mathbb{F}$; then an algebra with binary product λ, defined on vector space*

$$\mathfrak{A} = \mathfrak{A}_1 \oplus \mathfrak{A}_2 = \{(x_1, x_2) |\ x_1 \in \mathfrak{A}_1 \text{ and } x_2 \in \mathfrak{A}_2\}$$

as follows

$$\lambda((x_1, x_2), (y_1, y_2)) = (\lambda_1(x_1, y_1), \lambda_2(x_2, y_2))$$

is said to be the direct sum of the algebras $\mathfrak{A}_1$ and $\mathfrak{A}_2$. Note that $\mathfrak{A}_1$ and $\mathfrak{A}_2$ are ideals of $\mathfrak{A}$.

The definition above can be extended to the case of the direct sum

$$\mathfrak{A} = \mathfrak{A}_1 \oplus \mathfrak{A}_2 \oplus \cdots \oplus \mathfrak{A}_n$$

of n algebras $(\mathfrak{A}_1, \lambda_1)$, $(\mathfrak{A}_2, \lambda_2)$, $\ldots (\mathfrak{A}_n, \lambda_n)$ defining the product as follows

$$\lambda((x_1, x_2, \ldots, x_n), (y_1, y_2, \ldots, y_n)) = (\lambda_1(x_1, y_1), \lambda_2(x_2, y_2), \ldots, \lambda_n(x_n, y_n)).$$

Definition 1.5. *A linear transformation $d : \mathfrak{A} \longrightarrow \mathfrak{A}$ is said to be a derivation of an algebra $(\mathfrak{A}, \lambda)$ if*

$$d(\lambda(x, y)) = \lambda(d(x), y) + \lambda(x, d(y)) \text{ whenever } x, y \in \mathfrak{A}.$$

Proposition 1.1. *The set of all derivations of an algebra* $\mathfrak{A}$ *denoted by* $\mathrm{Der}(\mathfrak{A})$ *is a Lie algebra with respect to the bracket operation*

$$[d_1, d_2] = d_1 \circ d_2 - d_2 \circ d_1.$$

Proof. It is obvious that $[d_1, d_2]$ is a linear transformation, i.e.,

$$[d_1, d_2](\alpha x+\beta y) = \alpha[d_1, d_2](x)+\beta[d_1, d_2](y), \text{ for all } x, y \in \mathfrak{A} \text{ and } \alpha, \beta \in \mathbb{F}.$$

Let us show that $[d_1, d_2]$ is a derivation of $\mathfrak{A}$. Indeed,

$$\begin{aligned}
[d_1, d_2](\lambda(x, y)) &= (d_1 \circ d_2 - d_2 \circ d_1)(\lambda(x, y)) \\
&= (d_1 \circ d_2)(\lambda(x, y)) - (d_2 \circ d_1(\lambda(x, y)) \\
&= d_1(\lambda(d_2(x), y) + \lambda(x, d_2(y)) \\
&\quad -d_2(\lambda(d_1(x), y) + \lambda(x, d_1(y)) \\
&= \lambda((d_1 \circ d_2)(x), y) + \underline{\lambda(d_2(x), d_1(y))} \\
&\quad +\underline{\lambda(d_1(x), d_2(y))} + \lambda(x, (d_1 \circ d_2)(y)) \\
&\quad -\lambda((d_2 \circ d_1)(x), y) - \underline{\lambda(d_1(x), d_2(y))} \\
&\quad -\underline{\lambda(d_2(x), d_1(y))} - \lambda(x, (d_2 \circ d_1)(y)) \\
&= \lambda((d_1 \circ d_2 - d_2 \circ d_1)(x), y) \\
&\quad +\lambda(x, (d_1 \circ d_2 - d_2 \circ d_1)(y)) \\
&= \lambda([d_1, d_2](x), y) + \lambda(x, [d_1, d_2](y)).
\end{aligned}$$

□

Definition 1.6. *It is called that an algebra* $(\mathfrak{A}, \lambda)$ *is represented as a semidirect sum if there are subspaces* $\mathfrak{A}_1$ *and* $\mathfrak{A}_2$ *such that*

$$\mathfrak{A} = \mathfrak{A}_1 \oplus \mathfrak{A}_2 \ \textit{(the direct sum of vector spaces)},$$

$$\lambda(\mathfrak{A}_1, \mathfrak{A}_1) \subset \mathfrak{A}_1,\ \lambda(\mathfrak{A}_1, \mathfrak{A}_2) \subset \mathfrak{A}_1,\ \lambda(\mathfrak{A}_2, \mathfrak{A}_1) \subset \mathfrak{A}_1, \textit{ and } \lambda(\mathfrak{A}_2, \mathfrak{A}_2) \subset \mathfrak{A}_2.$$

Further the bilinear function $\lambda(x, y)$ we denote just by xy.
The associator in an algebra is the trilinear function

$$(x, y, z) = (xy)z - x(yz).$$

The Jacobian in an algebra is defined by

$$J(x, y, z) = (xy)z + (yz)x + (zx)y.$$

The left (right) Leibniz identity is defined by

$$x(yz) = (xy)z + y(xz) \quad ((xy)z = (xz)y + x(yz)).$$

Definition 1.7.

- *An algebra* $\mathfrak{A}$
  - *is said to be unital if there exists an element* $e \in \mathfrak{A}$ *such that*
  $$xe = ex = x \text{ for all } x \in \mathfrak{A}.$$
  - *is said to be a division algebra if any nonzero element of* $\mathfrak{A}$ *is invertible, i.e., for any* $x \neq 0 \in \mathfrak{A}$ *there exists* $y \in \mathfrak{A}$ *such that* $xy = yx = e$.
  - *is commutative if* $xy = yx$ *for all* $x, y \in \mathfrak{A}$.
  - *is anticommutative if* $x^2 = 0$ *for all* $x \in \mathfrak{A}$.
  - *is associative if* $(x, y, z) = 0$ *for all* $x, y, z \in \mathfrak{A}$.
- *A Jordan algebra is a commutative algebra satisfying the Jordan identity*
  $$(x^2, y, x) = 0 \text{ for all } x, y \in \mathfrak{A}.$$
- *An algebra* $\mathfrak{A}$ *is left (right) alternative if* $(y, x, x) = 0$ $((x, x, y) = 0)$ *for all* $x, y \in \mathfrak{A}$.
- *A Lie algebra is an anticommutative algebra* $\mathfrak{A}$ *satisfying the Jacobi identity*
  $$J(x, y, z) = 0 \text{ for all } x, y, z \in \mathfrak{A}.$$
- *A Malcev algebra is an anticommutative algebra* $\mathfrak{A}$ *satisfying the identity*
  $$J(x, y, xz) = J(x, y, z)x \text{ for all } x, y, z \in \mathfrak{A}.$$
- *A left (right) Leibniz algebra is an algebra* $\mathfrak{A}$ *satisfying the left (right) Leibniz identity*
  $$x(yz) = (xy)z + y(xz) \quad ((xy)z = (xz)y + x(yz)) \text{ for all } x, y, z \in \mathfrak{A}.$$

FACTS:

1. Every associative algebra is alternative.

2. Every Lie algebra is a Malcev algebra.

3. Every Lie algebra is a Leibniz algebra.

4. A Leibniz algebra $\mathfrak{A}$ is Lie if $x^2 = 0$ for all $x \in \mathfrak{A}$. Therefore, the intersection of classes of Malcev and Leibniz algebras is the class of Lie algebras.

5. The commutator in an algebra $\mathfrak{A}$ is the bilinear function

$$[x, y] = xy - yx.$$

The **minus algebra** $\mathfrak{A}^-$ of an algebra $\mathfrak{A}$ is the algebra with the same underlying vector space as $\mathfrak{A}$ but with $[x, y]$ as the multiplication.

Theorem 1.3. *(Poincaré-Birkhoff-Witt Theorem for Lie algebras) Every Lie algebra is isomorphic to a subalgebra of $\mathfrak{A}^-$ for some associative algebra $\mathfrak{A}$.*

6. If $\mathfrak{A}$ is an alternative algebra then $\mathfrak{A}^-$ is a Malcev algebra.

7. The **plus algebra** $\mathfrak{A}^+$ of an algebra $\mathfrak{A}$ is the algebra with the same underlying vector space as $\mathfrak{A}$ but with $x \circ y = \frac{1}{2}(xy + yx)$ as the multiplication. The algebra $\mathfrak{A}^+$ is always commutative. If $\mathfrak{A}$ is alternative then $\mathfrak{A}^+$ is a Jordan algebra. A Jordan algebra is called special if it is isomorphic to a subalgebra of $\mathfrak{A}^+$ for some associative algebra $\mathfrak{A}$; otherwise it is called exceptional. The analogue of the PBW theorem for Jordan algebras is false: not every Jordan algebra is special.

Example 1.4. *Let $\mathbb{A}$ be an algebra with an involution $\imath$. The set of all symmetric elements*

$$H(\mathfrak{A}, \imath) = \{a \in \mathfrak{A} \mid \imath(a) = a\}$$

is closed under the Jordan product $\circ$, therefore, it is a Jordan algebra.

- *If $\mathfrak{A}$ is an associative algebra then $H(\mathfrak{A}, \imath)$ is a special Jordan algebra.*

- *If* $\mathfrak{A} = \mathrm{O}_3$ *is an algebra of* 3×3 *hermitian matrices, i.e.,* $\iota : \mathrm{O}_3 \longrightarrow \mathrm{O}_3$ *defined by* $\iota(a_{ij}) = \overline{a}_{ij}$, *then* $H(\mathrm{O}_3)$ *is an example of exceptional simple Jordan algebra. A Jordan algebra* J *is said to be an Albert algebra if by extending the field of scalars* $\mathbb{F}$ *of* J *to* $\mathbb{K}$ *one gets an isomorphic to* $H(\mathrm{O}_3)$ *algebra, i.e., there exists an extension* $\mathbb{K}$ *of* $\mathbb{F}$ *such that* $J \bigotimes_{\mathbb{F}} \mathbb{K} \cong H(\mathrm{O}_3)$.

Definition 1.8.

- *An algebra* $\mathfrak{A}$ *is simple if* $\mathfrak{A}\mathfrak{A} \neq \{0\}$ *and* $\mathfrak{A}$ *has no ideals apart from* $\{0\}$ *and* $\mathfrak{A}$.

- *An algebra is semisimple if it is the direct sum of simple algebras.*

Set $\mathfrak{A}^1 = \mathfrak{A}^{(1)} = \mathfrak{A}$ and by induction we define

$$\mathfrak{A}^{n+1} = \sum_{i+j=n+1} \mathfrak{A}^i\mathfrak{A}^j \text{ and } \mathfrak{A}^{(n+1)} = \mathfrak{A}^{(n)}\mathfrak{A}^{(n)} \text{ for } n \geq 1.$$

Definition 1.9.

- *An algebra* $\mathfrak{A}$ *is called nilpotent if there exists k such that* $\mathfrak{A}^k = \{0\}$.

- *An algebra* $\mathfrak{A}$ *is called solvable if there exists s such that* $\mathfrak{A}^{(s)} = \{0\}$.

The smallest natural number k (respectively, s) with the properties above is called the nilpotency index (respectively solvability index) of $\mathfrak{A}$.

Remarks.

- If algebra $\mathfrak{A}$ is simple then $\mathfrak{A}\mathfrak{A} = \mathfrak{A}$.

- An algebra $\mathfrak{A}$ is nilpotent of index k if and only if any product of k elements (with any arrangement of parentheses) equals zero, and if there exists a nonzero product of $k - 1$ elements.

- Every nilpotent algebra is solvable but the converse is not true in general.

- Every solvable associative algebra is nilpotent.

Now we briefly recall some classical results concerning the theory of associative and Lie algebras. Then we review a few classes of algebras, which are closely related to the algebras considered in this book.

1.2 ASSOCIATIVE ALGEBRAS

An algebra A is said to be *associative* if its product " · " has the associativity property

$$(a \cdot b) \cdot c = a \cdot (b \cdot c) \text{ for all } a, b, c \in A. \tag{1.2}$$

Lemma 1.1. *An algebra A is associative if and only if*

$$(e_i \cdot e_j) \cdot e_k = e_i \cdot (e_j \cdot e_k), \text{ where } i, j, k = 1, 2, \ldots, n \tag{1.3}$$

for any basis $\{e_1, e_2, \ldots, e_n\}$ *of* A.

Proof. This is seen immediately combining (1.2) and (1.1). □

Note that the condition (1.3) gives polynomial identities for the structure constants $\{\gamma_{ij}^k\}$ $i, j, k = 1, 2, \ldots, n$ as follows

$$\sum_{r=1}^{n} \gamma_{ij}^r \gamma_{rk}^s = \sum_{r=1}^{n} \gamma_{jk}^r \gamma_{ir}^s \text{ for } 1 \leq i, j, k, s \leq n.$$

Therefore, the points corresponding to the associative algebra structures in $\mathbb{F}^{n^3}$ form a Zariski closed subset of $\mathbb{F}^{n^3}$.

Example 1.5.

1. The field of complex numbers $\mathbb{C}$ is a two-dimensional associative algebra over $\mathbb{R}$, but it is an infinite-dimensional algebra over $\mathbb{Q}$.

2. The space of square $n \times n$ matrices with the entries from a field $\mathbb{F}$ is an associative algebra of dimension n^2.

3. The space of endomorphisms $\mathrm{End}_{\mathbb{F}}(V)$ of an n-dimensional vector space V over a field $\mathbb{F}$ also is an n^2-dimensional associative algebra.

4. The space of polynomials with the coefficients from a field $\mathbb{F}$ is an infinite-dimensional associative algebra.

The theory of finite-dimensional associative algebras is one of the ancient areas of the modern algebra. It originates primarily from works of Hamilton, who discovered the famous quaternions, and Cayley, who developed the theory of matrices. Later the structural theory of finite dimensional associative algebras have been treated by a number of mathematicians, notably B. Pierce, C.S. Pierce, Clifford, Weierstrass, Dedekind, Jordan, Frobenius. At the end of the 19th century, T. Molien and E. Cartan described semisimple algebras over the fields of complex and real numbers, and they made first attempts in the study of non semisimple algebras.

A new era in the development of the theory of finite-dimensional associative algebras began due to works of Wedderburn, who obtained the fundamental results of this theory: description of the structure of semisimple algebras over a field, a theorem on the lifting of the quotient by the radical, the theorem on the commutativity of finite division rings, and others. Thanks to German school of algebraists, headed by Emmy Noether, E. Artin, R. Brouwer, the theory of semisimple algebras has found its modern form, and most of the results have been extended to some types of rings.

Further development of the theory of associative algebras occurred in the 1980s when many open problems, remaining unsolved since the 1930s were solved. The classification problem of associative algebras can be reduced to the classification problem of semisimple and nilpotent associative algebras.

Recall that an associative algebra (without unit) A is said to be *nilpotent* if $A^n = 0$ for some n (i.e., $a_1 \cdot a_2 \cdots \cdot a_n = 0$ for any $a_i \in A$). The representatives of low-dimensional nilpotent associative algebras over a field $\mathbb{F}$ are given by their structure constants on a basis $\{e_1, e_2, \ldots, e_n\}$ as follows (we give only non zero products)

$n = 1$. Trivial algebra.

$n = 2$.
- Trivial algebra.
- $A_1 := \{\mathrm{Span}_{\mathbb{F}}\{e_1, e_2\} \mid e_1e_1 = e_2\}$.

$n = 3$. Trivial algebra.

$$
\begin{array}{lll}
A_1 & := & \{\mathrm{Span}_{\mathbb{F}}\{e_1, e_2, e_3\} \mid e_1e_1 = e_2\}; \\
A_2(\alpha) & := & \{\mathrm{Span}_{\mathbb{F}}\{e_1, e_2, e_3\} \mid e_1e_1 = \alpha e_3,\ e_2e_2 = e_3\}, \\
 & & \quad \text{where } \alpha \in \mathbb{F}^*/(\mathbb{F}^*)^2; \\
A_3(\alpha) & := & \{\mathrm{Span}_{\mathbb{F}}\{e_1, e_2, e_3\} \mid e_1e_1 = \alpha e_3,\ e_2e_2 = e_3,\ e_2e_1 = e_3\}, \\
 & & \quad \text{where } \alpha \in \mathbb{F}^*; \\
A_4 & := & \{\mathrm{Span}_{\mathbb{F}}\{e_1, e_2, e_3\} \mid e_1e_2 = e_3,\ e_2e_1 = -e_3\}; \\
A_5 & := & \{\mathrm{Span}_{\mathbb{F}}\{e_1, e_2, e_3\} \mid e_1e_1 = e_2,\ e_1e_2 = e_3,\ e_2e_1 = e_3\}; \\
A_6 & := & \{\mathrm{Span}_{\mathbb{F}}\{e_1, e_2, e_3\} \mid e_1e_2 = e_3,\ e_2e_1 = -e_3,\ e_2e_2 = e_3\}.
\end{array}
$$

The lists above are taken from the paper [98], to which we refer the reader for the list in dimension four. Complete lists of low-dimensional complex associative algebras have been given in [157].

Radical Rad(A) of an algebra A is the largest nilpotent ideal of A (containing all nilpotent ideals of the algebra). The quotient $A/\text{Rad}(A)$ is *semisimple*, i.e., it has a zero radical.

The following two theorems form a foundation of the structural theory of semisimple associative algebras ([145]).

Theorem 1.4. *(Wedderburn – Artin) Any finite-dimensional semisimple associative algebra A is uniquely decomposed into a direct sum of simple algebras:*

$$A = B_1 \oplus B_2 \oplus \cdots \oplus B_k.$$

Recall that an algebra is *simple* if it has no nontrivial two-sided ideal.

Theorem 1.5. *Any finite-dimensional simple associative algebra A is isomorphic to the algebra of matrices $M_n(D)$ over a division ring D, the number n and the division ring D are uniquely determined by the algebra A.*

These theorems give a complete description of semisimple associative algebras. At the same time the structure of non-semisimple algebras (nilpotent part), is not sufficiently investigated, even over an algebraically closed field. So far there is no systematic way to study this case. Complex associative algebras in dimensions up to 5 were first classified by B. Pierce in 1870, initially in the form of manuscripts, which appeared later in [145]. There are classifications of unital 3, 4 and 5-dimensional associative algebras by Scorza [165], Gabriel [81], and Mazzola [127], respectively.

1.3 LIE ALGEBRAS

The notion of a Lie algebra (the old name is *Infinitesimal Lie group*) arose in the study of Lie groups, but later became an object of self-theory.

Lie algebra is a vector space L over a field $\mathbb{F}$, with a bilinear binary operation denoted by $[\cdot,\cdot]$, and the operation satisfies the following conditions:

- antisymmetry: $[x, x] = 0$ for all $x \in L$;
- Jacobi identity: $[[x, y], z] + [[y, z], x] + [[z, x], y] = 0$ for all $x, y, z \in L$.

Examples.

1. Let $V = \mathbb{R}^3$ and $\mathrm{x} = (x_1, x_2, x_3), \mathrm{y} = (y_1, y_2, y_3) \in \mathbb{R}^3$. Define
$$[\mathrm{x}, \mathrm{y}] = (x_2y_3 - x_3y_2, x_3y_1 - x_1y_3, x_1y_2 - x_2y_1).$$
Then $L = (\mathbb{R}^3, [\cdot, \cdot])$ is a Lie algebra.
2. Let A be an associative algebra and $x, y \in A$. Define
$$[x, y] = xy - yx.$$
Then $L = (A, [\cdot, \cdot])$ is a Lie algebra. Particularly, if $A = \mathrm{End}_{\mathbb{F}}(V)$ for the Lie algebra L there is a special notation $\mathfrak{gl}(V, \mathbb{F})$ (or $\mathfrak{gl}(V)$ if the field is clearly specified). If $A = M_n(\mathbb{F})$ the algebra of $n \times n$ matrices with the entries from $\mathbb{F}$ then the Lie algebra L is denoted by $\mathfrak{gl}(n, \mathbb{F})$ (sometimes the notation $\mathfrak{gl}_n(\mathbb{F})$ also is used).

Definition 1.10. *A subspace I of a Lie algebra L is said to be* ideal *if for $x \in I$ and $y \in L$ imply $[x, y] \in I$ (Since $[x, y] = -[y, x]$, in a Lie algebra all ideals are two-sided).*

Examples.

1. ***The center*** $\mathrm{Cent}(L) = \{x \in L \mid [x, y] = 0,$ *for all* $y \in L\}$ *of a Lie algebra L is an ideal of L.*
2. ***The derived algebra*** $[L, L] = \mathrm{Span}_{\mathbb{F}}\{[x, y] \in L$ *for all* $x, y \in L\}$ *is an ideal of L.*

Definition 1.11. *A Lie algebra L is said to be* **simple** *if $[L, L] \neq 0$ and L has no ideals except itself and zero.*

1.3.1 Simple Lie algebras

Every finite-dimensional complex simple Lie algebra is isomorphic to one of the following Lie algebras (the proof is based on works by Killing, Engel and Cartan):

1. Special Lie algebras

$$\mathfrak{sl}(l+1,\mathbb{C}) = \{A = (a_{ij}) \in \mathfrak{gl}(l+1,\mathbb{C}) \mid \mathrm{tr}(A) = \sum_{i=1}^{l+1} a_{ii} = 0\}.$$

Since $\mathrm{Tr}(AB) = \mathrm{Tr}(BA)$ and $\mathrm{Tr}(A+B) = \mathrm{Tr}(A) + \mathrm{Tr}(B)$ the subspace $\mathfrak{sl}(l+1,\mathbb{C})$ is a subalgebra of $\mathfrak{gl}(l+1,\mathbb{C})$. The dimension of $\mathfrak{sl}(l+1,\mathbb{C})$ is $l^2 + 2l$. Indeed, on the one hand $\mathfrak{sl}(l+1,\mathbb{C})$ is a proper subalgebra of $\mathfrak{gl}(l+1,\mathbb{C})$, hence its dimension at most $(l+1)^2 - 1 = l^2 + 2l$. On the other hand we can exhibit this number of linearly independent matrices of trace zero: take all E_{ij} $(i \neq j)$, along with all $H_i = E_{ii} - E_{i+1,i+1}$ $(1 \leq i \leq l)$, for total of $l + (l+1)^2 - (l+1) = l^2 + 2l$.

2. Orthogonal Lie algebras

   - $\mathfrak{o}(2l+1,\mathbb{C}) = \left\{A \in \mathfrak{gl}(2l+1,\mathbb{C}) \mid A^T J = -JA,\right\}$,
     where $J = \begin{pmatrix} 1 & 0 & 0 \\ 0 & 0 & I_l \\ 0 & I_l & 0 \end{pmatrix}$ and I_l is $l \times l$ identity matrix. Let us write elements of $\mathfrak{o}(2l+1,\mathbb{C})$ as blocks, of shapes adapted to the blocks of J. Calculations show that

     $$A = \begin{pmatrix} 0 & \mathrm{c}^T & -\mathrm{b}^T \\ \mathrm{b} & M & P \\ -\mathrm{c} & Q & -M^T \end{pmatrix} \quad \text{with } P = -P^T \text{ and } Q = -Q^T.$$

     For a basis,

     - take first the l diagonal matrices

       $$E_{ii} - E_{l+i,l+i} \quad (2 \leq i \leq l+1).$$

     - add the $2l$ matrices involving only row one and column one:

       $$E_{1,l+i+1} - E_{i+1,1} \text{ and } E_{i+1,1} - E_{1,l+i+1} \quad (1 \leq i \leq l).$$

– for the block M take

$$E_{i+1,j+1} - E_{l+j+1,l+i+1} \quad (1 \leq i \neq j \leq l).$$

– for the block P take

$$E_{i+1,l+j+1} - E_{j+1,l+i+1} \quad (1 \leq i < j \leq l).$$

– for the block Q take

$$E_{i+l+1,j+1} - E_{j+l+1,i+1} \quad (1 \leq j \neq i \leq l).$$

The total number of basis vectors is $2l^2 + l$, i.e.,

$$\dim \mathfrak{o}(2l+1, \mathbb{C}) = 2l^2 + l.$$

- $\mathfrak{o}(2l, \mathbb{C}) = \left\{ A \in \mathfrak{gl}(2l, \mathbb{C}) \mid A^T J = -JA, \right\}$
  where $J = \begin{pmatrix} 0 & I_l \\ I_l & 0 \end{pmatrix}$ and I_l is $l \times l$ identity matrix.

Write elements of $\mathfrak{o}(2l, \mathbb{C})$ as blocks, of shapes adapted to the blocks of J. Again calculations show that

$$A = \begin{pmatrix} M & P \\ Q & -M^T \end{pmatrix} \text{ with } P = -P^T \text{ and } Q = -Q^T.$$

For a basis,

- take first the l diagonal matrices $E_{ii} - E_{l+i,l+i}$ $(1 \leq i \leq l)$.
- for the block M take

$$E_{i,j} - E_{l+j,l+i} \quad (1 \leq i \neq j \leq l).$$

- for the block P take

$$E_{i,l+j} - E_{j,l+i} \quad (1 \leq i < j \leq l).$$

- for the block Q take

$$E_{i+l,j} - E_{j+l,i} \quad (1 \leq j < i \leq l).$$

The total number of basis vectors is $2l^2 - l$, i.e.,

$$\dim \mathfrak{o}(2l, \mathbb{C}) = 2l^2 - l.$$

3. Symplectic Lie algebras

$$\mathfrak{sp}(2l, \mathbb{C}) = \left\{ A \in \mathfrak{gl}(2l, \mathbb{C}) \mid A^T J = -JA, \right\}$$

where $J = \begin{pmatrix} 0 & I_l \\ -I_l & 0 \end{pmatrix}$ and I_l is the identity matrix. Write elements of $\mathfrak{sp}(2l, \mathbb{C})$ in a block form as follows

$$\begin{pmatrix} M & N \\ P & Q \end{pmatrix} \quad \text{where } M, N, P, Q \in \mathfrak{gl}(l, \mathbb{C}).$$

Then the condition of symplecticity $A^T J = -JA$ implies the following constraints $N^T = N, P^T = P$ and $M^T = -Q$ for $M, N, P, Q \in \mathfrak{gl}(l, \mathbb{C})$. As a basis of $\mathfrak{sp}(2l, \mathbb{C})$ we take

- l the diagonal matrices $E_{ii} - E_{l+i,l+j}$ $(1 \leq i \leq l)$;
- add to these $l^2 - l$ vectors $E_{ij} - E_{l+j,l+i}$ $(1 \leq i \neq j \leq l)$;
- use $l + \frac{1}{2}l(l-1)$ matrices $E_{i,l+i}$ $(1 \leq i \leq l)$ and $E_{i,l+j} + E_{j,l+i}$ $(1 \leq i \neq j \leq l)$ for the positions in N;
- and take $l + \frac{1}{2}l(l-1)$ matrices $E_{l+i,i}$ $(1 \leq i \leq l)$ and $E_{l+i,j} + E_{l+j,i}$ $(1 \leq i \neq j \leq l)$ for the positions in Q.

Summing up, we find $\dim \mathfrak{sp}(2l, \mathbb{C}) = 2l^2 + l$.

4. Exceptional simple Lie algebras. The constructions of the next simple Lie algebras are more delicate and we give them without details.

- $\mathrm{G}_2 = \mathrm{Der}(\mathrm{O})$, where O is the Cayley-Dickson matrix algebra (see Example 1.1). Note that $\dim \mathrm{G}_2 = 14$.

Here are Lie algebras given by some constructions over the Albert algebra J (see Example 1.4).

- $\mathrm{F}_4 = \mathrm{Der}\,\mathrm{J}$ and $\dim \mathrm{F}_4 = 52$.
- $\mathrm{E}_6 = \mathrm{Strl}_0\,\mathrm{J}$, where $\mathrm{Strl}_0\,\mathrm{J}$ is the reduced structure of the Jordan algebra J. It is a subalgebra of codim 1 of the structure algebra Strl J of J. $\dim \mathrm{E}_6 = 78$.
- $\mathrm{E}_7 = \mathrm{K}(\mathrm{J})$ (The Tits-Kantor-Koecher algebra of J) and $\dim \mathrm{E}_7 = 133$.

The simple Lie algebra E_8 is given as follows (Tits construction):

- $E_8 = \mathrm{Der}(O) \oplus O_0 \otimes J_0 \oplus \mathrm{Der}(J)$, where O_0 and J_0 are matrices with zero traces in O and J, respectively. The binary operation on the vector space $E_8 = \mathrm{Der}(O) \oplus O_0 \otimes J_0 \oplus \mathrm{Der}(J)$ is given as follows:
  - The part $\mathrm{Der}(O) \oplus \mathrm{Der}(J)$ is the direct sum of the Lie algebras;
  - $[a \otimes x, D] = aD \otimes x$, $[a \otimes x, E] = a \otimes xE$ for $D \in \mathrm{Der}\,(O)$, $E \in \mathrm{Der}\,(J)$, $a \in O_0$ and $x \in J_0$;
  - $[a \otimes x, b \otimes y] = \frac{1}{12}\mathrm{tr}(xy)D_{a,b} + (a * b) \otimes (x * y) + \frac{1}{2}t(ab)[R_x, R_y]$ for $a, b \in O_0$, $x, y \in J_0$ and $D_{a,b} = R_{[a,b]} - L_{[a,b]} - 3[L_a, R_b] \in \mathrm{Der}\ O$, where $t(a)$ is the trace of $a \in O$, $\mathrm{tr}(x)$ is the trace $x \in J$, $a * b = ab - \frac{1}{2}t(ab)$ and $x * y = xy - \frac{1}{3}\mathrm{tr}(xy)$.

  The dimension of E_8 is 248.

The list above exhausts all finite-dimensional simple Lie algebras over an algebraic closed field $\mathbb{F}$ of characteristic zero. They are pairwise not isomorphic except for the following cases:

- $\mathfrak{o}(3, \mathbb{F}) \cong \mathfrak{sp}(2, \mathbb{F}) \cong \mathfrak{sl}(2, \mathbb{F})$;
- $\mathfrak{o}(4, \mathbb{F}) \cong \mathfrak{sl}(2, \mathbb{F}) \oplus \mathfrak{sl}(2, \mathbb{F})$;
- $\mathfrak{o}(5, \mathbb{F}) \cong \mathfrak{sp}(4, \mathbb{F})$;
- $\mathfrak{o}(6, \mathbb{F}) \cong \mathfrak{sl}(4, \mathbb{F})$.

1.3.2 Solvable and nilpotent Lie algebras

Let $(L, [\cdot, \cdot])$ be a Lie algebra. The lower central series and derived series of L are defined as follows:

$$\begin{aligned} L^1 &= L, & L^{k+1} &= [L^k, L], \ k \geq 1, \\ L^{[1]} &= L, & L^{[s+1]} &= [L^{[s]}, L^{[s]}], \ s \geq 1. \end{aligned}$$

Obviously,

$$L^1 \supseteq L^2 \supseteq \cdots \supseteq L^i \supseteq \cdots,$$

and

$$L^{[1]} \supseteq L^{[2]} \supseteq \cdots \supseteq L^{[i]} \supseteq \cdots.$$

Note that $L^{[i]} \subseteq L^i$.

Definition 1.12. *A Lie algebra L is said to be nilpotent (respectively, solvable), if there exists $n \in \mathbb{N}$ ($m \in \mathbb{N}$) such that $L^n = 0$ (respectively, $L^{[m]} = 0$).*

The proofs of the following two lemmas are straightforward and can be found in any standard book on Lie algebras.

Lemma 1.2. *Let L be a Lie algebra, I and J be its ideals. Then*

- *If L is solvable, then so are all subalgebras and homomorphic images of L.*
- *If I and L/I are solvable, then L itself also is solvable.*
- *If I and J are solvable, so is $I + J$.*

As an application of the lemma for any Lie algebra L one can prove that there exists a unique maximal solvable ideal, called the *radical* of L and denoted $\mathrm{Rad}(L)$. If L is not trivial and $\mathrm{Rad}(L) = 0$, then L is said to be *semisimple*. The problem of classification of finite dimensional Lie algebras can be reduced to the following three separate tasks:

- Classification of nilpotent Lie algebras;
- Description of solvable algebras with given nilradical;
- Description Lie algebras with a given radical.

The latter two problems are the most studied part of the classification problem of Lie algebras and brought to fruition for the complex Lie algebras in the mid-20th century. The third problem is reduced to a description of semisimple subalgebras of the derivation algebras of a solvable algebra [124]. The problem of how to construct, by a given solvable algebra $\mathcal{R}$ and a semisimple algebra S, all the algebras L with the radical $\mathcal{R}$ and the quotient algebra $L/\mathcal{R}$ isomorphic to S, also has been solved. It turned out that such algebras L are finite in number, and they correspond one-to-one to semisimple subalgebras of the algebra of derivations of $\mathcal{R}$. Since semisimple algebras are completely described by the well-known theory of Cartan-Killing this problem reduces to the study of solvable algebras. The second problem is reduced to the description of the orbits of some nilpotent linear groups [125]. Thus the classification problem of Lie algebras is reduced to the study of nilpotent algebras. This problem is the most complicated and, unfortunately, still there is no any standard way to solve it. Recall that complete classification of complex Lie algebras is obtained only up to dimension six. As for nilpotent Lie algebras, then various methods of their classification have been implemented, and, accordingly, several lists of isomorphism classes have been presented. Recall just a few of them: the earliest results of the classification of complex nilpotent Lie algebras of dimension no greater than 6 were obtained in year 1891 by Umlauf in his thesis [177]. Umlauf's list was inaccurate, it contained many mutually isomorphic algebras. In 1958, Dixmier refined this list and published the list of nilpotent Lie algebras of dimensions at most 5 over an arbitrary field [61]. We mention as well results of Mubarakzjanov (1963, over the field of real numbers) [130], [131].

In dimension 6 there are several classification results: Morozov (1958, over a field of characteristic zero) [129], Shedler (1964, over an arbitrary field) [171]. Vergne (1966, over $\mathbb{C}$) [179], Skjelbred and Sund (1978, over $\mathbb{R}$) [172], Beck and Kolman (1981, over $\mathbb{R}$) [34].

In dimension 7, also there are several lists: Safiullina (1964, over $\mathbb{C}$) [163]), Romdani (1985, over $\mathbb{R}$ and $\mathbb{C}$) [162], Seeley (1988, over $\mathbb{C}$) [166], and Ancochea and Goze (1989, over $\mathbb{C}$) [20], [21]. All these results were obtained by using a variety of isomorphism invariants. In 1989, Carles, introducing a new type of invariants (the weights system), compared the results of Safiullina, Romdani and Seeley and in 1993 published his result in dimension 7 over $\mathbb{C}$ in [52]. In his PhD thesis Ming-Peng Gong [89] has applied the central extensions method to classify 7-dimensional nilpotent Lie algebras over algebraically closed fields and over $\mathbb{R}$. Moreover, M.-P. Gong compared the results obtained earlier by the authors above with results he obtained and purified the lists of isomorphism classes of 7-dimensional Lie algebras.

The classification of nilpotent Lie algebras in higher dimensions remains a vast open area. Again, as it was the case after Umlauf's pioneering work, the most efficient way of sorting through the mess of facts in higher dimensions is not obvious. These days computers offer some hope of being able to deal with the kind of messy details which haunt this subject.

There are results on classification of subclasses of nilpotent Lie algebras, such as Favre (1973, nilpotent Lie algebra of maximal rank) [67], Scheuneman (1967) [164], Gauger (1973) [83], Revoy (1980)

[158], and Galitssky and Timashev (1998) [82] (metabelian Lie algebra), Gómez, Jiménez-Merchán, Khakimdjanov (1998) [87] (filiform Lie algebras) and others.

1.3.3 Semisimple Lie algebras

In this section we shall give a short survey of the theory of complex semisimple Lie algebras (the reader interested in this subject is referred to the classical books on Lie algebras [105], [106], etc).

Definition 1.13. *A Lie algebra L is said to be semisimple if* $\mathrm{Rad}(L) = \{0\}$.

It is easy to show that for a Lie algebra L the quotient $L/\mathrm{Rad}(L)$ is semisimple. As a consequence of this we get that a semisimple Lie algebra does not have a nonzero abelian ideals.

Definition 1.14. *A bilinear form* $B : L \times L \longrightarrow \mathbb{C}$ *for a Lie algebra L is said to be invariant if*

$$B([x,y],z) + B(x,[y,z]) = 0 \text{ for all } x,y,z \in L.$$

The bilinear form defined by $B(x,y) = \mathrm{Tr}(\mathrm{ad}_x \circ \mathrm{ad}_y)$ on a Lie algebra L is an example of the invariant form. It is called the Killing form of L denoted by $K(x,y)$. The following proposition shows the importance of the Killing form.

Proposition 1.2. *(Cartan-Killing Criterion) A Lie algebra L is semisimple if and only if its Killing form* $K(x,y)$ *is non degenerate.*

Corollary 1.1. *Let L be a semisimple Lie algebra and* $\mathfrak{a}$ *its ideal. Then the set*

$$\mathfrak{a}^{\perp} = \{x \in L \mid K(x,y) = 0, \text{ for all } y \in \mathfrak{a}\}$$

is an ideal of L (called the kernel of the Killing form with respect to $\mathfrak{a}$*) and* $L = \mathfrak{a} \oplus \mathfrak{a}^{\perp}$ *(direct sum of ideals).*

Corollary 1.2. *A semisimple Lie algebra is represented as a direct sum of simple Lie algebras.*

Corollary 1.3. *If a Lie algebra L is semisimple, then* $L = [L,L]$.

Definition 1.15. *Let L be a Lie algebra over a field* $\mathbb{F}$.

i) *A representation of a Lie algebra L in a vector space V is a homomorphism of Lie algebras* $\rho : L \longrightarrow \mathfrak{gl}(V)$.

ii) *A vector space V over* $\mathbb{F}$ *endowed with an action* $L \times V \longrightarrow V$ *(denoted by* $x \cdot v$ *for* $x \in L$ *and* $v \in V$*) is said to be L-module if*

- $(\alpha x + \beta y) \cdot v = \alpha(x \cdot v) + \beta(y \cdot v)$ *for all* $\alpha, \beta \in \mathbb{F}$, $x, y \in L$ *and* $v \in V$;
- $x \cdot (\alpha v + \beta w) = \alpha(x \cdot v) + \beta(x \cdot w)$ *for all* $\alpha, \beta \in \mathbb{F}$, $x, y \in L$ *and* $v \in V$;
- $[x, y] \cdot v = x \cdot (y \cdot v) - y \cdot (x \cdot v)$ *for all* $x, y \in L$ *and* $v \in V$.

Definition 1.16. *A homomorphism of L-modules* V_1 *and* V_2 *is a linear function* $\varphi : V_1 \longrightarrow V_2$ *such that* $\varphi(x \cdot v) = x \cdot \varphi(v)$ *for all* $x \in L$ *and* $v \in V_1$.

The notions of the representation of a Lie algebra L in V and for V to be a L-module are equivalent. Indeed, if ρ is a representation of L in V then V is a L-module with the action $x \cdot v := \rho(x)(v)$. Conversely, if V is a L-module then the map $\rho : L \longrightarrow \mathfrak{gl}(V)$ given by $\rho(x)(v) := x \cdot v$ is a representation of L in V.

Examples. *Every Lie algebra L can be regarded as a L-module by setting* $x \cdot y := [x, y]$. *The corresponding representation is said to be adjoint representation of L. It is denoted by* ad_x, *for* $x \in L$, *i.e.,* $\mathrm{ad}_x(y) = [x, y]$.

Definition 1.17. *A L-module V is called simple if* $V \neq \{0\}$ *and V has no submodules other than* $\{0\}$ *and V. A L-module V is called semisimple if it is a direct sum of simple submodules.*

In terms of representations these notions correspond to the irreducibility and complete reducibility of representations. There is a fundamental result on finite-dimensional (i.e., V is a finite-dimensional) representations known as Weyl's Theorem. For the proof the reader is referred to [66] or [105].

Theorem 1.6. *Let L be a semisimple Lie algebra and let V be a finite-dimensional L-module. Then V is semisimple .*

Definition 1.18. *A Lie algebra is said to be reductive if its adjoint representation is completely reducible.*

Proposition 1.3. *Let L be a Lie algebra. The following conditions are equivalent:*

i) *L is reductive;*

ii) *$[L, L]$ is semisimple;*

iii) *$L = [L, L] \oplus \mathrm{Cent}(L)$;*

iv) *$\mathrm{Rad}(L) = \mathrm{Cent}(L)$.*

Here we give the following adapted version of well-known Schur's Lemma from [185, p.57, Corollary 3] which is needed to prove some results on conjugacy of Leibniz algebras.

Lemma 1.3. *Let L be a complex Lie algebra and let $V = V_1 \oplus \cdots \oplus V_m$ and $W = W_1 \oplus \cdots \oplus W_n$ be completely reducible L-modules, where $V_1, \dots, V_n, W_1, \dots, W_n$ are irreducible modules. Then any L-module homomorphism $\varphi\colon V \to W$ can be represented as*

$$\varphi = \sum_{i=1}^{m} \sum_{j=1}^{n} \lambda_{ij}\varphi_{ij},$$

where the operators $\varphi_{ij}\colon V_i \to W_j$ are fixed L-module homomorphisms and λ_{ij} are complex numbers. Furthermore, $\varphi_{ij} \neq 0$ if and only if φ_{ij} is an isomorphism.

Let us now discuss the following fundamental theorem on the structure of finite-dimensional Lie algebras.

Theorem 1.7. *(Levi's Theorem) Every Lie algebra L can be written as a semi-direct sum of its radical* $\mathrm{Rad}(L)$ *and a semisimple subalgebra $\mathcal{S}$ such that $L = \mathrm{Rad}(L) \dot{\oplus} \mathcal{S}$.*

The semisimple subalgebra $\mathcal{S}$ in the theorem above is called Levi's subalgebra (Levi's Factor) of L. The Levi's Factor is not unique. Note that if $\mathcal{S}_1$ and $\mathcal{S}_2$ are two Levi's Factors such that

$$L = \mathrm{Rad}(L) \dot{\oplus} \mathcal{S}_1 = \mathrm{Rad}(L) \dot{\oplus} \mathcal{S}_2,$$

then there exists an automorphism σ of L such that $\sigma(\mathcal{S}_1) = \mathcal{S}_2$. Moreover, the automorphism σ can be chosen of the form $\sigma = \exp(\mathrm{ad}_x)$, where x is in the nilradical of L. The result is due to A.I. Malcev.

Each element x of a complex semisimple Lie algebra is written uniquely as $x = d{+}n$, where $d, n \in L$ are such that ad_d is diagonalizable, ad_n is nilpotent and $[d, n] = 0$. Furthermore, if $y \in L$ commutes with x, then $[d, y] = 0$ and $[d, y] = 0$. If in $x = d + n$ the term $d = 0$, then x is said to be semisimple.

Let L be a semisimple Lie algebra. There exists a subalgebra H of L consisting entirely of semisimple elements of L such that L is decomposed into weight spaces for the action of $\mathrm{ad}H$. Indeed, the subalgebra H consists of all semisimple elements of L. The algebra L has a basis of common eigenvalues for the elements of $\mathrm{ad}H$. Given a common eigenvector $x \in L$, the eigenvalues are given by the associated weight, $\alpha : H \longrightarrow \mathbb{C}$, defined by

$$\mathrm{ad}_h x = \alpha(h)x, \text{ for all } h \in H.$$

Weights are elements of the dual space H^*. For each $\alpha \in H^*$, let

$$L_\alpha = \{x \in L \mid [h, x] = \alpha(h)x \text{ for all } h \in H\}$$

denote the corresponding weight space. One of these subspaces is the zero weight space:

$$L_0 = \{x \in L \mid [h, x] = 0 \text{ for all } h \in H\}.$$

The space L_0 is the same as the centralizer $C_L(H)$ of H in L. As H is abelian, one has $H \subseteq L_0$.

Let Φ denote the set of non-zero $\alpha \in L^*$ for which L_α is non-zero. The decomposition of L into the direct sum of weight spaces for H is written as follows:

$$L = L_0 \oplus \bigoplus_{\alpha\in\Phi} L_\alpha.$$

Since L is finite-dimensional Φ is finite. The elements of the set Φ are said to be roots of L with respect to the Cartan subalgebra H.

The following lemma expresses properties of weight spaces.

Lemma 1.4. *Suppose that $\alpha, \beta \in H^*$. Then*

i) $[L_\alpha, L_\beta] \subseteq L_{\alpha+\beta}$.

ii) *If $\alpha + \beta \neq 0$, then $K(L_\alpha, L_\beta) = 0$.*

iii) *The restriction of K to L_0 is non-degenerate. This property implies that for every $\alpha \in \Phi$ there is a vector $h \in H$ such that $K(h, h_\alpha) = \alpha(h)$ for all $h \in H$.*

iv) *If $\alpha \in \Phi$ then $-\alpha \in \Phi$ and we have $[L_\alpha, L_{-\alpha}] = \mathbb{C}h_\alpha$, $\alpha(h_\alpha) \neq 0$.*

Definition 1.19. *A subalgebra H of a Lie algebra L is said to be Cartan subalgebra if H is abelian and any element of H is semisimple and H is maximal with respect to these properties.*

Note that a Cartan subalgebra of a semisimple Lie algebra is not unique, all the Cartan subalgebras of a semisimple Lie algebra L are isomorphic. They also conjugate in the following sense: for any two subalgebras H_1 and H_2 of a semisimple Lie algebra L there is an automorphism σ of L of the form $\exp(\text{ad}_y)$ with $y \in L$ such that $H_2 = \sigma(H_1)$. The dimension of the Cartan subalgebra is called the rank of L.

Let r be the rank of the semisimple Lie algebra L with a Cartan subalgebra H. There is a basis $\{\alpha_1, \alpha_2, \dots, \alpha_r\}$ of H^* such that

1. $\alpha_i \in \Phi, \quad i = 1, 2, \dots, r.$
2. Every root α is decomposed as a linear combination of $\{\alpha_1, \alpha_2, \dots, \alpha_r\}$ with integer coefficients $k_1, k_2, \dots, k_r$:

$$\alpha = k_1\alpha_1 + k_2\alpha_2 + \cdots + k_r\alpha_r$$

such that the coefficients $k_1, k_2, \dots, k_r$ are all simultaneously positive or negative.

Definition 1.20.

- *The root system $S = \{\alpha_1, \alpha_2, \dots, \alpha_r\}$ satisfying the conditions above is said to be a system of simple roots.*
- *A root $\alpha \in \Phi$ is called positive (respectively, negative) if the coefficients $k_1, k_2, \dots, k_r$ in*

$$\alpha = k_1\alpha_1 + k_2\alpha_2 + \cdots + k_r\alpha_r$$

are positive (respectively, negative).

For a given root system Φ its subsets consisting of the positive and negative roots are denoted by Φ^+ and Φ^-, respectively. Thus one has

$$\Phi = \Phi^+ \cup \Phi^-, \text{ where } \Phi^- = -\Phi^+.$$

Note also that some texts use a "maximal toral subalgebra" instead of Cartan subalgebra.

Definition 1.21. *A Lie algebra L is called toral if it consists entirely of semisimple elements of L.*

Definition 1.22. *A maximal toral subalgebra of L is a subalgebra of L which is toral and is not contained in any larger toral subalgebra.*

The fact that the maximal toral subalgebras of a Lie algebra L are precisely the Cartan subalgebras is obtained from the statements below.

Lemma 1.5. *Any toral subalgebra of a semisimple Lie algebra is abelian.*

One immediately gets the following

Corollary 1.4. *A subalgebra of a semisimple Lie algebra L is maximal toral if and only if it is a Cartan subalgebra.*

1.3.4 More on finite-dimensional Lie algebras

In this section we review a few more results from [106] for further usage.

Lemma 1.6. *(Fitting's lemma) Let V be a vector space and $A : V \to V$ be a linear transformation. Then*

$$V = V_{0A} \oplus V_{1A},$$

where $A(V_{0A}) \subseteq V_{0A}$, $A(V_{1A}) \subseteq V_{1A}$ and $V_{0A} = \{v \in V \mid A^i(v) = 0$ for some $i\}$, $V_{1A} = \bigcap_{i=1}^{\infty} A^i(V)$. Moreover, $A_{|V_{0A}}$ is a nilpotent transformation and $A_{|V_{1A}}$ is an automorphism.

The spaces V_{0A} and V_{1A} are called respectively *Fitting null component* and *Fitting one component* of the space V with respect to the transformation A.

Lemma 1.7. *Let V be a vector space. And let A, B be linear transformations on V such that $[\dots[[B,\underbrace{A],A],\dots A}_{k \text{ times}}] = 0$. Then the Fitting components V_{0A}, V_{1A} of V with respect to A are invariant with respect to the transformation B.*

Further we make use the following results:

Theorem 1.8. *Let G be a nilpotent Lie algebra of linear transformations of a vector space V and $V_0 = \bigcap\limits_{A\in G} V_{0A}$, $V_1 = \bigcap\limits_{i=1}^{\infty} G^i(V)$. Then the subspaces V_0 and V_1 are invariant with respect to G (i.e., V_0 and V_1 are invariant with respect to every transformation B from G) and $V = V_0 \oplus V_1$. Moreover, $V_1 = \sum\limits_{A\in G} V_{1A}$.*

Theorem 1.9. *Let G be a split nilpotent Lie algebra of linear transformations of a vector space M. Then G has a finite number of different weights, weight subspaces are submodules of M, and M is decomposed into the direct sum of these modules. Moreover, if $M = M_1 \oplus M_2 \oplus \dots \oplus M_r$ is an arbitrary decomposition of M into the sum of subspaces M_i $(\neq 0)$, which are invariant with respect to G such that the following conditions hold:*

1) *for each i the restriction of $A \in G$ on M_i has only one characteristic root $\alpha_i(A)$ (of some multiplicity);*

2) *if $i \neq j$, then there exists $A \in G$ such that $\alpha_i(A) \neq \alpha_j(A)$;*

then the maps $A \to \alpha_i(A)$ are weights and M_i are weight subspaces.

Proposition 1.4. *Let L be a Lie algebra of linear transformations of a vector space over a field of zero characteristic, $\mathcal{R}$ be the radical of the algebra L, and let $\mathcal{N}$ be the radical of the associative algebra L^*, generated by L. Then $L \cap \mathcal{N}$ consists of all nilpotent elements of the radical $\mathcal{R}$ and $[\mathcal{R}, L] \subseteq \mathcal{N}$ ($\mathcal{N}$ is considered as a nilpotent radical of the associative algebra L^*).*

1.4 LODAY ALGEBRAS

In this section we briefly summarize basic facts about some classes of algebras which are closely related to Leibniz algebras. Any associative algebra gives rise to a Lie algebra by $[x, y] = xy - yx$. Leibniz algebras are non-commutative version of Lie algebras while associative dialgebras are generalization of associative algebras. One way to generalize the idea above, so as to obtain Leibniz algebra brackets that are not skew-symmetric, is to use two different associative algebra products. The notion of associative dialgebra was discovered by Loday while studying periodicity phenomena in algebraic K-theory. A few more classes of algebras and their relation with classical algebras are discussed.

1.4.1 Leibniz algebras

The concept of *Leibniz algebra* was introduced by Loday [117] in the study of Leibniz (co)homology as a noncommutative (to be more precise, as a non-anticommutative) analogue of Lie algebras (co)homology.

A Leibniz algebra over a field $\mathbb{F}$ is a vector space over $\mathbb{F}$, equipped with a $\mathbb{F}$-bilinear map $[\cdot,\cdot] : L \times L \longrightarrow L$ satisfying the Leibniz identity

$$[[x,y],z] = [[x,z],y] + [x,[y,z]], \tag{1.4}$$

for all $x, y, z \in L$.

In fact, the definition above is a definition of the right Leibniz algebra, whereas the identity for the left Leibniz algebra is as follows

$$[x,[z,y]] = [[x,z],y] + [z,[x,y]], \tag{1.5}$$

for all $x, y, z \in L$. The passage from the right to the left Leibniz algebra can be easily done by considering a new product "$[\cdot,\cdot]_{opp}$" on the same vector space defined by "$[x,y]_{opp} = [y,x]$". Therefore, the results proved for the right Leibniz algebras $(L, [\cdot.\cdot])$ can be easily reformulated for the left Leibniz algebras $(L, [\cdot,\cdot]_{opp})$ and vice versa.

Note that the versions

$$[x,[y,z]] = [[x,y],z] - [[x,z],y], \text{ for all } x,y,z \in L \tag{1.6}$$

and

$$[[x,z],y] = [x,[z,y]] - [z,[x,y]], \text{ for all } x,y,z \in L. \tag{1.7}$$

of the identities (1.4) and (1.5) are also often used.

Examples. 1. *Let $\mathfrak{g}$ be a Lie algebra and M be a $\mathfrak{g}$-module. Let $f : M \longrightarrow \mathfrak{g}$ be a $\mathfrak{g}$-equivariant linear map, i.e.,*

$$f(m \cdot x) = [f(m), x], \textit{ for all } m \in M \textit{ and } x \in \mathfrak{g}.$$

The bracket $[\cdot,\cdot]_M$ *defined by* $[m,n]_M := m\cdot f(n)$ *provides a Leibniz algebra structure on* M*. Indeed, for* $m,n,h \in M$ *we have*

$$\begin{aligned}
[[m,n]_M,h]_M &= [m\cdot f(n),h]_M = (m\cdot f(n))\cdot f(h)\\
&= (m\cdot f(h))\cdot f(n) - m\cdot [f(h),f(n)]\\
&= (m\cdot f(h))\cdot f(n) + m\cdot [f(n),f(h)]\\
&= [m\cdot f(h),n]_M + m\cdot f(n\cdot f(h))\\
&= [[m,h]_M,n]_M + [m,[n,h]_M]_M.
\end{aligned}$$

2. *Let* $\mathfrak{g}$ *be a Lie algebra and* M *be a* $\mathfrak{g}$*-module. Then the vector space* $Q=\mathfrak{g}\oplus M$ *equipped with the multiplication* $[x+m,y+n]=[x,y]+m\cdot y$*, where* $m,n\in M$, $x,y\in\mathfrak{g}$ *is a Leibniz algebra. Let us verify the Leibniz identity for the product. One has*

$$\begin{aligned}
&[[x+m,z+h]_Q,y+n]_Q + [x+m,[y+n,z+h]_Q]_Q\\
&\qquad = [[x,z]+m\cdot z,y+n]_Q + [[x+m,[y,z]+n\cdot z]_Q\\
&\qquad = [[x,z],y] + (m\cdot z)\cdot y + [x,[z,y]] + (m\cdot [y,z])\\
&\qquad = [[x,z],y] + [x,[z,y]] + (m\cdot z)\cdot y + (m\cdot [y,z])\\
&\qquad = [[x,y],z] + (m\cdot y)\cdot z\\
&\qquad = [[x,y]+m\cdot y,z+h]_Q\\
&\qquad = [[x+m,y+n]_Q,z+h]_Q.
\end{aligned}$$

3. *A vector space* $L=\mathfrak{g}\otimes\mathfrak{g}$*, equipped with the product*

$$[x\otimes y,a\otimes b]_L = [x,[a,b]]\otimes y + x\otimes [y,[a,b]]$$

is a Leibniz algebra, where $\mathfrak{g}$ *is a Lie algebra, i.e.,*

$$[[x\otimes y,a\otimes b]_L,s\otimes t]_L = [[x\otimes y,s\otimes t]_L,a\otimes b]_L + [x\otimes y,[a\otimes b,s\otimes t]_L]_L.$$

Indeed,

$$\begin{aligned}
&[[x\otimes y,a\otimes b]_L,s\otimes t]_L\\
&\qquad = \underbrace{[[x,[a,b]],[s,t]]\otimes y} + \underset{\sim\sim\sim}{[x,[a,b]]\otimes [y,[s,t]]}\\
&\qquad\quad \underline{+[x,[s,t]]\otimes [y,[a,b]]}\ \overbrace{+x\otimes [[y,[a,b]],[s,t]]}
\end{aligned}$$

and

$$\begin{aligned}
&[[x\otimes y,s\otimes t]_L,a\otimes b]_L + [x\otimes y,[a\otimes b,s\otimes t]_L]_L\\
&= \underbrace{([[x,[a,b]],[s,t]] + [x,[[a,[s,t]],b]] + [x,[a,[b,[s,t]]]])\otimes y}
\end{aligned}$$

$$+\underbrace{[x,[a,b]]\otimes[y,[s,t]]}+\underline{[x,[s,t]]\otimes[y,[a,b]]}$$

$$\overbrace{+x\otimes([[y,[s,t]],[a,b]]+[y,[[a,[s,t]],b]]+[y,[a,[b,[s,t]]]])}.$$

By using anti-symmetricity and the Jacobi identity of the Lie bracket. It it is easy to see that

$$([[x,[a,b]],[s,t]]+[x,[[a,[s,t]],b]]+[x,[a,[b,[s,t]]]])$$

and

$$([[y,[s,t]],[a,b]]+[y,[[a,[s,t]],b]]+[y,[a,[b,[s,t]]]])$$

are reduced to

$$[[x,[a,b]],[s,t]] \text{ and } [[y,[a,b]],[s,t]],$$

respectively.

4. *Let A be an associative algebra and D be a linear map $D : A \to A$ satisfying the condition*

$$D(aD(b)) = D(a)D(b) = D(D(a)b), \forall a,b \in A.$$

Then A equipped with the multiplication $[a,b]_D = aD(b) - D(b)a$ is a Leibniz algebra (denoted by A_D).

Remark. (Examples of the map D)

- Let $A = A_+ \oplus A_-$ be a $\mathbb{Z}_2$-graded associative algebra and $D(x) := x_+$;
- A derivation D of an associative algebra with $D^2 = 0$;
- An endomorphism D such that $D^2 = D$.

Clearly, a Lie algebra is a Leibniz algebra, and conversely, a Leibniz algebra L with property $[x,y] = -[y,x]$, for all $x,y \in L$, is a Lie algebra. The inherent properties of non-Lie Leibniz algebras imply that the subspace spanned by squares of elements of the algebra L is a non-trivial ideal (further denoted by $\mathcal{I}$.) Moreover, ideal $\mathcal{I}$ is abelian.

There is an inclusion functor i : **Lie** $\hookrightarrow$ **Leib**. This functor has a left adjoint p : **Leib** $\longrightarrow$ **Lie**, which is defined on the objects by

$L_{Lie} = L/\mathcal{I}$. That is any Leibniz algebra L gives rise to the Lie algebra L_{Lie}, obtained as the quotient of L by relation $[x, x] = 0$. The ideal $\mathcal{I}$ is the minimal ideal with respect to the property that $\mathfrak{g} := L/\mathcal{I}$ is a Lie algebra.

The quotient mapping $\pi : L \longrightarrow \mathfrak{g}$ is a homomorphism of Leibniz algebras. One has an exact sequence of Leibniz algebras:

$$0 \longrightarrow \mathcal{I} \longrightarrow L \longrightarrow L_{Lie} \longrightarrow 0.$$

We consider finite-dimensional algebras L over a field $\mathbb{F}$ of characteristic zero (in fact, it is sufficient that this characteristic is not equal to 2).

A linear transformation d of a Leibniz algebra L is said to be a derivation if

$$d([x, y]) = [d(x), y] + [x, d(y)],$$

for all $x, y \in L$.

Note that for $x \in L$ the operator R_x of right multiplication, $R_x \colon L \to L, R_x(y) = [y, x], y \in L$, is a derivation (for a left Leibniz algebra L, the operator L_x left multiplication, $L_x \colon L \to L$, $L_x(y) = [x, y], y \in L$, also is a derivation). Such kind of derivations are called inner derivations on L. The set of all the inner derivations of a Leibniz algebra L is denoted by $\mathrm{Inn}(L)$.

In other words, the right Leibniz algebra is characterized by this property, i.e., any right multiplication operator is a derivation of L. For the left Leibniz algebras, all left multiplication operators are derivations.

The set of all derivations of L (further denoted by $\mathrm{Der}(L)$) equipped with the commutator operation forms a Lie algebra. The set $\mathrm{Inn}(L)$ is an ideal of $\mathrm{Der}(L)$. The automorphism group $\mathrm{Aut}(L)$ of an algebra L also is naturally defined. If the field $\mathbb{F}$ is $\mathbb{R}$ or $\mathbb{C}$, then the automorphism group is a Lie group and the $\mathrm{Der}(L)$ is its Lie algebra. One can consider $\mathrm{Aut}(L)$ as an algebraic group (or as a group of $\mathbb{F}$-points of an algebraic group defined over the field $\mathbb{F}$).

The motivation to study the structural properties of Loday algebras (in fact, in [117] Loday introduced a few classes of algebras: diassociative, dendriform and Zinbiel algebras, and gave functorial diagram connecting these classes emphasizing their Koszulean duality, see Section 1.4.6) was provoked due to private discussions between J.L. Loday and Sh.A. Ayupov held in 1994 (Strasbourg).

The appearance of Leibniz algebras was motivated by Loday as follows.

The Chevalley-Eilenberg chain complex of a Lie algebra $\mathfrak{g}$ is the sequence of chain modules given by the exterior powers of $\mathfrak{g}$ as follows

$$\bigwedge * \mathfrak{g} : \quad \cdots \longrightarrow \wedge^{n+1}\mathfrak{g} \xrightarrow{d_{n+1}} \wedge^{n}\mathfrak{g} \xrightarrow{d_n} \wedge^{n-1}\mathfrak{g} \xrightarrow{d_{n-1}} \dots$$

and boundary operators $d_n : \wedge^n\mathfrak{g} \longrightarrow \wedge^{n-1}\mathfrak{g}$ classically defined by

$$d_n(x_1 \wedge x_2 \wedge \cdots \wedge x_n) := \sum_{i<j} (-1)^{i+j+1}[x_i, x_j] \wedge x_1 \wedge \cdots \wedge \hat{x}_i \wedge \cdots \wedge \hat{x}_j \wedge \cdots \wedge x_n.$$

The property $d_n \circ d_{n+1} = 0$, which makes this sequence a chain complex, is proved by using the antisymmetry $x \wedge y = -y \wedge x$ of the exterior product, the Jacobi identity $[[x, y], z] + [[y, z], x] + [[z, x], y] = 0$ and the antisymmetry $[x, y] = -[y, x]$ of the Lie bracket given on $\mathfrak{g}$.

Loday noticed that, if one replaces the exterior product with the tensor product and rewrites the Chevalley-Eilenberg boundary operator as

$$d_n(x_1 \otimes x_2 \otimes \cdots \otimes x_n) = \sum_{i<j} (-1)^{n-j} x_1 \otimes \cdots \otimes [x_i, x_j] \otimes \cdots \otimes \hat{x}_j \otimes \cdots \otimes x_n,$$

for the chain complex

$$\cdots \longrightarrow L^{\otimes(n+1)} \xrightarrow{d_{n+1}} L^{\otimes n} \xrightarrow{d_n} L^{\otimes(n-1)} \xrightarrow{d_{n-1}} \dots$$

of an algebra L, then the property $d_n \circ d_{n+1} = 0$ is proved without making use of the antisymmetry properties of both the exterior product and the bracket on the algebra L. It suffices only that the bracket $[\cdot, \cdot]$ on L satisfies the so-called Leibniz identity

$$[[x, y], z] = [[x, z], y] + [x, [y, z]].$$

Later on, besides this purely algebraic motivation to introduce the Leibniz algebras, some of their relationships with classical geometry, non-commutative geometry and as well as physics have been discovered.

During the last 25 years the theory of Leibniz algebras has been a topic of active research. Some (co)homology and deformations properties; relations with R-matrices and Yang-Baxter equations; results on various types of decompositions; structure of semisimple, solvable and nilpotent Leibniz algebras; classifications of some classes of graded nilpotent Leibniz algebras were obtained in numerous papers devoted to Leibniz algebras. In fact, many results of theory of Lie algebras have been extended to the Leibniz algebras case. For instance, the classical results on Cartan subalgebras, Levi's decomposition, properties of solvable algebras with a given nilradical and others from the theory of Lie algebras are also true for Leibniz algebras.

Recently, D. Barnes [32] has proved an analogue of Levi-Malcev Theorem for Leibniz algebras. He proved that a Leibniz algebra is decomposed into a semidirect sum of its solvable radical and a semisimple Lie algebra. Therefore, the main problem of the description of finite-dimensional Leibniz algebras consists of the study of solvable Leibniz algebras. D. Barnes also noted that the non-uniqueness of the semisimple subalgebra S may appear in Levi-Malcev theorem (the minimum dimension of Leibniz algebra where this phenomena appears is six). It is known that in the case of Lie algebras the semisimple Levi quotient is unique up to conjugation via an inner automorphism. The conjugation in the Leibniz algebras case is not true, in general. Recently, in [115] the authors studied the conditions for the semisimple parts of Leibniz algebras in the decomposition to be conjugated.

With a given Leibniz algebra we can associate a few Lie algebras, namely: the quotient algebra by the ideal $\mathcal{I}$; the algebra of all right multiplication operators with the commutator operation; the quotient algebra by the ideal of right annihilators etc. Some of properties of the Leibniz algebra can be stated in terms of these algebras. For instance, the solvability of the Leibniz algebra is equivalent to the solvability of each of the Lie algebras mentioned.

Note that the nilradical of a Leibniz algebra (the maximal nilpotent ideal) is not radical in the sense of Kurosh, because the quotient algebra by the nilradical may contain a nilpotent ideal. Nevertheless, similarly to the case of Lie algebras, it plays a crucial role in the description of solvable Leibniz algebras. The fact that the square of a solvable Leibniz algebra is contained in the nilradical also motivates the study of nilpotent algebras.

In the case of Lie algebras to classify the solvable part a result of the paper [131] has been used (see [17], [49], [132], [173] and [181] , etc). In the case of Leibniz algebras the analogue of Mubarakzjanov's

[131] results has been also successfully implemented in [50], [54], [55], [111] and [169].

A linear representation (sometimes referred as module) of a Leibniz algebra L is a vector space M, equipped with two actions (left and right) on L :

$$[\cdot,\cdot] : L \times M \longrightarrow M \text{ and } [\cdot,\cdot] : M \times L \longrightarrow M,$$

such that the identity

$$[[x,y],z] = [[x,z],y] + [x,[y,z]]$$

is true whenever one (any) of the variables is in V, and the other two are in L, i.e.,

$$[[x,y],m] = [[x,m],y] + [x,[y,m]];$$

$$[[y,m],x] = [[y,x],m] + [y,[m,x]];$$

$$[[m,x],y] = [[m,y],x] + [m,[x,y]],$$

for $x, y \in L$ and $m \in M$.

Note that to take the conditions above is predicted by the fact that on the direct sum $L \oplus M$ of vectors spaces the bracket

$$[[l+m, l'+m']] := [l,l']+[m,l']+[l,m'], \text{ where } l,l' \in L \text{ and } m,m' \in M.$$

defines a Leibniz algebra structure.

Note that the concepts of representations for Lie and Leibniz algebras are different. Therefore, such an important theorem in the theory of Lie algebras, as the Ado theorem on the existence of faithful representation in the case of Leibniz algebras is proved much easier and gives a stronger result. Because the kernel of the Leibniz algebra representation is the intersection of the kernels (in general, different ones) of right and left actions, in contrast to the representations of Lie algebras, where these kernels are the same. Therefore, a faithful representation of Leibniz algebras can be obtained easier than a faithful representation in the case of Lie algebras (see [31]). The (co)homology theory, the representations and related problems for Leibniz algebras were studied by Frabetti, [80], Loday, and Pirashvili, [119] and others. A good survey of all these and related problems is the book [118].

The problems related to the group theoretical realizations and integrability problems of Leibniz algebras are studied by Kinyon, Weinstein, [112] and others.

Different generalizations of Leibniz algebras as Leibniz superalgebras and n-ary Leibniz algebras also have been introduced and studied (see [9], [26], [147]).

Deformation theory of Leibniz algebras and related physical applications are initiated by Fialowski, Mandal, Mukherjee [72].

The notion of simple Leibniz algebra has been introduced by Dzhumadil'daev in [63].

1.4.2 Associative dialgebras

Here we present a generalization of associative algebras named by Loday as an *associative dialgebra.*

A **diassociative algebra** D is a vector space equipped with two bilinear binary associative operations:

$$\dashv : D \times D \to D \text{ and } \vdash : D \times D \to D$$

satisfying the axioms

$$\begin{aligned}(x \vdash y) \vdash z &= (x \dashv y) \vdash z,\\ (x \vdash y) \dashv z &= x \vdash (y \dashv z),\\ x \dashv (y \vdash z) &= x \dashv (y \dashv z).\end{aligned}$$

for all $x, y, z \in D$ (the binary operations $\vdash$ and $\dashv$ are said to be left and right products in D, respectively).

Example 1.6. *Let (A, d) be a differential associative algebra with $d^2 = 0$. Define the left and right products on A as follows*

$$x \dashv y := xd(y) \text{ and } x \vdash y := d(x)y.$$

Then $(A, \dashv, \vdash)$ is a diassociative algebra.

Example 1.7. *Let A be an associative algebra, M be an A-bimodule and let $f : M \longrightarrow A$ be an A-bimodule map. Define the left and right products on M as follows*

$$m \dashv m' := mf(m') \text{ and } m \vdash m' := f(m)m'.$$

Then $(M, \dashv, \vdash)$ is a diassociative algebra.

Example 1.8. *Let A be an associative algebra and n be a positive integer. The left and right products defined by*

$$(x \dashv y)_i := x_i \left(\sum_{j=1}^{n} y_j \right) \text{ for } 1 \leq i \leq n$$

and

$$(x \vdash y)_i := \left(\sum_{j=1}^{n} x_j \right) y_i \text{ for } 1 \leq i \leq n$$

on the A-module $D = A^n$ make it an associative dialgebra.

Example 1.9. *Let $D = \mathbb{F}[x, y]$ be a polynomial algebra at two indeterminates x, y over a field $\mathbb{F}$ of characteristic zero. Define the left and right products $\dashv, \vdash$ on D by*

$$f(x, y) \dashv g(x, y) := f(x, y)g(y, y)$$

and

$$f(x, y) \vdash g(x, y) := f(x, x)g(x, y).$$

Then $(D, \dashv, \vdash)$ is an associative dialgebra.

Proposition 1.5. *Let D be an associative dialgebra. Then the bracket*

$$[x, y] = x \dashv y - y \vdash x$$

turns D into a Leibniz algebra.

Proof. It is a straightforward checking since

$$[[x, y].z] = (x \dashv y) \dashv z - (y \vdash x) \dashv z - z \vdash (x \dashv y) + z \vdash (y \vdash x),$$

$$[[x, z].y] = (x \dashv z) \dashv y - (z \vdash x) \dashv y - y \vdash (x \dashv z) + y \vdash (z \vdash x),$$

$$[x, [y.z]] = x \dashv (y \dashv z) - x \dashv (z \vdash y) - (y \dashv z) \vdash x + (z \vdash y) \vdash x.$$

□

This construction defines a functor

$$\mathbf{Dias} \xrightarrow{-} \mathbf{Leib}$$

from category **Dias** of associative dialgebras to the category **Leib** of Leibniz algebras.

1.4.3 Universal enveloping algebra (Poincaré-Birkhoff-Witt Theorem)

The universal enveloping algebra for a Leibniz algebra has been constructed by Loday and Pirashvili in [119] (also see [118]). Loday showed that such an algebra comes in as an algebra with two bilinear binary associative operations satisfying three axioms (the algebra has been called *associative dialgebra* by Loday).

The enveloping algebra of a Leibniz algebra L was constructed by using the concept of free diassociative algebra. In algebras over fields this construction is interpreted as follows. Let V be a vector space over a field $\mathbb{F}$. By definition the free dialgebra structure on V is the dialgebra $D(V)$ equipped with a $\mathbb{F}$-linear map $i : V \longrightarrow D$ such that for any $\mathbb{F}$-linear map $f : V \longrightarrow D'$, where D' is a dialgebra over $\mathbb{F}$, there exists a unique factorization

$$f : V \xrightarrow{i} D(V) \xrightarrow{h} D',$$

where h is a dialgebra morphism. The authors proved the existence of $D(V)$ giving it as the tensor module

$$D(V) = T(V) \otimes V \otimes T(V),$$

where

$$T(V) := \mathbb{F} \oplus V \oplus V^{\otimes 2} \oplus \cdots \oplus V^{\otimes n} \oplus \cdots$$

with the following two products inductively defined by

$$(v_{-n} \cdots v_{-1} \otimes v_0 \otimes v_1 \cdots v_m) \dashv (w_{-p} \cdots w_{-1} \otimes w_0 \otimes w_1 \cdots w_q)$$
$$= v_{-n} \cdots v_{-1} \otimes v_0 \otimes v_1 \cdots v_m w_{-p} \cdots w_q,$$

$$(v_{-n} \cdots v_{-1} \otimes v_0 \otimes v_1 \cdots v_m) \vdash (w_{-p} \cdots w_{-1} \otimes w_0 \otimes w_1 \cdots w_q)$$
$$= v_{-n} \cdots\cdots v_m w_{-p} \cdots w_{-1} \otimes w_0 \otimes w_1 \cdots w_q,$$

where $v_i, w_j \in V$.

Let L^l and L^r be two copies of the Leibniz algebra L which is supposed to be free as $\mathbb{F}$-module. We denote by l_x and r_x the elements of L^l and L^r corresponding to $x \in L$. Consider the tensor $\mathbb{F}$-algebra $T(L^l \oplus L^r)$, which is associative and unital.

Let J be the two-sided ideal corresponding to the relations

$$r_{[x,y]} = r_x r_y - r_y r_x, \quad l_{[x,y]} = l_x r_y - r_y l_x, \quad (r_y + l_y) l_x = 0, \text{ for any } x, y \in L.$$

Definition 1.23. *The universal enveloping algebra of the Leibniz algebra L is the associative and unital algebra*

$$UL(L) := T(L^l \oplus L^r)/J.$$

Let $\tau : V \to W$ be an epimorphism of $\mathbb{F}$-modules. Define the associative algebra $SL(\tau)$ as the quotient of $S(W) \otimes T(V)$ by the 2-sided ideal generated by $1 \otimes xy + \tau(x) \otimes y$, for all $x, y \in V$.

Note that $UL(L)$ is a filtered algebra, the filtration being induced by the filtration of $T(L^l \oplus L^r)$, that is,

$$F_n UL(L) = \{\text{image of } \mathbb{F} \oplus E \oplus \cdots \oplus E^{\otimes n} \text{ in } UL(L)\},$$

where $E = L^l \oplus L^r$.

The associated graded algebra is denoted $grUL(L) := \bigoplus_{n \geq 0} gr_n UL(L)$. Then Poincaré-Birkhoff-Witt Theorem for Leibniz algebras is given as follows.

Theorem 1.10. *For any Leibniz $\mathbb{F}$-algebra L such that L and L_{Lie} are free as $\mathbb{F}$-modules, there is an isomorphism of graded associative $\mathbb{F}$-algebras*

$$grUL(L) \cong SL(L \to L_{Lie}).$$

Classification Results. The classification of the associative dialgebras in low-dimensions has been given by using the structure constants and a computer program in Maple [33], [152], [153].

1.4.4 Zinbiel algebras

The result intertwining diassociative algebras and dendriform algebras are best expressed in the framework of algebraic operads. In order to illustrate this point, Loday defined a class of Zinbiel algebras, which is Koszul dual to the category of Leibniz algebras.

Definition 1.24. *A Zinbiel algebra Z is an algebra with a binary operation $\cdot : Z \times Z \to Z$, satisfying the condition:*

$$(a \cdot b) \cdot c = a \cdot (b \cdot c) + a \cdot (c \cdot b) \textit{ for any } a, b, c \in Z.$$

Example 1.10. *Let V be the vector space of integrable functions. Let a product operation $\cdot$ on V be defined as follows*

$$(f \cdot g)(x) = f(x) \int_0^x g(t)dt \ \ for \ \ f, g \in V.$$

It is immediate to verify that $(V, \cdot)$ is a Zinbiel algebra.

Proposition 1.6. *Let Z be a Zinbiel algebra. Then the product*

$$xy = x \cdot y + y \cdot x$$

makes Z into a commutative algebra.

Proof. Obvious. □

Note that according to V.V. Gorbatsevich (see [93]) J.-L. Loday wrote as an author under a pseudonym "Guillaume William Zinbiel", here Zinbiel is reading of Leibniz in the reverse order.

Due to Proposition 1.6 we have a functor

$$\mathbf{Zinb} \xrightarrow{+} \mathbf{Com}$$

from category **Zinb** of Zinbiel algebras to the category **Com** of commutative algebras.

Classification Results. A.S. Dzhumadildaev et al. proved that any finite-dimensional Zinbiel algebra is nilpotent and gave the list of isomorphism classes of Zinbiel algebras in dimension 3 [64]. By the way, Example 1.10 above is taken from [64] and it is an example of non-nilpotent infinite-dimensional Zinbiel algebra. Earlier 2-dimensional case has been classified by B. Omirov [137]. Further classifications have been carried out by using the isomorphism invariant called the characteristic sequence which was used successfully in the case of Lie algebras [2], [4], [5], [43].

1.4.5 Dendriform algebras: Koszul dual of associative dialgebras

Definition 1.25. *Dendriform algebra E is an algebra with two binary operations*

$$\succ: E \times E \to E, \prec: E \times E \to E$$

satisfying the following identities:

$$(a \prec b) \prec c = (a \prec c) \prec b + a \prec (b \succ c),$$

$$(a \succ b) \prec c = a \succ (b \prec c), \tag{1.8}$$

$$(a \prec b) \succ c + (a \succ b) \succ c = a \succ (b \succ c).$$

Recall that an *associative Rota-Baxter algebra* (over a field $\mathbb{F}$) is an associative algebra $(A, \cdot)$ endowed with a linear operator $R : A \longrightarrow A$ subject to the following relation:

$$R(x) \cdot R(y) = R(R(x) \cdot y + x \cdot R(y) + \alpha x \cdot y),$$

$\alpha \in K$. The operator $\tilde{R} := -\alpha \mathrm{id} - R$ is called a Rota-Baxter operator of weight α. The Rota-Baxter operator $\tilde{R}$ generates the following dendriform algebra structure on A.

Example 1.11. *Let (A, R, α) be a Rota-Baxter algebra of weight α and define $\prec$ and $\succ$ as follows:*

$$x \prec y := -x \cdot \tilde{R}(y) \ \ and \ \ x \succ y := \tilde{R}(x) \cdot y.$$

It is immediate to check that $(A, \prec, \succ)$ is a dendriform algebra.

Proposition 1.7. *Let E be a dendriform algebra. Then with the product defined by*

$$x * y = x \prec y - x \succ y$$

E becomes an associative algebra.

Proof. It is an immediate checking if add up both sides of the equalities in (1.8). □

Proposition 1.7 provides a functor

$$\mathbf{Dend} \xrightarrow{+} \mathbf{As}$$

from the category **Dend** of dendriform algebras to the category **As** of associative algebras.

Classification Results. The essential results on Dendriform algebras have been obtained by Aguiar, Guo and Ebrahimi-Fard [8], [65], [101]. There is a classification of two-dimensional dendriform algebras [161].

1.4.6 Loday diagram and Koszul duality

The results intertwining Loday algebras are best expressed in the framework of algebraic operads. The notion of diassociative algebra defines an algebraic operad **Dias**, which is binary and quadratic. According to the theory of Ginzburg and Kapranov there is a well-defined "dual operad" **Dias**$^!$. Loday has showed that this is exactly the operad **Dend** of the dendriform algebras, in other words a dual diassociative algebra is nothing but a dendriform algebra. A similar duality can be established between the algebraic operads **Leib** defined by the notion of Leibniz algebra and the algebraic operads **Zinb** defined by the notion of Zinbiel algebra. Operadic dualities **Com**$^!$ = **Lie**, **As**$^!$ = **As** have been proved in [86], **Dias**$^!$ = **Dend** is in [118] and **Leib**$^!$ = **Zinb** is in [120]. The categories of algebras over these operads assemble into the following commutative diagram of functors (see Figure 1.1) which reflects the Koszul duality (see [118]).

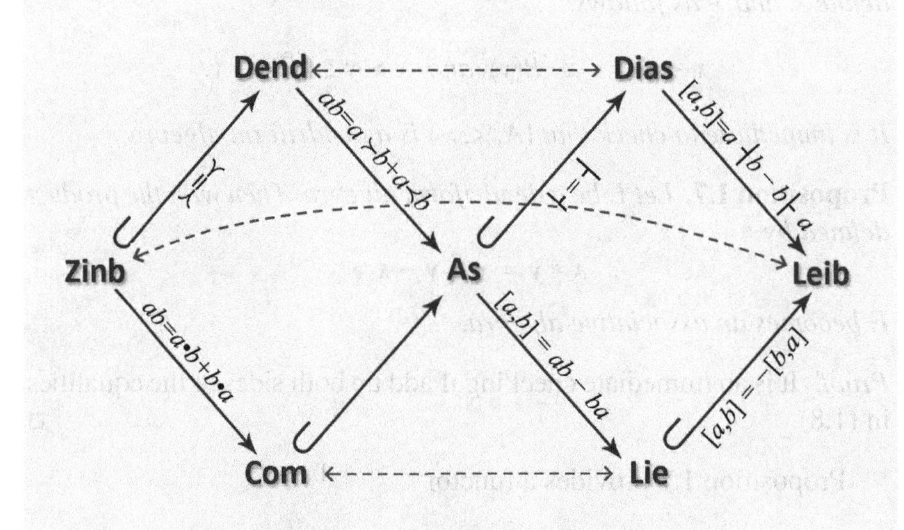

Figure 1.1 Loday diagram and Koszul duality

CHAPTER 2

STRUCTURE OF LEIBNIZ ALGEBRAS

2.1 SOME PROPERTIES OF LEIBNIZ ALGEBRAS

In this section we review extensions of main fundamental theorems of Lie algebras to Leibniz algebras. For those results which can not be extended counterexamples are provided.

Let L be a (right) Leibniz algebra. The right and left multiplication operators for a fixed element a of L are denoted by R_a and L_a, respectively, i.e., $R_a(x) = [x, a]$ and $L_a(x) = [a, x]$ for $x \in L$. The set of right multiplication operators $\Re(L) = \{R_a \mid a \in L\}$ forms an ideal of the Lie algebra $\mathrm{Der}(L)$.

In a right Leibniz algebra the following identities

$$R_{[b,a]} = R_a \circ R_b - R_b \circ R_a, \quad L_{[a,b]} = R_b \circ L_a + L_a \circ L_b, \quad L_{[a,b]} = R_b \circ L_a - L_a \circ R_b$$

hold true.

Recall that $\mathcal{I} = \mathrm{Span}_{\mathbb{F}}\{[x, x] \mid x \in L\}$ (Loday has denoted it by L^{ann} and another term "liezator" has been used by Gorbatsevich [93]). In fact, $\mathcal{I}$ is a two-sided ideal of L. Indeed, it follows from the equalities

$$[[x, x], y] = [[y, y] + x, [y, y] + x] - [x, x],$$

$$[y, [x, x]] = [[y, x], x] - [[y, x], x] = 0.$$

For a non-Lie Leibniz algebra L the ideal $\mathcal{I}$ is non trivial. The quotient algebra $L/\mathcal{I}$ is a Lie algebra and $\mathcal{I}$ is the smallest ideal in L, satisfying this condition. Therefore, $\mathcal{I}$ can be regarded as the "non Lie core" of the Leibniz algebra L. The ideal $\mathcal{I}$ can also be described as the linear span of all elements of the form $[x, y] + [y, x]$. The quotient algebra $L/\mathcal{I}$ is called liezation of L and denoted by L_{Lie}.

For a Leibniz algebra L we define right and left annihilators as follows

$$\mathrm{Ann}_r(L) = \{x \in L \mid [L, x] = 0\} \quad \text{and} \quad \mathrm{Ann}_l(L) = \{x \in L \mid [x, L] = 0\},$$

respectively.

These annihilators can be considered for both left and for right Leibniz algebras. For a right Leibniz algebra L the right annihilator $\mathrm{Ann}_r(L)$ is a two-sided ideal (since $[L, [x, y]] = -[L, [y, x]]$), while $\mathrm{Ann}_l(L)$ might not be even a subalgebra. For a left Leibniz algebra it is exactly the opposite. In general, the left and right annihilators are different, even they might have different dimensions. Obviously, $\mathcal{I} \subseteq \mathrm{Ann}_r(L)$ for a right Leibniz algebra L, whereas $\mathcal{I} \subseteq \mathrm{Ann}_l(L)$ for a left Leibniz algebra L. Therefore $L/\mathrm{Ann}_r(L)$ (respectively, $L/\mathrm{Ann}_r(L)$) is a Lie algebra in the case of right (respectively, left) Leibniz algebra, which is anti-isomorphic to the Lie subalgebra $\mathfrak{R}(L)$ (respectively, $\mathfrak{L}(L) = \{L_a \mid a \in L\}$) of $\mathrm{Der}(L)$, consisting of all right (left) multiplication operators.

We denote by $\mathrm{Cent}(L) = \mathrm{Ann}_r(L) \cap \mathrm{Ann}_l(L)$ the center of a Leibniz algebra L.

2.2 NILPOTENT AND SOLVABLE LEIBNIZ ALGEBRAS

The concepts of nilpotency and solvability for Leibniz algebras are defined similarly to those of Lie algebras.

For a given Leibniz algebra $(L, [\cdot, \cdot])$ the lower central series and derived series are defined, respectively, as follows:

$$L^1 = L, \; L^{k+1} = [L^k, L], \; k \geq 1,$$

$$L^{[1]} = L, \; L^{[s+1]} = [L^{[s]}, L^{[s]}], \; s \geq 1.$$

Obviously,

$$L^1 \supseteq L^2 \supseteq \cdots \supseteq L^i \supseteq \cdots,$$

and

$$L^{[1]} \supseteq L^{[2]} \supseteq \cdots \supseteq L^{[i]} \supseteq \cdots.$$

Moreover, $L^{[i]} \subseteq L^i$.

Definition 2.1. *A Leibniz algebra L is said to be nilpotent (respectively, solvable), if there exists $n \in \mathbb{N}$ $(m \in \mathbb{N})$ such that $L^n = 0$ (respectively, $L^{[m]} = 0$). The minimal number n (respectively, m) with such property is said to be the index of nilpotency (respectively, of solvability) of the algebra L.*

Example 2.1.

- *A nilpotent Lie algebra is nilpotent as a Leibniz algebra;*
- *The sum of nilpotent Leibniz algebras is nilpotent;*
- *On the space of quadratic $n \times n$ matrices with complex coefficients $M_n(\mathbb{C})$ we consider a linear map $D : M_n(\mathbb{C}) \to M_n(\mathbb{C})$ satisfying the condition (see Example in Section 1.4.1)*

$$D(AD(B)) = D(A)D(B) = D(D(A)B), \text{ for all } A, B \in M_n(\mathbb{C}).$$

Then the space $M_n(\mathbb{C})$ equipped with the multiplication $[A, B]_D = AD(B) - D(B)A$ becomes a Leibniz algebra (denoted by $M_n(\mathbb{C})_D$). Note that $M_n(\mathbb{C})_D$ is nilpotent if and only if $Card(\mathrm{Spec}X) = 1$ for every $X \in \mathrm{Im}D$;

- *The algebras with the table of multiplications below on a basis $\{e_1, e_2, \ldots, e_n\}$ are nilpotent Leibniz algebras:*

i)

$$\begin{cases} [e_i, e_1] & = e_{i+1}, \quad 1 \leq i \leq n-3, \\ [e_1, e_{n-1}] & = e_2 + e_n, \\ [e_i, e_{n-1}] & = e_{i+1}, \quad 2 \leq i \leq n-3, \end{cases}$$

ii)

$$\begin{cases} [e_i, e_1] & = e_{i+1}, \quad 1 \leq i \leq n-3, \\ [e_1, e_{n-1}] & = e_n. \end{cases}$$

Here are examples of solvable Leibniz algebras.

Example 2.2.

- *A solvable Lie algebra is solvable as a Leibniz algebra;*
- *A nilpotent Leibniz algebra is solvable;*

- *The sum of solvable Leibniz algebras is solvable;*
- *A right Leibniz algebra is solvable if and only if the Lie algebra* $\mathfrak{R}(L)$ *is solvable;*
- *A right Leibniz algebra is solvable if and only if the quotient Lie algebra* $L/\mathrm{Ann}_r(L)$ *is solvable;*
- *An algebra with the following table of multiplications on a basis* $\{e_1, e_2, \ldots, e_n, x\}$

$$\begin{cases} [e_i, e_1] = e_{i+1}, & 1 \leq i \leq n-1, \\ [x, e_1] = e_1, & \\ [e_i, x] = -ie_i, & 1 \leq i \leq n. \end{cases}$$

is a solvable Leibniz algebra.

One of the classical results on finite-dimensional nilpotent Lie algebras is Engel's theorem on criterion of nilpotency of the Lie algebra in terms of nilpotency of the right multiplications operators. An analogue of Engel's theorem for Leibniz algebras also holds. The various versions of the theorem have been proved in [23], [93] and [143]. Here we give the following version from [93].

Theorem 2.1. *If all operators* R_x *of right multiplication for the right Leibniz algebra* L *are nilpotent, then the algebra* L *is nilpotent. In particular, for the right multiplications there is a common eigenvector with zero eigenvalue.*

Proof. Since $L/\mathcal{I}$ is a Lie algebra the assertion of the proposition for $L/\mathcal{I}$ holds true. Note that $\mathcal{I}$ is the central ideal, its action on L by right multiplications is trivial. Since the Leibniz algebra L is a central extension of the nilpotent Lie algebra $L/\mathcal{I}$, it is also nilpotent. □

Corollary 2.1. *If for a right Leibniz algebra all right multiplication operators are nilpotent, then all left multiplication operators also are nilpotent.*

The proof of this corollary is based on the identity $(L_x)^n = (-1)^{n-1}L_x(R_x)^{n-1}$ to prove that we use induction on n, the case $n = 2$ being obvious from the identity $[x, [x, z]] = -[x, [z, x]]$. Assume that $(L_x)^{n-1} = (-1)^{n-2}L_x(R_x)^{n-2}$.

Then

$$\begin{aligned}(L_x)^n(z) &= [x,(L_x)^{n-1}(z)] = [x,(-1)^{n-2}L_x(R_x)^{n-2}(z)]\\ &= (-1)^{n-2}[x,[x,(R_x)^{n-2}(z)]] = (-1)^{n-1}[x,[(R_x)^{n-2}(z),x]]\\ &= (-1)^{n-1}L_x(R_x)^{n-1}(z).\end{aligned}$$

For the other nilpotency properties of Leibniz algebras we refer the reader to [143].

Proposition 2.1. *Any Leibniz algebra L has a maximal nilpotent ideal containing all nilpotent ideals of L.*

Proof. Let J be a nilpotent ideal in L. Then the sum $J+\mathcal{I}$ is nilpotent, it follows immediately from the right centrality of $\mathcal{I}$. Therefore, in discussing the question of the existence of maximal nilpotent ideal it is sufficient to consider only nilpotent ideals containing $\mathcal{I}$. Consider the epimorphism of liezation $\pi : L \to L/\mathcal{I} = L_{Lie}$. For the Lie algebra L_{Lie} the existence of a maximal nilpotent ideal is known. Its inverse image under the mapping π is also nilpotent. This completes the proof. □

The maximal nilpotent ideal of a Leibniz algebra is said to be *the nilradical* of the algebra (further denoted by $\mathcal{N}$).

The existence of solvable radical or just radical (the maximal solvable ideal), denoted by $\mathcal{R}$ is carried out in a similar way as in the case of Lie algebras.

The following analogue of the well-known Lie Theorem for solvable Leibniz algebras also holds true (see [93], [144]).

Theorem 2.2. *A right complex Leibniz algebra L has a complete flag of subspaces which is invariant under the right multiplication operator.*

In other words, there is a basis with respect to that all right multiplication operators R_x, $x \in L$ can be simultaneously given in the triangular form.

Proof. For Lie algebras this theorem is well known (in one of the variants of its formulation). Now let L be an arbitrary solvable Leibniz algebra. Since the liezator $\mathcal{I}$ is central, then the solvability of the Leibniz algebra L is equivalent to the solvability of the Lie algebra $L/\mathcal{I}$. By the classical Lie theorem $L/\mathcal{I}$ has a complete flag, which is invariant under the multiplication (both left and right). Let x be an element of L. Its right action on $\mathcal{I}$ is trivial and therefore in $\mathcal{I}$ there exists a complete flag, which is invariant under right multiplication (for the left multiplication on the right Leibniz algebra such reasoning does not pass).

Extending it to the full invariant flag in the quotient space until the full flag in L, we obtain a complete flag in L, which is invariant under the right multiplication operator. □

Proposition 2.2. *Let $\mathcal{R}$ be the radical of a Leibniz algebra L and let $\mathcal{N}$ be the nilradical of L. Then $[L, \mathcal{R}] \subseteq \mathcal{N}$.*

Proof. For Lie algebras the statement above is true. Now let L be an arbitrary Leibniz algebra. Then $\mathcal{I} \subseteq \mathcal{N} \subseteq \mathcal{R}$ and $I \subseteq Ann_r(L)$. Consider $L^* := L_{Lie}$ the liezation of L. From the definitions of the radical and nilradical it follows that $\mathcal{R}^* = \mathcal{R}/\mathcal{I}$ and $\mathcal{N}^* = \mathcal{N}/\mathcal{I}$ are radical and nilradical of the Lie algebra L^*, respectively. Since L^* is a Lie algebra we get $[L^*, \mathcal{R}^*] \subset \mathcal{N}^*$. But then, by the fact that $\mathcal{I} \subseteq \mathcal{N} \subseteq \mathcal{R}$, we obtain $[L, \mathcal{R}] \subseteq N$. □

Corollary 2.2. $[\mathcal{R}, \mathcal{R}] \subset \mathcal{N}$. *In particular,* $[\mathcal{R}, \mathcal{R}]$ *is nilpotent.*

We can give the characterization of the solvability of a Leibniz algebra in terms of the nilpotence of its square.

Corollary 2.3. *A Leibniz algebra L is solvable if and only if $[L, L]$ is nilpotent.*

Proof. In one direction the assertion of Corollary 2.3 is proved in Corollary 2.2. The converse is proved as follows. Let $[L, L]$ be nilpotent and $\mathcal{R}$ be the radical of L. Consider the quotient Lie algebra $L/\mathcal{R}$. Since $L/\mathcal{R}$ does not contain a solvable ideal it is a semisimple Lie algebra. Therefore, the quotient algebra coincides with its commutator. Since $[L, L]$ is nilpotent the algebra $L/\mathcal{R}$ must be trivial, i.e., L is solvable. □

2.3 ON LEVI'S THEOREM FOR LEIBNIZ ALGEBRAS

In the theory of Lie algebras there is well-known Levi's Theorem, which asserts that any finite-dimensional Lie algebra over a field of characteristic zero is decomposed into a semidirect sum of the solvable radical and a semisimple subalgebra. Moreover, all complements to solvable radical are conjugated. In this section we show that Levi's Theorem can be extended to Leibniz algebras, but the conjugacy property does not hold (see [32]).

Theorem 2.3. *(Levi's theorem for Leibniz algebras) For arbitrary finite-dimensional Leibniz algebra L over a field of characteristic zero there exists a subalgebra $\mathcal{S}$ (which is a semisimple Lie algebra), such that $L = \mathcal{S} \dot{+} \mathcal{R}$, where $\mathcal{R}$ is the radical of L.*

Proof. Let us consider the Lie algebra $L^* := L_{Lie}$. By the classical Levi's theorem there is a semisimple algebra $\mathcal{S}^* \subseteq L^*$, which gives the semidirect sum decomposition $L^* = \mathcal{S}^* + \mathcal{R}^*$. Here $\mathcal{R}^*$ is the radical of L^*. Let $\mathcal{R}$ be the radical of the Leibniz algebra L and let $\pi : L \longrightarrow L^*$ be the natural epimorphism. Then it is clear that $\pi(\mathcal{R}) = \mathcal{R}^*$. Let $F = \pi^{-1}(\mathcal{S}^*)$. Then F is a subalgebra of L containing $\mathcal{I}$, and the quotient algebra $F/\mathcal{I}$ is isomorphic to the semisimple Lie algebra $\mathcal{S}^*$.

For the Lie algebra $\mathcal{S}^*$ the abelian Leibniz algebra $\mathcal{I}$ is an $\mathcal{S}^*$- module. For the semisimple Lie algebra $\mathcal{S}^*$, it follows that the subalgebra F, considered as an extension of $\mathcal{S}^*$ by means of $\mathcal{I}$, is split (by Whitehead's Lemma for semisimple Lie algebras). But this means that in F there exists a subalgebra (semisimple Lie algebra) which is complementary to the abelian ideal $\mathcal{I}$. So one has $\mathcal{S} \dot{+} \mathcal{R} = L$. □

The subalgebra $\mathcal{S}$ in the theorem above, similarly to the case of Lie algebras is called a *Levi subalgebra* of the Leibniz algebra L.

In the splitting theorem for Lie algebras over a field of characteristic zero all the semisimple parts are mutually conjugate. But such a conjugacy for Leibniz algebras fails to be true which is illustrated by the following example of D. Barnes [32].

Example 2.3. *Let S be a simple Lie algebra and K be isomorphic to S as a right S-module. Denote by x' the element of K corresponding to $x \in S$ under this isomorphism. One can make K into a Leibniz module by defining the left action to be 0. Let L be the split extension of K by S. Then L is a Leibniz algebra and $\mathcal{I} = K$. The space $S_1 = \{(s, s')| \ s \in S\}$ is another subalgebra complementing K since, using the module isomorphism, one has*

$$(t, t')(s, s') = (ts, t's) = (ts, (ts)').$$

For any $x = (s, k) \in L$, the inner derivation $d_x = d_s$, $d_x(S) \subseteq S$ and so $\exp(d_x)(S) \subseteq S$. Thus, S and S_1 are not conjugate.

Corollary 2.4. *Let L be a Leibniz algebra such that its left annihilator lies in the right annihilator (i.e., $[L, \mathrm{Ann}_l(L)] = \{0\}$). If the Lie algebra*

$L/\mathrm{Ann}_l(L)$ is semisimple, then L is a Lie algebra and its radical is central.

Proof. Let us consider the Levi decomposition $L = \mathcal{S} \dot{+} \mathcal{R}$. Since, by the hypothesis, the algebra $L/\mathrm{Ann}_r(L)$ is semisimple, we have $\mathcal{R} = \mathrm{Ann}_l(L)$. The right action of the subalgebra $\mathcal{S}$ on $\mathrm{Ann}_l(L)$ is always trivial. But due to $[\mathrm{Ann}_l(L), L] = \{0\}$, the right action of $\mathcal{S}$ on $\mathrm{Ann}_l(L)$ also is trivial. Therefore, $\mathcal{S}$ commutes with the radical and then Levi's decomposition of L is a direct sum. The radical of $\mathcal{R} = \mathrm{Ann}_l(L)$ is abelian. Such Lie algebras have been called reductive Lie algebras. □

The statement of Corollary 2.4 was proved in [23] for the case where $\mathcal{S}$ is a classical simple Lie algebra.

2.3.1 On conjugacy of Levi subalgebras of Leibniz algebras

Here we provide conditions for the Levi-Malcev theorem to hold or not (i.e., for two Levi subalgebras to be conjugate or not by an inner automorphism) in the context of finite-dimensional Leibniz algebras over a field of characteristic zero. Particularly, in the case of the field of complex numbers, we consider all possible cases for Levi subalgebras to be conjugate or not. We make use an adapted version of Schur's Lemma (see Lemma 1.3, Section 1.4.2) given earlier.

Let $\mathcal{N}$ be the nilradical of a Leibniz algebra L, $a \in \mathcal{N}$ and R_a be the right multiplication operator by a. Since R_a is nilpotent there exists n depending on a such that $R_a^{n+1} = 0$.

We set

$$\exp(R_a)(x) = x + R_a(x) + \frac{R_a^2(x)}{2!} + \cdots + \frac{R_a^n(x)}{n!}.$$

Then $\exp(R_a)$ is an automorphism of L. Similarly to the case of Lie algebras $\exp(R_a)$ is called an *inner* automorphism. It is easy to see that the composition of inner automorphisms again is an inner automorphism and they form a subgroup of the group of all automorphisms of L.

Let us start with the following elementary result on invariantness of the ideals $\mathcal{I}$, $\mathcal{N}$ and $\mathcal{R}$ under automorphisms of a Leibniz algebra L.

Proposition 2.3. *For any $\varphi \in \mathrm{Aut}(L)$ we have*

$$\varphi(\mathcal{I}) = \mathcal{I}, \quad \varphi(\mathcal{N}) = \mathcal{N}, \quad \varphi(\mathcal{R}) = \mathcal{R}.$$

Proof. Let $\varphi \in \mathrm{Aut}(L)$. The inclusion $\varphi(\mathcal{I}) \subseteq \mathcal{I}$ is obvious. Let us prove that $\mathcal{I} \subseteq \varphi(\mathcal{I})$. Let $x = \sum \alpha_i[x_i, x_i] \in \mathcal{I}$. For each $x_i \in L$ there exists $y_i \in L$ such that $\varphi(y_i) = x_i$. Due to $x = \sum \alpha_i[x_i, x_i] = \sum \alpha_i[\varphi(y_i), \varphi(y_i)] = \varphi(\sum \alpha_i[y_i, y_i])$, we conclude $x = \varphi(y)$, where $y = \sum \alpha_i[y_i, y_i] \in \mathcal{I}$ i.e., $\mathcal{I} \subseteq \varphi(\mathcal{I})$. Hence, $\varphi(\mathcal{I}) = \mathcal{I}$.

Due to isomorphism properties we easily get that for any $k \in \mathbb{N}$ and an ideal $\mathcal{J}$ of L the following equalities are true:

$$\varphi(\mathcal{J})^k = \varphi(\mathcal{J}^k), \qquad \varphi(\mathcal{J})^{[k]} = \varphi(\mathcal{J}^{[k]}).$$

Consequently, $\varphi(\mathcal{N})$ and $\varphi(\mathcal{R})$ are nilpotent and solvable radicals, respectively. □

Let L be a Leibniz algebra, $L = \mathcal{S} \dot{+} \mathcal{R}$ its Levi decomposition and let φ be an automorphism of L. Then due to Proposition 2.3 we get $\varphi = \varphi_{\mathcal{S},\mathcal{S}} + \varphi_{\mathcal{S},\mathcal{R}} + \varphi_{\mathcal{R},\mathcal{R}}$, where

$$\varphi_{\mathcal{S},\mathcal{S}} = \mathrm{Proj}_{\mathcal{S}}(\varphi_{|\mathcal{S}}), \quad \varphi_{\mathcal{S},\mathcal{R}} = \mathrm{Proj}_{\mathcal{R}}(\varphi_{|\mathcal{S}}), \quad \varphi_{\mathcal{R},\mathcal{R}} = \varphi_{|\mathcal{R}}.$$

Denote by $\mathrm{Hom}_{\mathcal{S}}(\mathcal{S}, \mathcal{I})$ the set of all $\mathcal{S}$-module homomorphisms from $\mathcal{S}$ into $\mathcal{I}$.

We set

$$\mathcal{S}_\theta = \{x + \theta(x) : x \in \mathcal{S}, \theta \in \mathrm{Hom}_{\mathcal{S}}(\mathcal{S}, \mathcal{I})\}. \tag{2.1}$$

Since

$$[x + \theta(x), y + \theta(y)] = [x, y] + [\theta(x), y] = [x, y] + \theta([x, y]),$$

it follows that the mapping $x \mapsto x + \theta(x) \in \mathcal{S}_\theta$, $x \in \mathcal{S}$, is an isomorphism from $\mathcal{S}$ onto $\mathcal{S}_\theta$.

Remark. Further we shall consider only automorphisms of the form $\exp(R_a)$ since results can be easily extended for their compositions (inner automorphisms).

Theorem 2.4. *Let $\mathcal{S}$ and $\mathcal{S}_1$ be two Levi subalgebras of a Leibniz algebra L. Then there exist $\tau \in \mathrm{Hom}_{\mathcal{S}}(\mathcal{S}, \mathcal{I})$ and an element $a \in \mathcal{N}$ such that* $\exp(R_a)(\mathcal{S}_\tau) = \mathcal{S}_1$.

Proof. Let $\widetilde{L} = \mathcal{S} \dot{+} \widetilde{\mathcal{R}}$, where $\widetilde{\mathcal{R}} = \mathcal{R}/\mathcal{I}$. Due to Levi–Malcev theorem for Lie algebras (see [123]), there exists an element $\widetilde{a} \in \widetilde{\mathcal{N}} = \mathcal{N}/\mathcal{I}$ such that $\exp(R_{\widetilde{a}})(\mathcal{S}) = \mathcal{S}_1$. For an element a from the class $\widetilde{a}$, we consider the automorphism $\exp(R_a)$ of L. Let $\pi\colon L \to \widetilde{L}$ be the quotient map. Since any automorphism maps $\mathcal{I}$ into itself, the quotient map $\widetilde{\exp(R_a)}$ is well defined, i.e.,

$$\widetilde{\exp(R_a)}(\pi(x)) = \pi\left(\exp(R_a)(x)\right), \; x \in L.$$

Then $\widetilde{\exp(R_a)}$ is an automorphism of $\widetilde{L}$. For $x \in S$ we have

$$\widetilde{\exp(R_a)}(x) = \widetilde{\exp(R_a)}(\pi(x)) = \pi\left(\exp(R_a)(x)\right) = \pi\left(\sum_{k=0}^{n} \frac{R_a^k(x)}{k!}\right)$$
$$= \sum_{k=0}^{n} \frac{R_{\pi(a)}^k(\pi(x))}{k!} = \sum_{k=0}^{n} \frac{R_{\widetilde{a}}^k(x)}{k!} = \exp(R_{\widetilde{a}})(x).$$

Therefore, $\widetilde{\exp(R_a)}\big|_{\mathcal{S}} = \exp(R_{\widetilde{a}})\big|_{\mathcal{S}}$.

Thus, for any $x \in \mathcal{S}$ there exists an element $\tau(x) \in \mathcal{I}$ such that

$$\exp(R_a)^{-1}\left(\exp(R_{\widetilde{a}})(x)\right) = x + \tau(x).$$

For $x, y \in \mathcal{S}$ we consider

$$\begin{aligned}
[x, y] + \tau([x, y]) &= \exp(R_a)^{-1}\left(\exp(R_{\widetilde{a}})([x, y])\right) \\
&= \left[\exp(R_a)^{-1}\left(\exp(R_{\widetilde{a}})(x)\right), \exp(R_a)^{-1}\left(\exp(R_{\widetilde{a}})(y)\right)\right] \\
&= [x + \tau(x), y + \tau(y)] = [x, y] + [\tau(x), y],
\end{aligned}$$

i.e., $\tau \in \mathrm{Hom}_{\mathcal{S}}(\mathcal{S}, \mathcal{I})$.

Then,

$$\exp(R_{\widetilde{a}})(x) = \exp(R_a)(x + \tau(x)) \text{ for all } x \in \mathcal{S}$$

implies $\mathcal{S}_1 = \exp(R_a)(\mathcal{S}_\tau)$. □

The following result follows from Theorem 2.4 and it asserts that the question on the conjugacy of Levi subalgebras is reduced to verification of the conjugacy of Levi subalgebras $\mathcal{S}$ and $\mathcal{S}_\theta$, where $\mathcal{S}_\theta$ is the algebra form (2.1).

Corollary 2.5. *A Levi subalgebra $\mathcal{S}$ is unique up to conjugation by an inner automorphism of L if and only if for every $\tau \in \mathrm{Hom}_{\mathcal{S}}(\mathcal{S}, \mathcal{I})$ there exists an element $b \in \mathcal{N}$ such that* $\exp(R_b)\big|_{\mathcal{S}} = \mathrm{id}_{\mathcal{S}} + \tau$.

Proof. *Sufficiency.* Suppose that $\mathcal{S}$ is unique up to conjugation by an inner automorphism of L. Take an arbitrary $\tau \in \mathrm{Hom}_{\mathcal{S}}(\mathcal{S}, \mathcal{I})$ and consider the Levi subalgebra $\mathcal{S}_\tau$. Since $\mathcal{S}$ is unique up to conjugation by an inner automorphism, there exists an element an element $a \in \mathcal{N}$ such that $\exp(R_b)(\mathcal{S}) = \mathcal{S}_\tau$ and therefore $\exp(R_b)\big|_{\mathcal{S}} = \mathrm{id}_{\mathcal{S}} + \tau$.

Necessity. Let $\mathcal{S}$ and $\mathcal{S}_1$ be two Levi subalgebras. By Theorem 2.4 there exist $\tau \in \mathrm{Hom}_{\mathcal{S}}(\mathcal{S}, \mathcal{I})$ and an element $a \in \mathcal{N}$ such that $\exp(R_a)(\mathcal{S}_\tau) = \mathcal{S}_1$. By the assumption there exists an element $b \in \mathcal{N}$ such that $\exp(R_b)\big|_{\mathcal{S}} = \mathrm{id}_{\mathcal{S}} + \tau$, i.e., $\exp(R_b)(\mathcal{S}) = \mathcal{S}_\tau$. This implies that $\exp(R_a) \circ \exp(R_b)(\mathcal{S}) = \mathcal{S}_1$. □

In the next two propositions we give sufficient conditions of the conjugacy of Levi subalgebras.

Proposition 2.4. *Let $\mathcal{S}$ be a Levi subalgebra of a Leibniz algebra L and* $\mathrm{Hom}_{\mathcal{S}}(\mathcal{S}, \mathcal{I}) = \{0\}$. *Then $\mathcal{S}$ is unique up to conjugation by an inner automorphism of L.*

Proof. Taking into account the condition $\mathrm{Hom}_{\mathcal{S}}(\mathcal{S}, \mathcal{I}) = \{0\}$ in Theorem 2.4, we obtain $\mathcal{S}_\tau = \mathcal{S}$ and therefore, $\exp(R_a)(\mathcal{S}) = \mathcal{S}_1$. □

Remark. Note that the equality $[\mathcal{I}, \mathcal{S}] = \{0\}$ implies $\mathrm{Hom}_{\mathcal{S}}(\mathcal{S}, \mathcal{I}) = \{0\}$. Therefore, due to Proposition 2.4 we conclude that Levi subalgebras of a Leibniz algebra $L = \mathcal{S} \dot{+} \mathcal{R}$ satisfying the condition $[\mathcal{I}, \mathcal{S}] = \{0\}$ are conjugate via an inner automorphism. Note that an analogue of this result also has been obtained by G. Mason and G. Yamskulma in [126].

Thanks to Proposition 2.4, henceforth we shall consider only the case when $\mathrm{Hom}_{\mathcal{S}}(\mathcal{S}, \mathcal{I}) \neq \{0\}$.

Let $L = \mathcal{S} \dot{+} \mathcal{R}$ be a Levi decomposition of a Leibniz algebra L. A Levi subalgebra $\mathcal{S}$ and the $\mathcal{S}$-module $\mathcal{I}$ can be uniquely represented as follows

$$\mathcal{S} = G \oplus Q \quad \text{and} \quad \mathcal{I} = J \oplus K,$$

where

$$J = \{i \in \mathcal{I} \mid \exists\, \varphi \in \mathrm{Hom}_{\mathcal{S}}(\mathcal{S}, \mathcal{I}) \text{ such that } i \in \mathrm{Im}\varphi\},$$

$$K = \{i \in \mathcal{I} \mid \forall \varphi \in \mathrm{Hom}_{\mathcal{S}}(\mathcal{S}, \mathcal{I}) \text{ such that } i \notin \mathrm{Im}\varphi\},$$

$$G = \{x \in \mathcal{S} \mid \exists\, \varphi \in \mathrm{Hom}_{\mathcal{S}}(\mathcal{S}, \mathcal{I}) \text{ such that } \varphi(x) \in J\},$$

$$\mathcal{Q} = \{x \in \mathcal{S} \mid \forall \varphi \in \mathrm{Hom}_{\mathcal{S}}(\mathcal{S}, \mathcal{I}) \text{ such that } \varphi(x) = 0\}.$$

Clearly, for any $\varphi \in \mathrm{Hom}_{\mathcal{S}}(\mathcal{S}, \mathcal{I})$ we have $\varphi = \mathrm{proj}_J(\varphi|_G)$. Therefore, we can identify $\mathrm{Hom}_{\mathcal{S}}(\mathcal{S}, \mathcal{I})$ with $\mathrm{Hom}_{\mathcal{S}}(G, J)$ (we denote it further by $\mathrm{Hom}_{\mathcal{S}}(\mathcal{S}, \mathcal{I}) \equiv \mathrm{Hom}_{\mathcal{S}}(G, J)$).

Let now assume that $[J, \mathcal{R}] = \{0\}$. For $\theta \in \mathrm{Hom}_{\mathcal{S}}(\mathcal{S}, \mathcal{I})$ we set

$$\delta_\theta(x_{\mathcal{S}} + x_{\mathcal{R}}) = \theta(x_{\mathcal{S}}), \quad x = x_{\mathcal{S}} + x_{\mathcal{R}} \in \mathcal{S} \dot{+} \mathcal{R}.$$

Then δ_θ is a derivation of L. Indeed,

$$\begin{aligned}
[\delta_\theta(x), y] + [x, \delta_\theta(y)] &= [\theta(x_{\mathcal{S}}), y_{\mathcal{S}} + y_{\mathcal{R}}] + [x, \theta(y_{\mathcal{S}})] \\
&= [\theta(x_{\mathcal{S}}), y_{\mathcal{S}}] + [\theta(x_{\mathcal{S}}), y_{\mathcal{R}}] = [\theta(x_{\mathcal{S}}), y_{\mathcal{S}}] \\
&= \theta([x_{\mathcal{S}}, y_{\mathcal{S}}]) = \delta_\theta([x, y]).
\end{aligned}$$

It is clear that $\delta_\theta^2 = 0$. Hence, $\exp(\delta_\theta)(x) = x + \delta_\theta(x)$, $x \in L$ and $\exp(\delta_\theta)(x) = x + \theta(x)$, $x \in \mathcal{S}$. Thus, we obtain

$$\exp(\delta_\theta)(\mathcal{S}) = \mathcal{S}_\theta.$$

Let us define a nilpotent derivation D as follows

$$D = R_a + \delta_\theta, \quad a \in \mathcal{N}, \quad \theta \in \mathrm{Hom}_{\mathcal{S}}(\mathcal{S}, \mathcal{I}). \tag{2.2}$$

Then one has

Proposition 2.5. *Let $L = \mathcal{S} \dot{+} \mathcal{R}$ be a Levi decomposition of a Leibniz algebra L such that* $[J, \mathcal{R}] = \{0\}$. *Then any two Levi subalgebras $\mathcal{S}$ and $\mathcal{S}_1$ are conjugate by* $\exp(D)$, *where D has the form* (2.2).

Proof. By Theorem 2.4 there exist $\tau \in \mathrm{Hom}_{\mathcal{S}}(\mathcal{S}, \mathcal{I})$ and an element $a \in \mathcal{N}$ such that $\exp(R_a)(\mathcal{S}_\tau) = \mathcal{S}_1$. Since $[J, \mathcal{R}] = \{0\}$, it follows that δ_τ is a derivation with $\delta_\tau^2 = 0$. For the derivation $D = R_a + \delta_\tau$, we have

$$\begin{aligned}
\exp(D)(\mathcal{S}) &= \exp(R_a + \delta_\tau)(\mathcal{S}) = (\exp(R_a) \circ \exp(\delta_\tau))(\mathcal{S}) \\
&= \exp(R_a)(\exp(\delta_\tau)(\mathcal{S})) = \exp(R_a)(\mathcal{S}_\tau) = \mathcal{S}_1.
\end{aligned}$$

This completes the proof. □

A version (when $J = \mathfrak{I}$) of Proposition 2.5 above has been proved in [126].

Note that Example 2.3 satisfies the conditions of Proposition 2.5.

Now we consider the cases of Leibniz algebras with Levi subalgebras which are not conjugate via automorphisms.

Proposition 2.6. *Let $L = \mathcal{S} \dot{+} \mathcal{R}$ be a Levi decomposition with conditions $[\mathcal{S}, \mathcal{R}] = \{0\}$ and $[J, \mathcal{R}] \neq \{0\}$. Then there exists a Levi subalgebra $\mathcal{S}_1$ such that the algebras $\mathcal{S}$ and $\mathcal{S}_1$ are not conjugate via any automorphism.*

Proof. Due to $[J, \mathcal{R}] \neq \{0\}$ we have the existence of elements $i \in J$ and $y \in \mathcal{R}$ such that $[i, y] \neq 0$. Consider an $\mathcal{S}$-module homomorphism θ from $\mathcal{S}$ to J such that there exists $x \in \mathcal{S}$ with the condition $\theta(x) = i$.

We set

$$\mathcal{S}_1 = \{x + \theta(x) : x \in \mathcal{S}\}.$$

Let assume the existence of an automorphism $\varphi = \varphi_{\mathcal{S},\mathcal{S}} + \varphi_{\mathcal{S},\mathcal{R}} + \varphi_{\mathcal{R},\mathcal{R}} \in \mathrm{Aut}(L)$ such that $\varphi(\mathcal{S}) = \mathcal{S}_1$. Then for $x \in \mathcal{S}$ we have $\varphi_{\mathcal{S},\mathcal{S}}(x) + \varphi_{\mathcal{S},\mathcal{R}}(x) = \varphi(x) \in \mathcal{S}_1 \subset \mathcal{S} + J$. Hence $\varphi_{\mathcal{S},\mathcal{R}}(x) \in J$ and $\varphi_{\mathcal{S},\mathcal{R}} = \varphi_{\mathcal{S},J}$.

Since $\varphi(\mathcal{R}) = \mathcal{R}$ there exists $z \in \mathcal{R}$ such that $\varphi_{\mathcal{R},\mathcal{R}}(z) = y$. Due to $[\mathcal{S}, \mathcal{R}] = \{0\}$, we have

$$\begin{aligned} 0 = \varphi([x, z]) &= [\varphi(x), \varphi(z)] = [\varphi_{\mathcal{S},\mathcal{S}}(x) + \varphi_{\mathcal{S},J}(x), \varphi_{\mathcal{R},\mathcal{R}}(z)] \\ &= [\varphi_{\mathcal{S},J}(x), \varphi_{\mathcal{R},\mathcal{R}}(z)] = [i, y] \neq 0. \end{aligned}$$

This contradiction shows that there is no automorphism of L which maps $\mathcal{S}$ onto $\mathcal{S}_1$. □

Let $K = \mathcal{S} \dot{+} \mathcal{R}$ be a Levi decomposition of a Lie algebra K satisfying the following conditions:

$$\mathcal{R} = \mathcal{N} + \mathrm{Span}\{p\}, \quad \mathcal{N} = [\mathcal{R}, \mathcal{R}], \quad [\mathcal{S}, p] = \{0\}, \quad [\mathcal{S}, x] \neq \{0\},$$

for any $0 \neq x \in \mathcal{N}$.

We set

$$\mathcal{I} = \mathcal{S}, \qquad \theta = \mathrm{id}|_{\mathcal{S}} : \mathcal{S} \to \mathcal{I}.$$

On the space $L = \mathcal{S} \dot{+} \mathcal{R} \dot{+} \mathcal{I}$, we define products as follows

$$[L, \mathcal{I}] = [\mathcal{I}, \mathcal{N}] = 0, \quad [\theta(x), y] = \theta([x, y])$$

for $x, y \in \mathcal{S}$, $[i, p] = i$, for $i \in \mathcal{I}$.

A straightforward verification shows that L is a Leibniz algebra. Note that $\mathcal{S}$ and $\mathcal{I}$ are isomorphic $\mathcal{S}$-modules via the isomorphism θ. This phenomenon is not accidental.

Proposition 2.7. *Leibniz algebra L constructed above admits two Levi subalgebras which are non conjugate by any automorphism of L.*

Proof. Consider the Levi subalgebra $\mathcal{S}_\theta = \{x + \theta(x) : x \in \mathcal{S}\}$. Suppose that there is $\varphi = \varphi_{\mathcal{S},\mathcal{S}} + \varphi_{\mathcal{S},\mathcal{R}+\mathcal{I}} + \varphi_{\mathcal{R}+\mathcal{I},\mathcal{R}+\mathcal{I}} \in \mathrm{Aut}(L)$ such that $\varphi(\mathcal{S}) = \mathcal{S}_\theta$. Similarly to that of the proof of Proposition 2.6 we obtain $\varphi_{\mathcal{S},\mathcal{R}+\mathcal{I}} = \varphi_{\mathcal{S},\mathcal{I}} \neq 0$. Taking into account $\varphi(\mathcal{N}) = \mathcal{N}$ we get $\varphi(p) = b + c$, where $b = \alpha p \neq 0$ and $c \in \mathcal{N}$. For an element $0 \neq x \in S$ we have

$$\begin{aligned} 0 = \varphi(0) &= \varphi([x, p]) = [\varphi(x), \varphi(p)] \\ &= [\varphi_{\mathcal{S},\mathcal{S}}(x), \varphi_{\mathcal{R}+\mathcal{I},\mathcal{R}+\mathcal{I}}(p)] + [\varphi_{\mathcal{S},\mathcal{I}}(x), \varphi_{\mathcal{R}+\mathcal{I},\mathcal{R}+\mathcal{I}}(p)] \\ &= [\varphi_{\mathcal{S},\mathcal{S}}(x), b + c] + [\varphi_{\mathcal{S},\mathcal{I}}(x), b + c] = [\varphi_{\mathcal{S},\mathcal{S}}(x), c] + [\varphi_{\mathcal{S},\mathcal{I}}(x), b] \\ &= [\varphi_{\mathcal{S},\mathcal{S}}(x), c] + \alpha\varphi_{\mathcal{S},\mathcal{I}}(x). \end{aligned}$$

But $[\varphi_{\mathcal{S},\mathcal{S}}(x), c] + \alpha\varphi_{\mathcal{S},\mathcal{I}}(x) \neq 0$, because $[\varphi_{\mathcal{S},\mathcal{S}}(x), c] \in \mathcal{N}$ and $0 \neq \varphi_{\mathcal{S},\mathcal{I}}(x) \in \mathcal{I}$.

Thus, we get a contradiction with the existence of an automorphism of L mapping $\mathcal{S}$ onto $\mathcal{S}_\theta$. □

Example 2.4. *Let us consider the following Leibniz algebra* $L = \mathrm{Span}\{e_1, e_2, e_3, y_4, y_5, y_6, x_7, x_8, x_9\}$ *with the multiplication table:*

$$\begin{aligned} [e_1, e_2] &= -[e_2, e_1] = 2e_2, [e_1, e_3] = -[e_3, e_1] = -2e_3, \\ [e_2, e_3] &= -[e_3, e_2] = e_1, \\ [e_1, y_4] &= -[y_4, e_1] = y_4, [e_1, y_5] = -[y_5, e_1] = -y_5, \\ [e_2, y_5] &= -[y_5, e_2] = y_4, [e_3, y_4] = -[y_4, e_3] = y_5, \\ [y_4, y_6] &= -[y_6, y_4] = y_4, [y_5, y_6] = -[y_6, y_5] = y_5. \end{aligned}$$

$$\begin{array}{lll} [x_7, e_1] = -2x_7, & [x_7, e_3] = x_8, & [x_8, e_2] = 2x_7, \\ [x_8, e_3] = -2x_9, & [x_9, e_1] = 2x_9, & \\ [x_9, e_2] = -x_8, & [x_7, y_6] = x_7, & \\ [x_8, y_6] = x_8, & [x_9, y_6] = x_9. & \end{array}$$

Then $L = \mathcal{S} \dot{+} (\mathcal{R} \dot{+} \mathcal{I})$, *where* $\mathcal{S} = \mathrm{Span}\{e_1, e_2, e_3\} \cong \mathfrak{sl}_2$, $\mathcal{R} = \mathrm{Span}\{y_4, y_5, y_6\}$ *and* $\mathcal{I} = \mathrm{Span}\{x_7, x_8, x_9\}$. *Note that* $\mathcal{N} = \mathrm{Span}\{x_7, x_8, x_9, y_4, y_5\}$ *and* $[\mathcal{S}, y_6] = \{0\}$. *The Leibniz algebra L satisfies the conditions of Proposition* 2.7.

2.3.2 On conjugacy of Levi subalgebras of complex Leibniz algebras

In this subsection we shall discuss conditions for conjugacy of Levi subalgebras of Leibniz algebras over the field $\mathbb{C}$. First we prove the following auxiliary lemma.

Lemma 2.1. *Let b be an element of $\mathcal{R}$ such that $[\mathcal{S}, b] \subseteq \mathcal{I}$. Then $R_b^n\big|_{\mathcal{S}} \in \mathrm{Hom}_{\mathcal{S}}(\mathcal{S}, \mathcal{I})$ for all $n \in \mathbb{N}$.*

Proof. Since $[y, b] \in \mathcal{I}$ and $[L, \mathcal{I}] = \{0\}$, for any $x, y \in \mathcal{S}$ we have

$$R_b([x,y]) = [[x,y],b] = [[x,b],y] + [x,[y,b]] = [[x,b],y] = [R_b(x),y].$$

Now assume that R_b^{n-1} is an $\mathcal{S}$-module homomorphism, $n \geq 2$. Then

$$R_b^n([x,y]) = R_b(R_b^{n-1}([x,y])) = R_b([R_b^{n-1}(x),y]) = [R_b^n(x),y].$$

This completes the proof. □

Proposition 2.8. *Let $L = \mathcal{S} \dotplus \mathcal{R}$ be a Levi decomposition of a complex Leibniz algebra L such that $[\mathcal{S}, \mathcal{E}] = \mathcal{I}$, where $\mathcal{E} = \{b \in \mathcal{N} : [\mathcal{S}, b] \subseteq \mathcal{I}\}$. Then $[\mathcal{I}, \mathcal{E}] = 0$.*

Proof. Let us first consider the case when $\mathcal{S}$ and $\mathcal{I}$ are simple $\mathcal{S}$-modules. Fix a non-zero element $b \in \mathcal{E}$ such that $[\mathcal{S}, b] \neq 0$. Taking into account that $\mathcal{S}$ and $\mathcal{I}$ are simple $\mathcal{S}$-modules, we get a non-zero module homomorphism $R_b|_{\mathcal{S}}$. Applying Schur's Lemma (see Lemma 1.3) we derive that the modules $\mathcal{S}$ and $\mathcal{I}$ are isomorphic and the isomorphism $R_b|_{\mathcal{S}}$ is unique up to a constant. Take a non-trivial $\tau \in \mathrm{Hom}_{\mathcal{S}}(\mathcal{S}, \mathcal{I})$. Then by Schur's Lemma we have $\tau = \alpha\, R_b|_{\mathcal{S}}, \alpha \in \mathbb{C}$. By Lemma 2.1, $R_b^2\big|_{\mathcal{S}}$ is also an $\mathcal{S}$-module isomorphism and again applying Schur's lemma we get $R_b^2\big|_{\mathcal{S}} = \lambda\, R_b|_{\mathcal{S}}$ for some $\lambda \in \mathbb{C}$. Since $b \in \mathcal{N}$, it follows that R_b is nilpotent, and therefore there exists $n \in \mathbb{N}$ such that $R_b^n = 0$. Then $0 = R_b^n\big|_{\mathcal{S}} = \lambda^{n-1}\, R_b|_{\mathcal{S}}$, which implies $\lambda = 0$, due to $R_b|_{\mathcal{S}} \neq 0$. Thus $R_b^2 = 0$. Taking an arbitrary $y \in \mathcal{I}$, we conclude that there exists an element $x \in \mathcal{S}$ such that $R_b(x) = [x, b] = y$. Moreover,

$$[y,b] = [[x,b],b] = R_b^2(x) = 0.$$

This means that $[\mathcal{I}, \mathcal{E}] = 0$.

Let us now consider the case when $\mathcal{S}$ is semisimple and $\mathcal{I}$ is a simple $\mathcal{S}$-module. Suppose that we have the decomposition $\mathcal{S}$ into the sum of simple Lie ideals: $\mathcal{S} = \bigoplus_{i=1}^{n} \mathcal{S}_i$. Since $\mathcal{I}$ is irreducible, it follows that there exists an index i such that $[\mathcal{S}_i, \mathcal{E}] = \mathcal{I}$. By the above case we obtain $[\mathcal{I}, \mathcal{E}] = 0$.

Finally, we consider the general case. Let $\mathcal{I} = \bigoplus_{j=1}^{m} \mathcal{I}_j$ be a decomposition into sum of irreducible $\mathcal{S}$-modules. Set $L_i = \mathcal{S} \dot{+} \mathcal{R}_i$, where $\mathcal{R}_i = \mathcal{R}/J_i$, $J_i = \bigoplus_{j \neq i} \mathcal{I}_j$ and $i = 1, \dots, m$. Then $[\mathcal{S}, \mathcal{E}_i] = \mathcal{I}_i$, for all $i = 1, \dots, m$, where $\mathcal{E}_i = \mathcal{E}/J_i$. By the above case we have $[\mathcal{I}_i, \mathcal{E}_i] = 0$, for $i = 1, \dots, m$. Thus

$$[\mathcal{I}_i, \mathcal{E}] = [\mathcal{I}_i, \mathcal{E}_i + J_i] = [\mathcal{I}_i, \mathcal{E}_i] = 0$$

and

$$[\mathcal{I}, \mathcal{E}] = \left[\bigoplus_{j=1}^{m} \mathcal{I}_j, \mathcal{E}\right] = \bigoplus_{j=1}^{m} \left[\mathcal{I}_j, \mathcal{E}\right] = 0.$$

□

Now we provide an example which satisfies the conditions of Proposition 2.8.

Example 2.5. *Let L be a Leibniz algebra given by the following table of multiplication on a basis* $\{e, f, h, x_0, x_1, x_2, y_1, y_2\}$ *(see [44]):*

$$\begin{array}{lll}
[e,h] = 2e, & [h,f] = 2f, & [e,f] = h, \\
[h,e] = -2e, & [f,h] = -2f, & [f,e] = -h, \\
[x_1,e] = -2x_0, & [x_2,e] = -2x_1, & [x_0,f] = x_1, \\
[x_1,f] = x_2, & [x_0,h] = 2x_0, & [x_2,h] = -2x_2, \\
[e,y_1] = 2x_0, & [f,y_1] = x_2, & [h,y_1] = 2x_1, \\
[y_1,y_2] = y_1, & [y_2,y_1] = -y_1, & \\
[x_0,y_2] = x_0, & [x_1,y_2] = x_1, & [x_2,y_2] = x_2.
\end{array}$$

It is easy to see that $L = \mathcal{S} \dot{+} \mathcal{R}$*, where* $\mathcal{S} = \mathrm{Span}\{e, f, h\}$, $\mathcal{R} = \mathrm{Span}\{x_0, x_1, x_2, y_1, y_2\}$, $\mathcal{I} = \mathrm{Span}\{x_0, x_1, x_2\}$ *and* $[\mathcal{S}, \mathcal{E}] = \mathcal{I}$*, with* $\mathcal{E} = \mathrm{Span}\{x_0, x_1, x_2, y_1\}$.

Let $L = \mathcal{S}\dot{+}\mathcal{R}$ be a Leibniz algebra and

$$\mathcal{S} = G \oplus Q \qquad \text{and} \qquad \mathcal{I} = J \oplus K \tag{2.3}$$

be the decompositions from Section 2.3.1.

Let θ be an $\mathcal{S}$-module homomorphism such that $\theta(G) = J$. Suppose that

$$G = \bigoplus_{i=1}^{n} G_i$$

is a decomposition into the sum of simple Lie ideals.

Set

$$J_i = \theta(G_i) \text{ for all } i \in \{1, \ldots, n\}.$$

Since G_i is an ideal in G, it follows that J_i is a submodule of J. Take $z = \theta(x_i) = \theta(x_j) \in J_i \cap J_j$, where $i \neq j$. Let $y_k \in G_k$.

If $k \neq i$, then

$$[z, y_k] = [\theta(x_i), y_k] = \theta([x_i, y_k]) = 0.$$

If $k \neq j$, then

$$[z, y_k] = [\theta(x_j), y_k] = \theta([x_j, y_k]) = 0.$$

Thus $\left[J_i \cap J_j, G\right] = 0$, and therefore the intersections $J_i \cap J_j$ are trivial submodules. Since θ maps G onto J, it follows that

$$J = \bigoplus_{i=1}^{n} J_i,$$

and each submodule is decomposed as

$$J_i = \bigoplus_{j=1}^{n_i} J_j^{(i)},$$

where $J_j^{(i)}$ is a simple G_i-module. Since $J_i = \theta(G_i)$, it follows that $G_i \cong J_j^{(i)}$ for all $j \in \{1, \ldots, n_i\}$ and $i \in \{1, \ldots, n\}$.

Proposition 2.9. *Let $L = \mathcal{S}\dot{+}\mathcal{R}$ be the Levi decomposition of a complex Leibniz algebra L such that $[\mathcal{S}, \mathcal{E}] = J$. Then any two Levi subalgebras of L are conjugate via an inner automorphism.*

Proof. Let $\tau \in \mathrm{Hom}_{\mathcal{S}}(\mathcal{S}, J)$. By Lemma 1.3 we have $\tau = \sum\limits_{i=1}^{n}\sum\limits_{j=1}^{n_i} \tau_{ij}$, where $\tau_{ij}\colon G_i \to J_j^{(i)}$ are module homomorphisms. For i, j take an element $b_{i,j} \in \mathcal{N}$ such that $0 \neq [G_i, b_{i,j}] \subseteq J_j^{(i)}$. Then there exists $\lambda_{ij} \in \mathbb{C}$ such that $\tau_{i,j} = \lambda_{i,j} R_{b_{i,j}}$. By Proposition 2.8 we get $R^2_{b_{i,j}}\big|_{\mathcal{S}} = 0$.

Set $b = \sum\limits_{i=1}^{n}\sum\limits_{j=1}^{n_i} \lambda_{i,j} b_{ij}$. Then

$$\exp(R_b)\big|_{\mathcal{S}} = \prod_{i=1}^{n}\prod_{j=1}^{n_i} \exp(R_{\lambda_{i,j}b_{ij}})\big|_{\mathcal{S}} = \mathrm{id}|_{\mathcal{S}} + \sum_{i=1}^{n}\sum_{j=1}^{n_i} R_{\lambda_{i,j}b_{ij}}\big|_{\mathcal{S}} = \mathrm{id}|_{\mathcal{S}} + \tau.$$

Finally, to get the proof we need just to apply Corollary 2.5. □

Now we give the criterion on conjugacy of the Levi subalgebras.

Theorem 2.5. *Let L be a complex Leibniz and $L = \mathcal{S} \dot{+} \mathcal{R}$ be its Levi decomposition. Then the following are equivalent:*

i) *A Levi subalgebra $\mathcal{S}$ is unique up to conjugacy via an inner automorphism;*

ii) $[\mathcal{S}, \mathcal{E}] = J$.

Proof. The implication (ii) ⇒ (i) follows from Proposition 2.9.

Now we prove the implication (i) ⇒ (ii). Let $\mathcal{S} = G \oplus Q$ be a decomposition of the form (2.3). We first consider the case where G is a simple Lie algebra. Let $J = \bigoplus\limits_{i=1}^{n} J_i$ be a decomposition into simple modules and $G \cong J_i$ for all $i = 1, \dots, n$. Take an arbitrary module isomorphism $\tau_j\colon G \to J_j$. Since any two subalgebras of L are conjugate, there exists an element $b_j \in \mathcal{N}$ such that

$$\exp(R_{b_j})\big|_G = \mathrm{id}|_G + \tau_j.$$

Similarly as in the proof of Proposition 2.8, we obtain $R^2_{b_j}\big|_G = 0$, i.e., $\tau_j = R_{b_j}\big|_G$, and therefore $[G, b_j] = J_j$. Thus

$$J \supseteq [G, \mathcal{E}] \supseteq [G, b_1] + \cdots + [G, b_n] = J.$$

Hence $[G, \mathcal{E}] = J$.

Now let us consider the general case. Let G be a semisimple Lie algebra and $J = \bigoplus_{j=1}^{m} J_j$ be the decomposition into the sum of irreducible G-modules. Set $L_i = \mathcal{S} \dotplus \mathcal{R}_i$, where $\mathcal{R}_i = \mathcal{R}/\mathcal{I}_i$ and $\mathcal{I}_i = \bigoplus_{j \neq i} J_j$ for $i = 1, \ldots, m$. Then

$$\mathrm{Hom}_{\mathcal{S}}(\mathcal{S}, J_i) \equiv \mathrm{Hom}_{\mathcal{S}}(G_i, J_i). \tag{2.4}$$

Since a Levi subalgebra S of L is unique up to conjugacy via an inner automorphism and the quotient of inner automorphism is also inner, it follows that $\mathcal{S}$ is unique up to conjugacy via an inner automorphism as Levi subalgebra of L_i, for all $i = 1, \ldots, m$. Taking into account (2.4) and the fact that G_i is a simple algebra, by the above case we obtain $[G, \mathcal{E}_i] = J_i$, for all $i = 1, \ldots, m$, where $\mathcal{E}_i = \mathcal{E}/J_i$. Thus

$$[\mathcal{S}_i, \mathcal{E}] = [\mathcal{S}_i, \mathcal{E}_i + \mathcal{I}_i] = J_i.$$

Further

$$[\mathcal{S}, \mathcal{E}] = \left[\bigoplus_{j=1}^{m} \mathcal{S}_j, \mathcal{E}\right] = \bigoplus_{j=1}^{m} \left[\mathcal{S}_j, \mathcal{E}\right] = \bigoplus_{j=1}^{m} J_j = J.$$

□

Remark. Note that in Proposition 2.7 and Example 2.4, we have $\mathcal{E} = 0$. Therefore, $0 = [\mathcal{S}, \mathcal{E}] \neq J$.

Corollary 2.6. *Let $L = \mathcal{S} \dotplus \mathcal{R}$ be a Levi decomposition of a complex Leibniz algebra L such that $[\mathcal{S}, \mathcal{N}] = J$. Then any two Levi subalgebras of L are conjugate via an inner automorphism.*

Recall that J is the maximal submodule of $\mathcal{I}$ such that $\mathrm{Hom}_{\mathcal{S}}(\mathcal{S}, \mathcal{I}) \equiv \mathrm{Hom}_{\mathcal{S}}(\mathcal{S}, J)$.

In Figure 2.1 we describe the cases that may occur:

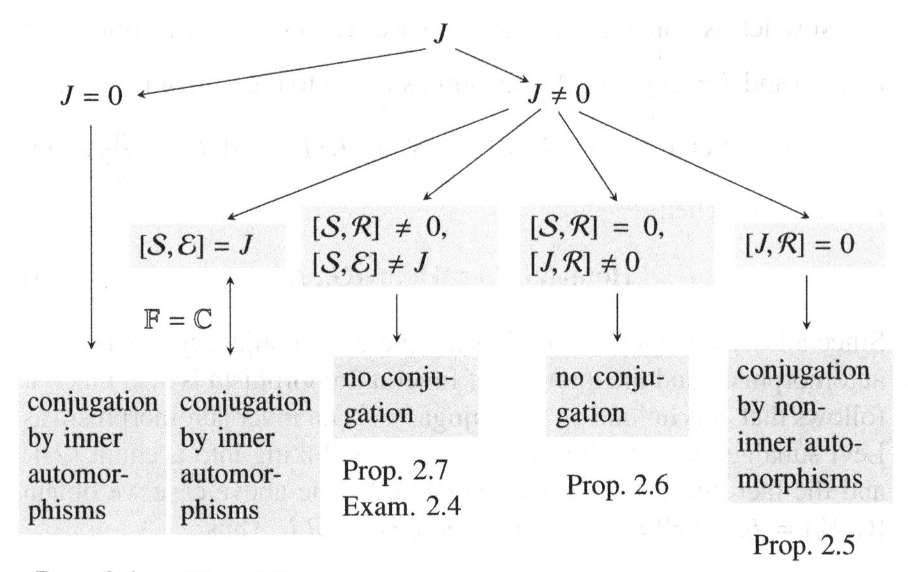

Figure 2.1 Conjugacy of Levi's subalgebras

2.4 SEMISIMPLE LEIBNIZ ALGEBRAS

Although Theorem 2.3 reduces the study of finite-dimensional Leibniz algebras to semisimple Lie and solvable Leibniz algebras the study of "simple" and "semisimple" Leibniz algebras are of interest in their own right. In fact, in a non-Lie Leibniz algebra the ideal $\mathcal{I}$ is non-trivial. That is why usual notion of simplicity is not applicable to the case of Leibniz algebras. In [63] the authors have proposed the notion of simple Leibniz algebra as follows.

Definition 2.2. *A Leibniz algebra L is said to be simple if it contains only the following ideals:* $\{0\}$, $\mathcal{I}$, L *and* $L^2 \neq \mathcal{I}$.

Note that the definition of simplicity of Leibniz algebra above agrees with that of simplicity of Lie algebra, since in this case $\mathcal{I} = 0$. A simple application of Levi's decomposition for finite-dimensional Leibniz algebras implies that the above definition can be rephrased in terms of quotient Lie algebra as follows: a Leibniz algebra L is simple if and only if its Lie algebra L_{Lie} is simple and the ideal $\mathcal{I}$ is an irreducible right L_{Lie}-module.

Example 2.6.

1. *Let* $L := sl_2(\mathbb{C}) \times L(1)$, *where* $sl_2(\mathbb{C})$ *is the Lie algebra of traceless complex* 2×2 *matrices and* $L(1)$ *is the two-dimensional left* $sl_2(\mathbb{C})$*-module. Then* L *with the multiplication defined by* $(X, a)(Y, b) := ([X, Y], X \cdot b)$ *for any* $X, Y \in sl_2(\mathbb{C})$ *and any* $a, b \in L(1)$ *is a simple left Leibniz algebra (see [68]).*

2. *The example above is generalized as follows. Let* G *be a (simple) Lie algebra and* V *a right (irreducible)* G*-module. Endow the vector space* $L = G \oplus V$ *with the bracket product as follows:*

$$[(g_1, v_1), (g_2, v_2)] := ([g_1, g_2], v_1 \cdot g_2),$$

where $v \cdot g$ *(sometimes denoted as* $[v, g]$*) is an action of an element* g *of* G *on* $v \in V$*. Then straightforward verification of the Leibniz identity shows that* $L = G \dot{+} V$ *is a (simple) Leibniz algebra.*

Note that from Levi's theorem it is immediate that any simple Leibniz algebra can be obtained in the manner of part 2 of Example 2.6.

The following definition and results on Lie algebras from [105], given there as Proposition 19.1 and Corollary 21.2, will be applied in the description of semisimple Leibniz algebras.

Definition 2.3. *A Lie algebra* G *is said to be reductive, if* $\mathcal{R} = \mathrm{Cent}(G)$, *where* $\mathcal{R}$ *is the solvable radical of* G.

Lemma 2.2.

i) *Let* G *be a reductive Lie algebra. Then* $G = [G, G] \oplus \mathrm{Cent}(G)$, *and* $[G, G]$ *is either semisimple or* 0.

ii) *Let* V *be a finite-dimensional space and* $G \subset \mathfrak{gl}(V)$ *be a non zero Lie algebra acting irreducibly on* V*. Then* G *is reductive with* $\dim \mathrm{Cent}(G) \leq 1$. *If, in addition,* $G \subset \mathfrak{sl}(V)$, *then* G *is semisimple.*

We shall also need the following result from [180, Page 143, Theorem 6].

Theorem 2.6. *Let* $G = G_1 \oplus G_2$ *be a semisimple Lie algebra. Let a Cartan subalgebra* H *of* G *be respectively decomposed as* $H = H_1 \oplus H_2$, *where* H_i *is a Cartan subalgebra of* G_i, $(i = 1, 2)$*. Suppose that the simple root system* Π *of* G *with respect to* H *is decomposed into* $\Pi = \Pi_1 \cup \Pi_2$, *where* $\Pi_1 = \{\alpha_1, \ldots, \alpha_n\}$ *and* $\Pi_2 = \{\beta_1, \ldots, \beta_m\}$ *are the fundamental root system of* G_i *with respect to* H_i, $i = 1, 2$.

a) *Suppose that* ϱ_i *is an irreducible representation of* G_i *with the highest weight* ω_i *and the representation space* V_i, $i = 1, 2$. *If*

$$\varrho(x)(v_1 \otimes v_2) = \varrho_1(x_1)v_1 \otimes v_2 + v_1 \otimes \varrho_2(x_2)v_2,$$

$$x = x_1 + x_2, \ \ x_i \in G_i, \ \ i = 1, 2,$$

then ϱ *is an irreducible representation of* G *with the highest weight* $\omega_1 + \omega_2$, *where*

$$(\omega_1 + \omega_2)(h) = \omega_1(h_1) + \omega_2(h_2), \ \ h = h_1 + h_2, \ \ h_i \in H_i, \ i = 1, 2.$$

b) *Conversely, every irreducible representation of G is obtained as above.*

The concept of semisimple Leibniz algebra is defined as follows.

Definition 2.4. *A Leibniz algebra L is called semisimple if its radical coincides with* $\mathcal{I}$.

In other words, a Leibniz algebra L is semisimple if its liezation L_{Lie} is a semisimple Lie algebra. Note that the definition of semisimple Leibniz algebra agrees with that of Lie algebra (since in that case $\mathcal{I} = 0$). Clearly, a simple Leibniz algebra is semisimple.

Application of Theorem 2.3 gives the following

Proposition 2.10. *Let L be a finite-dimensional semisimple Leibniz algebra. Then* $L = \mathcal{S} \dot{+} \mathcal{I}$, *where* $\mathcal{S}$ *is a semisimple Lie subalgebra of L. Moreover* $[\mathcal{I}, \mathcal{S}] = \mathcal{I}$.

Now the structure of a simple Leibniz algebra is clearly seen. By Example 2.6 a simple Leibniz algebra is a semidirect sum of a simple Lie algebra $\mathcal{S}$ and an irreducible right module over $\mathcal{S}$.

Note that a semisimple Leibniz algebra need not necessarily be presented as a direct sum of simple Leibniz algebras as justified by the following example.

Example 2.7. *Let L be a* 10*-dimensional Leibniz algebra given on a basis* $\{e_1, h_1, f_1, e_2, h_2, f_2, x_1, x_2, x_3, x_4\}$ *by the multiplication table as follows*

$$[\mathfrak{sl}_2^i, \mathfrak{sl}_2^i]: \begin{array}{lll} [e_i, h_i] = 2e_i, & [f_i, h_i] = -2f_i, & [e_i, f_i] = h_i, \\ [h_i, e_i] = -2e_i & [h_i, f_i] = 2f_i, & [f_i, e_i] = -h_i, \ i = 1, 2, \end{array}$$

$$[\mathcal{I}, \mathfrak{sl}_2^1]: \begin{array}{llll} [x_1, f_1] = x_2, & [x_1, h_1] = x_1, & [x_2, e_1] = -x_1, & [x_2, h_1] = -x_2, \\ [x_3, f_1] = x_4, & [x_3, h_1] = x_3, & [x_4, e_1] = -x_3, & [x_4, h_1] = -x_4, \end{array}$$

$$[\mathcal{I}, \mathfrak{sl}_2^2]: \begin{array}{llll} [x_1, f_2] = x_3, & [x_1, h_2] = x_1, & [x_3, e_2] = -x_1, & [x_3, h_2] = -x_3, \\ [x_2, f_2] = x_4, & [x_2, h_2] = x_2, & [x_4, e_2] = -x_2, & [x_4, h_2] = -x_4, \end{array}$$

(omitted products of the basis vectors are equal to zero).

It is easy to verify that L is a semisimple Leibniz algebra and $\mathcal{I} = \mathrm{Span}\{x_1, x_2, x_3, x_4\}$. *From the table of multiplications we have* $[\mathcal{I}, \mathfrak{sl}_2^1] = [\mathcal{I}, \mathfrak{sl}_2^2] = \mathcal{I}$. *Moreover,* $\mathcal{I}$ *splits over* $\mathfrak{sl}_2^1$ *as* $\mathcal{I} = \mathrm{Span}\{x_1, x_2\} \oplus \mathrm{Span}\{x_3, x_4\}$ *and over* $\mathfrak{sl}_2^2$ *as* $\mathcal{I} = \mathrm{Span}\{x_1, x_3\} \oplus \mathrm{Span}\{x_2, x_4\}$. *In fact,* $\mathcal{I}$ *is a simple ideal (an irreducible* $(\mathfrak{sl}_2^1 \oplus \mathfrak{sl}_2^2)$*-module). Therefore,* $L = (\mathfrak{sl}_2^1 \oplus \mathfrak{sl}_2^2) \dotplus \mathcal{I}$ *cannot be a direct sum of two non zero ideals.*

We give a generalization of Theorem 2.6, which is of independent interest.

Theorem 2.7. *Let* $G = G_1 \oplus G_2$ *be a direct sum of complex Lie algebras and suppose that* V *is a finite-dimensional irreducible* G*-module with* $dim_{\mathbb{C}} V > 1$. *Then there are finite-dimensional irreducible* G_i*-modules* V_i, $i = 1, 2$ *such that* $V \cong V_1 \otimes V_2$.

Proof. Consider $\mathrm{Ann}(V) = \{x \in G : V \cdot x = 0\}$ which is an ideal of G. Then V is a faithful irreducible module over $\overline{G} = G/\mathrm{Ann}(V)$, in particular, for an element $x \in \overline{G}$ from $v \cdot x = 0$ for all $v \in V$ we get $x = 0$. Thus we have an injective linear map $\rho : \overline{G} \to \mathfrak{gl}(V)$ given by $\rho(\bar{g})(v) = v \cdot \bar{g}$ for any $v \in V$ and $\bar{g} \in \overline{G}$ (a representation of $\overline{G}$ on V. Therefore, $\overline{G}$ is also finite-dimensional.

Applying part b) of Lemma 2.2 to $\overline{G}$ we conclude that $\overline{G}$ is reductive, $\dim \mathrm{Cent}(\overline{G}) \leq 1$, and $\overline{G} = [\overline{G}, \overline{G}] \oplus \mathrm{Cent}(\overline{G})$. Denote the natural projections by $\rho_1 : \overline{G} \to [\overline{G}, \overline{G}]$ and $\rho_2 : \overline{G} \to \mathrm{Cent}(\overline{G})$, and define $\pi = \rho_1 \circ \rho$.

Assume that $\mathrm{Cent}(\overline{G}) = \mathbb{C}z \neq 0$. Since $[[v, x], z] = [[v, z], x]$ for all $x \in \overline{G}$ and $v \in V$, it follows that the mapping

$$v \in V \hookrightarrow [v, z] \in V$$

is a $\overline{G}$-module homomorphism of V. Taking into account that V is an irreducible $\overline{G}$-module, by Schur's Lemma, there exists $\lambda \in \mathbb{C}$ such that $[v, z] = \lambda v$ for all $v \in V$. Note that V is a simple module over

$\pi(G) = [\overline{G}, \overline{G}]$. If $\pi(G_1) \cap \pi(G_2) \neq \{0\}$, it has to be semisimple. We note that

$$\pi(G_1) \cap \pi(G_2) = [\pi(G_1) \cap \pi(G_2), \pi(G_1) \cap \pi(G_2)\}] \subset [\pi(G_1), \pi(G_2)\}] = 0,$$

which is a contradiction. Thus $\pi(G_1) \cap \pi(G_2) = \overline{0}$, and $\pi(G) = \pi(G_1) \oplus \pi(G_2)$. Since V is also an irreducible module over $\pi(G_1) \oplus \pi(G_2)$, both $\pi(G_1)$ and $\pi(G_2)$ are semisimple. By Theorem 2.6, we deduce that there are finite-dimensional irreducible $\pi(G_i)$-modules V_i so that $V \cong V_1 \otimes V_2$. We can easily extend $\pi(G_i)$-module V_i into a G_i-module. This completes the proof. □

We say that a semisimple Leibniz algebra $L = \mathcal{S} \dot{+} \mathcal{I}$ is *decomposable*, if $L = (\mathcal{S}_1 \dot{+} \mathcal{I}_1) \oplus (\mathcal{S}_2 \dot{+} \mathcal{I}_2)$, where $\mathcal{S}_1 \dot{+} \mathcal{I}_1$ and $\mathcal{S}_2 \dot{+} \mathcal{I}_2$ are non-trivial semisimple Leibniz algebras. Otherwise, we say that L is *indecomposable.*

Lemma 2.3. *Any semisimple Leibniz algebra has the form:*

$$L = \bigoplus_{i=1}^{n} (\mathcal{S}_i \dot{+} \mathcal{I}_i),$$

where $\mathcal{S}_i \dot{+} \mathcal{I}_i$ is an indecomposable semisimple Leibniz algebra for all $i \in \{1, \dots, n\}$.

Proof. Let $L = \mathcal{S} \dot{+} \mathcal{I}$ be a semisimple Leibniz algebra and $\mathcal{S} = \bigoplus_{i=1}^{n} \mathcal{S}_i$ be a decomposition of simple Lie ideals $\mathcal{S}_i$. Note that $[\mathcal{I}, \mathcal{S}] = \mathcal{I}$. We proceed the proof by induction on n.

Let $n = 1$, i.e., $\mathcal{S}$ is a simple Lie algebra. Then $L = \mathcal{S} \dot{+} \mathcal{I}$ is an indecomposable semisimple Leibniz algebra.

Suppose that the assertion of the theorem is true for all numbers less than n and $\mathcal{S} = \bigoplus_{i=1}^{n} \mathcal{S}_i$. Consider a partition of the set $\{1, 2, \dots, n\} = A \cup B$ into union of disjoint subsets A and B. Set

$$\mathcal{I}_A = \left[\mathcal{I}, \bigoplus_{i \in A} \mathcal{S}_i\right] \text{ and } \mathcal{I}_B = \left[\mathcal{I}, \bigoplus_{j \in B} \mathcal{S}_j\right].$$

Case 1. Let $\mathcal{I}_A \cap \mathcal{I}_B \neq \{0\}$ for any non trivial partition $A \cup B$ of the set $\{1, 2, \dots, n\}$. In this case L is indecomposable.

Case 2. Let $\mathcal{I}_A \cap \mathcal{I}_B = \{0\}$ for some partition $A \cup B$ of the set $\{1, 2, \ldots, n\}$. Then L is decomposable and

$$L = \left(\bigoplus_{i \in A} \mathcal{S}_i \dot{+} \mathcal{I}_A \right) \oplus \left(\bigoplus_{j \in B} \mathcal{S}_j \dot{+} \mathcal{I}_B \right).$$

By the hypothesis of the induction, the algebras $\bigoplus_{i \in A} \mathcal{S}_i \dot{+} \mathcal{I}_A$ and $\bigoplus_{j \in B} \mathcal{S}_j \dot{+} \mathcal{I}_B$ can be represented as a direct sum of indecomposable algebras. □

Further we need the following auxiliary result.

Lemma 2.4. *Let $\mathcal{S}$ be a semisimple Lie algebra and $\mathcal{I}$ be an irreducible $\mathcal{S}$-module. Then $[\mathcal{I}, \mathcal{S}] = 0$ if and only if* $\dim \mathcal{I} = 1$.

Proof. Let ρ be the representation of $\mathcal{S}$ on $\mathcal{I}$ given by $\rho(g)(v) = [v, g]$ for any $v \in \mathcal{I}$ and $g \in \mathcal{S}$. Then we have the Lie algebra homomorphism $\rho : \mathcal{S} \to \mathfrak{gl}(\mathcal{I})$. So $\rho(\mathcal{S})$ is a subalgebra of $\mathfrak{gl}(\mathcal{I})$. Then the result follows from Lemma 2.2. □

Remark. From Lemma 2.4 we conclude that for an indecomposable semisimple Leibniz algebra $L = \mathcal{S} \dot{+} \mathcal{I}$ the inequality $\dim \mathcal{I} \geq 2$ holds true if $\mathcal{I} \neq 0$.

From Lemma 2.3, we need only to determine the structure of indecomposable semisimple Leibniz algebras. Suppose that L is an indecomposable semisimple Leibniz algebra with a given Levi decomposition $L = \mathcal{S} \dot{+} \mathcal{I}$ (see Theorem 2.3). We may assume that $\mathcal{S} = \oplus_{i=1}^m \mathcal{S}_i$ and $\mathcal{I} = \oplus_{i=1}^n \mathcal{I}_i$, where each $\mathcal{S}_i$ is a simple Lie algebra and each $\mathcal{I}_i$ is an irreducible $\mathcal{S}$-module. We say that $\mathcal{S}_i$ and $\mathcal{S}_j$ are *adjacent* if there exists $\mathcal{I}_k$ such that $[\mathcal{I}_k, \mathcal{S}_i] = [\mathcal{I}_k, \mathcal{S}_j] = \mathcal{I}_k$. We say that $\mathcal{S}_i$ and $\mathcal{S}_j$ are *connected* if there exist $\mathcal{S}_{k_1} = \mathcal{S}_i, \mathcal{S}_{k_2}, \cdots, \mathcal{S}_{k_r} = \mathcal{S}_j$ such that $\mathcal{S}_{k_l}$ and $\mathcal{S}_{k_{l+1}}$ are adjacent.

Now we can obtain our main result on indecomposable semisimple Leibniz algebras.

Theorem 2.8. *Let $L = \mathcal{S} \dot{+} \mathcal{I}$ be an indecomposable semisimple Leibniz algebra with $\mathcal{I} \neq \{0\}$. Then*

(a) $\mathcal{S} = \oplus_{i=1}^m \mathcal{S}_i$ *where each $\mathcal{S}_i$ is a simple Lie algebra;*

(b) $\mathcal{I} = \oplus_{i=1}^{n} \mathcal{I}_i$ *where each* $\mathcal{I}_i$ *is an irreducible* $\mathcal{S}$*-module with* $[\mathcal{I}_i, \mathcal{S}] = \mathcal{I}_i$;

(c) *For any* $1 \leq i \leq m$ *and* $1 \leq j \leq n$ *there is an irreducible* $\mathcal{S}_i$*-module* $\mathcal{I}_{ij}$ *such that* $\mathcal{I}_i = \otimes_{j=1}^{m} \mathcal{I}_{ij}$;

(d) *Any two* $\mathcal{S}_i$ *and* $\mathcal{S}_j$ *are connected.*

Proof. Parts (a) and (b) are clear, (c) follows from Theorem 2.6.

(d) The proof is carried out by induction on n as follows. Let $n = 1$. Then $\mathcal{I}$ is an irreducible $\mathcal{S}$-module. By Theorem 2.6 for each $j \in \{1, \dots, m\}$ there is an irreducible $\mathcal{S}_j$-module J_j such that $\mathcal{I} = \otimes_{j=1}^{m} J_j$. Since $\mathcal{S} \dotplus \mathcal{I}$ is indecomposable, Lemma 2.4 implies that $[J_j, \mathcal{S}_j] = J_j$ for all j. Thus $\left[\mathcal{I}, \mathcal{S}_j\right] = \mathcal{I}$ for all j. This means that $\mathcal{S}_i$ and $\mathcal{S}_j$ are adjacent.

Suppose that the assertion (e) is true for all numbers less than n ($n > 1$). Consider the Leibniz algebra $\mathcal{S} \dotplus J$, where $J = \oplus_{i=1}^{n-1} \mathcal{I}_i$. Assume that $\mathcal{S} \dotplus J$ is indecomposable. By the hypothesis of the induction, $\mathcal{S}_i$ and $\mathcal{S}_j$ are already connected.

Now assume that $\mathcal{S} \dotplus J$ is decomposable. Let us consider the decomposition of $\mathcal{S} \dotplus J$ into a direct sum of indecomposable semisimple Leibniz algebras: $\bigoplus_{t=1}^{p} (\oplus_{i \in A_t} \mathcal{S}_i \dotplus \oplus_{s \in B_t} \mathcal{I}_s)$, where $A_1, \dots, A_p$ and $B_1, \dots, B_p$ are partitions of $\{1, \dots, m\}$ and $\{1, \dots, n-1\}$, respectively. Since $L = \mathcal{S} \dotplus \mathcal{I}$ is indecomposable, for every $s \in \{1, \dots, p\}$ there exists an index $i_s \in A_s$ such that $[\mathcal{I}_n, \mathcal{S}_{i_s}] = \mathcal{I}_n$. Thus $\mathcal{S}_{i_1}, \dots, \mathcal{S}_{i_p}$ are mutually adjacent. On the other hand, since $\oplus_{i \in A_t} \mathcal{S}_i \dotplus \oplus_{s \in B_t} \mathcal{I}_s$ is indecomposable, by the hypothesis of the induction, for each $i, j \in A_s$ ($1 \leq s \leq p$) the algebras $\mathcal{S}_i$ and $\mathcal{S}_j$ are connected. Thus $\mathcal{S}_i$ and $\mathcal{S}_j$ are connected for every $i, j \in \{1, \dots, m\}$. □

Here we present an example that generalizes Example 2.7 (see [45, Theorem 4.2]) of a semisimple Leibniz algebra which can not be decomposed into the direct sum of simple ideals.

Example 2.8. *Let* $m, n \in \mathbb{N}$. *Let* $L = \left(\mathfrak{sl}_2^1 \bigoplus \mathfrak{sl}_2^2\right) \dotplus \mathcal{I}$ *be a semisimple Leibniz algebra with a basis* $\{e_1, h_1, f_1, e_2, h_2, f_2, x_i \otimes y_j \, : \, i = 0, 1, \cdots, m; j = 0, 1, \cdots, n\}$ *and the table of multiplication given as* *follows*

$$\begin{aligned}
[e_i, h_i] &= -[h_i, e_i] = 2e_i,\\
[h_i, f_i] &= -[f_i, h_i] = 2f_i,\\
[e_i, f_i] &= -[f_i, e_i] = h_i,\\
[x_i \otimes y_j, h_1] &= (m-2i)x_i \otimes y_j,\\
[x_i \otimes y_j, h_2] &= (n-2j)x_i \otimes y_j,\\
[x_i \otimes y_j, e_1] &= ix_{i+1} \otimes y_j,\\
[x_i \otimes y_j, e_2] &= jx_i \otimes y_{j+1},\\
[x_i \otimes y_j, f_1] &= (m-i)x_{i-1} \otimes y_j,\\
[x_i \otimes y_j, f_2] &= (n-j)x_i \otimes y_{j-1}.
\end{aligned}$$

Consider $V_1 = \mathrm{span}\{x_0, x_1, \ldots, x_m\}$ *and* $V_2 = \mathrm{span}\{y_0, y_1, \ldots, y_n\}$. *Then* V_1 *and* V_2 *are irreducible* $(m+1)$*-dimensional and* $(n+1)$*-dimensional modules over* $\mathfrak{sl}_2^1$ *and* $\mathfrak{sl}_2^2$, *respectively. It is obvious that*

$$L = \left(\mathfrak{sl}_2^1 \bigoplus \mathfrak{sl}_2^2\right) \dot{+} (V_1 \otimes V_2).$$

2.5 ON CARTAN SUBALGEBRAS OF FINITE-DIMENSIONAL LEIBNIZ ALGEBRAS

The subsets

$$\mathrm{Nor}_l(H) = \{x \in L \mid [x, H] \subseteq H\}$$

and

$$\mathrm{Nor}_r(H) = \{x \in L \mid [H, x] \subseteq H\}$$

of L are said to be *left* and *right normalizers*, respectively, of the subalgebra H in L.

Definition 2.5. *A subalgebra* H *of a Leibniz algebra* L *is called a Cartan subalgebra if the following two conditions are satisfied:*

a) *H is nilpotent;*

b) *H coincides with the left normalizer of H in L.*

It is obvious that if the antisymmetricity property is supposed the sets $\mathrm{Nor}_l(H)$ and $\mathrm{Nor}_r(H)$ coincide. For a Cartan subalgebra H of a Leibniz algebra L we have the inclusion $\mathrm{Nor}_l(H) \subseteq \mathrm{Nor}_r(H)$. It is easy to see that if H contains the ideal generated by squares of elements of

L, then we have $\mathrm{Nor}_l(H) = \mathrm{Nor}_r(H)$. In general, for non Lie Leibniz algebras the equality does not hold, which can be confirmed by the following example.

Example 2.9. *Let L be three-dimensional Leibniz algebra defined by the following table of multiplications:*

$$[x, z] = x, \ [z, y] = y, \ [y, z] = -y, \ [z, z] = x,$$

where $\{x, y, z\}$ is a basis of L and the omitted products are supposed to be zero. Then $H = \mathrm{Span}\{x - z\} = \mathrm{Nor}_l(H)$ is the Cartan subalgebra of L and $\mathrm{Nor}_r(H) = \mathrm{Span}\{x, z\}$, i.e., $\mathrm{Nor}_l(H) \neq \mathrm{Nor}_r(H)$.

Let H be a nilpotent subalgebra of a Leibniz algebra L and $L = L_0 \oplus L_1$ be the Fitting decomposition of the algebra L with respect to the nilpotent Lie algebra $\Re(H)$ of transformations of the vector space L according to Theorem 1.8. Similar to the case of Lie algebras each Cartan subalgebra H of a Leibniz algebra has a close relation with the Fitting null component with respect to the Lie algebra $\Re(H)$.

Proposition 2.11. *A nilpotent subalgebra H of a Leibniz algebra L is a Cartan subalgebra if and only if H coincides with L_0 in the Fitting decomposition of L with respect to $\Re(H)$.*

Proof. Firstly, we note that $\mathrm{Nor}_l(H) \subseteq L_0$. Indeed, if $x \in \mathrm{Nor}_l(H)$ then $[x, h] \in H$ for any $h \in H$. Since the subalgebra H is nilpotent, there exists $k \in \mathbb{N}$ such that $[\ldots[[x, \underbrace{h], h], \ldots, h}_{k\ times}] = R_h^k(x) = 0$, this implies that $x \in L_0$, i.e., $\mathrm{Nor}_l(H) \subseteq L_0$. Since $H \subseteq \mathrm{Nor}_l(H)$ we have $H \subseteq L_0$. Suppose that the inclusion $H \subset L_0$ is strict, i.e., $H \neq L_0$. By Theorem 1.8, the space L_0 is invariant with respect to $\Re(H)$ and for $h \in H$ the restriction of the operator R_h on L_0 is nilpotent. Moreover, H is an invariant subspace of L_0 with respect to $\Re(H)$. Thus, we obtain the induced Lie algebra $\overline{H}$ of linear transformations which acts on the non null space L_0/H. Since these transformations are nilpotent by a version of Engel's theorem we have $\overline{H}(x + H) = \overline{0}$, where $x + H$ is a non zero vector, i.e., $[x, h] \in H$ for any $h \in H$. Hence, $x \in \mathrm{Nor}_l(H)$, $x \notin H$ and $H \neq \mathrm{Nor}_l(H)$. Thus, $H \subset \mathrm{Nor}_l(H)$ if and only if $H \subset L_0$, and the assertion is proved. □

Proposition 2.12. *Let H be a nilpotent subalgebra of a Leibniz algebra L and $L = L_0 \oplus L_1$ be the Fitting decomposition of L with respect to $\Re(H)$. Then L_0 is a subalgebra and $[L_1, L_0] \subseteq L_1$.*

Proof. Let $h \in H$ and $a \in L_0$. Then there exists $k \in \mathbb{N}$ such that

$$[\dots[a, \underbrace{h], h], \dots, h}_{k \text{ times}}] = 0.$$

This means that

$$[\dots[[R_a, R_h], R_h], \dots, R_h] = (-1)^k R_{[\dots[a,h],h],\dots,h]} = 0.$$

Applying the relation above and Lemma 1.7 we derive that the Fitting subspaces L_{0R_h} and L_{1R_h} of the algebra L corresponding to the endomorphism R_h are invariant subspaces with respect to R_a. Since $L_0 = \bigcap\limits_{h \in H} L_{0R_h}$ and $L_1 = \bigcap\limits_{h \in H} L_{1R_h}$ we have $R_a(L_0) \subseteq L_0$ and $R_a(L_1) \subseteq L_1$. Since a is an arbitrary element in L_0, we obtain $[L_0, L_0] \subseteq L_0$ and $[L_1, L_0] \subseteq L_1$. □

Similar to that of Lie algebras case an element h of a Leibniz algebra L is said to be *regular* if the dimension of the Fitting null component of L with respect to R_h is minimal. This dimension is called *the rank* of L.

Recall that the dimension of the Fitting null component of a linear transformation A is equal to the multiplicity of zero root of the characteristic polynomial of A.

Therefore an element h is regular if and only if the multiplicity of zero root of R_h is minimal.

Note that in the Lie algebras case the linear transformation R_h is degenerated since $[h, h] = 0$ for any h, and therefore the rank of the Lie algebra is positive.

The following proposition shows that in the case of Leibniz algebras the rank also is positive.

Proposition 2.13. *Multiplications (right and left) in the Leibniz algebra are degenerate linear operators.*

Proof. For Lie algebras this is obvious, since $[x, x] = 0$ for any $x \in L$, i.e., the vector x is the root vector for the zero eigenvalue. Thus for any non Lie but Leibniz algebra L the quotient Lie algebra $L/\mathcal{I}$ has positive dimension. It is clear that the subspace $\mathcal{I}$ is invariant under operators L_x and R_x, where $0 \neq x \in L$. On the quotient space which is a Lie algebra, the linear operators induced by these multiplications, are degenerated (their determinants equal to zero). Therefore these multiplications operators are degenerate in the whole space L. □

The following theorem establishes relations between regular elements of a Leibniz algebra and its Cartan subalgebras.

Theorem 2.9. *Let L be a Leibniz algebra over an infinite field $\mathbb{F}$ and a be a regular element of L. Then the Fitting null component H of L with respect to R_a is a Cartan subalgebra.*

Proof. Let $L = H \oplus V_1$ be the Fitting decomposition of L with respect to R_a. By Proposition 2.12 the subspace H is a subalgebra and $[V_1, H] \subseteq H$. Now we prove that for any $b \in H$ the transformation $R_{b|H}$ is nilpotent. Indeed, let $b \in H$ be an element such that $R_{b|H}$ is not nilpotent. Let us consider a basis of L consisting of bases of H and V_1. The matrix of R_h with $h \in H$, in this basis has the form $\begin{pmatrix} (\rho_1) & 0 \\ 0 & (\rho_2) \end{pmatrix}$, where (ρ_1) and (ρ_2) are the matrices $R_{h|H}$ and $R_{h|V_1}$, respectively.

Let $A = \begin{pmatrix} (\alpha_1) & 0 \\ 0 & (\alpha_2) \end{pmatrix}$ and $B = \begin{pmatrix} (\beta_1) & 0 \\ 0 & (\beta_2) \end{pmatrix}$ be the matrices of R_a and R_b, respectively. Since (α_2) is not singular, that is, $\det(\alpha_2) \neq 0$, by hypothesis, the matrix (β_1) is not nilpotent. Therefore if $l = \operatorname{rank}L$, then $\dim H = l$ and the characteristic polynomial of the matrix (β_1) is not divisible by λ^l. Let λ, μ, ν be algebraic independent variables and let $P(\lambda, \mu, \nu)$ be the characteristic polynomial, i.e. $P(\lambda, \mu, \nu) = \det(\lambda I - \mu A - \nu B) = \det(\lambda I - (\mu A + \nu B))$. Then the equality $P(\lambda, \mu, \nu) = P_1(\lambda, \mu, \nu)P_2(\lambda, \mu, \nu)$, where $P_i(\lambda, \mu, \nu) = \det(\lambda_1 - \mu(\alpha_i) - \nu(\beta_i)) = \det(\lambda I - (\mu(\alpha_i) + \nu(\beta_i)))$ holds. As it was noted above the polynomial $P_2(\lambda, I, 0) = \det(\lambda I - (\alpha_2))$ is not divisible by λ and the polynomial $P_1(\lambda, 0, I) = \det(\lambda I - (\beta_1))$ is not divisible by λ^l. Therefore the greatest degree of λ on which the polynomial $P(\lambda, \mu, \nu)$ can be divided is $\lambda^{l'}$, where $l' < l$. Since the field $\mathbb{F}$ is infinite, we can choose μ_0 and ν_0 such that $P(\lambda, \mu_0, \nu_0)$ is not divisible by $\lambda^{l'+1}$. Put $c := \mu_0 a + \nu_0 b$, then the characteristic polynomial $\det(\lambda I - R_c) = \det(\lambda I - \mu_0 A - \nu_0 B) = P(\lambda, \mu_0, \nu_0)$ is not divisible by $\lambda^{l'+1}$. Therefore the multiplicity of zero root for R_c is equal to l', $(l' < l)$. But this contradicts the condition that a is a regular element. Therefore, for any $b \in H$ the operator $R_{b|H}$ is nilpotent. By Engel's theorem (see Theorem 2.1) we obtain that H is a nilpotent Leibniz algebra. Let now L_0 be a Fitting's null component of L with respect to $R(H)$. Since H is the Fitting null component of the transformation R_a, then $L_0 \subseteq H$. Indeed, when $a \in H$ we have that $L_0 = \bigcap\limits_{b \in H} L_{0R_b} \subseteq L_{0R_a}$.

Let $a \notin H$. Then one has $a^k \neq 0$ for any $k \in \mathbb{N}$. Consider the following set of vectors $\{a_1 := a, a_2 := [a,a], \ldots, a_{n+1} := \underbrace{[[[a,a],a],\ldots,a]}_{(n+1)\ times}\}$, where $n = \dim L$. This set of vectors is linearly dependent, i.e., there exists a non trivial linear combination:

$$\alpha_1 a_1 + \alpha_2 a_2 + \cdots + \alpha_{n+1} a_{n+1} = 0.$$

Let $\alpha_1 \neq 0$, then using the fact that $a_i \in \mathcal{I}$ for any $2 \leq i \leq n+1$, we obtain $a_1 \in \mathcal{I} \subseteq \mathrm{Ann}_r(L)$ and consequently $a_2 = 0$, i.e., we have a contradiction with the condition $a_2 \neq 0$. Thus, $\alpha_1 = 0$.

Let k be the minimal number, such that $\alpha_k \neq 0$. Then $\alpha_k a_k + \cdots + \alpha_{n+1} a_{n+1} = 0$, therefore for the element $t = \alpha_k a_1 + \cdots + \alpha_{n+1} a_{n+1-k}$ we have $\underbrace{[[[t,t],t],\ldots,t]}_{k\ times} = 0$ and $R_t = \alpha_k R_a$, i.e., t is a regular element and $L_{0R_a} = L_{0R_t} = H$. On the other hand $L_0 \supseteq H$ for any nilpotent subalgebra H. In fact, $L_0 = \bigcap\limits_{b \in H} L_{0R_b}$ and if $h \in H$ then $\underbrace{[[[h,b],b],\ldots,b]}_{s\ times} = 0$ for any $b \in H$ (here s is the index of nilpotence of the algebra H) thus $h \in L_0$ and $L_0 = H$. Due to Proposition 2.11 we conclude that H is a Cartan subalgebra, which completes the proof of the theorem. □

Another useful remark about regular elements and Cartan subalgebras is the following

Corollary 2.7. *If a Cartan subalgebra H of a Leibniz algebra L contains a regular element a, then H is uniquely defined by the element a as a Fitting null component of the L with respect to R_a, i.e., $H = L_{0R_a}$.*

Proof. Let us denote L_{0R_a} by V. It is clear that $H \subseteq V$, since H is nilpotent (if $h \in H$ then using the nilpotency of H and $a \in H$ we have $[[[h,a],a],\ldots,a] = 0$ which implies $h \in L_{0R_a} = V$).

On the other hand, due to Theorem 2.9 V is nilpotent. And if $H \neq V$ then there exists $z \in V \setminus H$. If $[z,L] \subseteq H$, then since H is a Cartan subalgebra, we have $z \in H$ which is a contradiction. Therefore, for any $z \in V \setminus H$ we have $[z,L] \not\subseteq H$. Then there exist $l_1, l_2, \ldots, l_k \in H$ such that $[[[z,l_1],l_2],\ldots,l_k] \in V \setminus H$ and $[[[z,l_1],l_2],\ldots,l_k] \neq 0$, i.e., V is not nilpotent which is also a contradiction. Hence, $H = V$. We conclude that two Cartan subalgebras having the same regular element coincide. □

Proposition 2.14. *Let L be a complex finite-dimensional Leibniz algebra. Then the image of a regular element of L under $\pi : L \to L_{Lie}$ is a regular element of the Lie algebra L_{Lie}.*

Proof. Let a be a regular element of L. We prove that the element $\overline{a} = a + \mathcal{I}$ is a regular element of the Lie algebra L_{Lie}. Suppose the contrary, i.e., $\overline{a} = a + \mathcal{I}$ is not a regular element of L_{Lie}. Let $\overline{b} = b + \mathcal{I}$ be any regular element of L_{Lie}. For any $x \in L$ one has $R_x(\mathcal{I}) \subseteq \mathcal{I}$. Let $\{i_1, i_2, \ldots, i_n\}$ be a basis of $\mathcal{I}$. We extend the basis $\{i_1, i_2, \ldots, i_n\}$ to the basis $\{e_1, e_2, \ldots, e_m, i_1, i_2, \ldots, i_n\}$ of L. Then the matrix of the transformation R_x with respect to this basis has the following block form

$$R_x = \begin{pmatrix} X, & 0 \\ Z_X, & I_X \end{pmatrix},$$

where X is the matrix of the linear transformation $R_x|_{\mathrm{Span}\{e_1,e_2,\ldots,e_m\}}$ and I_X is the matrix of the transformation $R_x|_{\mathcal{I}}$.

Let

$$R_a = \begin{pmatrix} A, & 0 \\ Z_a, & I_a \end{pmatrix}, \quad R_b = \begin{pmatrix} B, & 0 \\ Z_b, & I_b \end{pmatrix}$$

be the matrices of the transformations R_a and R_b, respectively.

Let k (respectively k') and s (respectively s') be the multiplicities of the root 0 of the characteristic polynomials of A (respectively B) and I_a (respectively I_b). Obviously, we have $k' < k$, $s < s'$.

Put

$$U = \left\{ y \in L \setminus \mathcal{I} \;\middle|\; R_y = \begin{pmatrix} Y, & 0 \\ Z_y, & I_y \end{pmatrix} \text{ and } Y \text{ has the multiplicity of the root 0 of the characteristic polynomial less than } k \right\},$$

$$V = \left\{ y \in L \setminus \mathcal{I} \;\middle|\; R_y = \begin{pmatrix} Y, & 0 \\ Z_y, & I_y \end{pmatrix} \text{ and } I_y \text{ has the multiplicity of the root 0 of the characteristic polynomial less than } s+1 \right\}.$$

Since $b \in U$ and $a \in V$ these sets are not empty. Let us show that the set U is an open subset of the set $L \setminus \mathcal{I}$ with respect to the Zariski topology.

The set Y has the multiplicity of the root 0 of the characteristic polynomial less than k, hence the matrix Y^k has the rank greater than $n - k$. This means that there exists a non-zero minor of the order $n - k + 1$. In other words, there exists a non zero polynomial of structure constants of the algebra L, hence the set U is an open subset of $L \setminus \mathcal{I}$. Similarly, one can prove that the set V is open in $L \setminus \mathcal{I}$. It is not difficult to check that the sets U and V are dense in $L \setminus \mathcal{I}$. Therefore, there exists an element $y \in U \cap V$ such that Y has the multiplicity of the root 0 of the characteristic polynomial less than k and I_y has the multiplicity of the root 0 of the characteristic polynomial less than $s+1$. Thus, for this element y the multiplicity of the root 0 of the characteristic polynomial is not greater than $k + s - 1$, i.e., the rank of the algebra L is less than $k + s$ and we obtain a contradiction to the assumption that $\overline{a}$ is not a regular element of the Lie algebra L_{Lie}. □

Let L be a Leibniz algebra with a basis $\{e_1, e_2, \dots, e_n\}$ over a field $\mathbb{F}$. Let $\xi_1, \xi_2, \dots, \xi_n$ be independent variables and $\mathbb{P} = \mathbb{F}(\xi_1, \xi_2, \dots, \xi_n)$ be the field of rational functions at ξ_i. Consider the extension $L_{\mathbb{P}} = \mathbb{P}e_1 + \mathbb{P}e_2 + \cdots + \mathbb{P}e_n$ of $\mathbb{P}$.

The following definition and its comments are step by step modifications of the Lie algebras case and they are included for the sake of completeness.

Definition 2.6. *An element* $x = \sum_{i=1}^{n} \xi_i e_i$ *of the algebra* $L_{\mathbb{P}}$ *is called a generic element of the algebra* L *and the characteristic polynomial* $f_x(\lambda)$ *of the transformation* R_x *in* $L_{\mathbb{P}}$ *is called characteristic polynomial of the Leibniz algebra* L.

If we take the basis $\{e_1, e_2, \dots, e_n\}$ of $L_{\mathbb{P}}$ then $[e_i, x] = \sum_{j=1}^{n} \rho_{ij} e_j$, where $i = 1, \dots, n$ and ρ_{ij} are homogeneous functions of degree 1 with respect to ξ_k. Then

$$\begin{aligned} f_x(\lambda) &= \det(\lambda I - R_x) \\ &= \lambda^n - \tau_1(\xi)\lambda^{n-1} + \tau_2(\xi)\lambda^{n-2} - \cdots + (-1)^l \tau_{n-l}(\xi)\lambda^l, \end{aligned} \quad (2.5)$$

where $\tau_i(\xi)$ are homogeneous polynomials of degree i in the variables ξ_i, $\xi = \{\xi_1, \cdots, \xi_n\}$ and $\tau_{n-l}(\xi) \neq 0$, where $\tau_{n-l+k}(\xi) = 0$ for $k > 0$. Since $x \neq 0$ and R_x is a degenerate operator, it follows that $l > 0$ and $\det R_x = 0$.

The value of the characteristic polynomial on an arbitrary element $a = \sum_{i=1}^{n} \alpha_i e_i$ of the algebra L is obtained by specialization of $\xi_i = \alpha_i$, $i = 1, \ldots, n$ in (2.5). Therefore it is evident that the order of zero root of the characteristic polynomial of R_a is not less than l. On the other hand, if $\mathbb{F}$ is an infinite field, the polynomial $\tau_{n-l}(x)$ is a non zero element of $\mathbb{F}[\xi_1, \xi_2, \ldots, \xi_n]$, we can choose $\xi_i = \alpha_i$ such that $\tau_{n-l}(\alpha) \neq 0$. Then the transformation R_a for the element $a = \sum_{i=1}^{n} \alpha_i e_i$ has exactly l characteristic roots which are equal to zero and therefore a is regular. Thus in the case of an infinite field the element a is regular if and only if $\tau_{n-l}(\alpha) \neq 0$. In this sense "almost all" elements of the algebra L are regular (i.e., they form an open set in the Zariski topology).

The above statement depends on the choice of the basis $\{e_1, e_2, \ldots, e_n\}$. However, it is easy to observe what happens when we pass to another basis $\{f_1, f_2, \ldots, f_n\}$, where $f_i = \sum_{j=1}^{n} \mu_{ij} e_j$. If $\eta_1, \eta_2, \ldots, \eta_n$ are independent variables then $y = \sum_{i=1}^{n} \eta_i f_i = \sum_{i=1}^{n} \eta_i \mu_{ij} e_j$. Therefore the characteristic polynomial $f_y(\lambda)$ is obtained from polynomial $f_x(\lambda)$ by substitution $\xi_j \to \sum_{i=1}^{n} \eta_i \mu_{ij}$ in its coefficients.

2.6 SOME PROPERTIES OF WEIGHT SPACES OF LEIBNIZ ALGEBRAS AND CARTAN'S CRITERION OF SOLVABILITY

In order to define a weight module over a Leibniz algebra we need the definition of the right representation for Leibniz algebras.

Definition 2.7. *A vector space M is said to be a right representation of a Leibniz algebra L if there is an action:* $[\cdot, \cdot] : M \times L \to M$ *such that*

$$[m, [x, y]] = [[m, x], y] - [[m, y], x]$$

for all $x, y \in L$, $m \in M$.

The definition agrees with that of symmetric representation in [117].

Observe that M has natural right L–module structure (in the Lie sense) and below we shall consider M in this sense. Let M be a

L-module and $\alpha : L \to \mathbb{F}$ be linear function. We define:

$$M_\alpha = \{a \in L \mid \exists x \in M \text{ such that } (a - \alpha(a)I)^k(x) = 0 \text{ for some } k \in \mathbb{N}\}.$$

The map $\alpha : L \to \mathbb{F}$ defined above is said to be a *weight* and M_α is called *weight space* corresponding to the weight α.

Let L be a Leibniz algebra and M be a weight subspace over the algebra L with respect to the weight α. Then for each element $x \in M$ one has $[x, (a - \alpha(a)I)^k] = 0$ if k is sufficiently large. Moreover, if $\dim M = n$, then the polynomial $(\lambda - \alpha(a))^n$ is the characteristic polynomial of the α. Therefore $[x, (a - \alpha(a)I)^n] = 0$ for any $x \in M$.

Consider the contradredient (conjugated) right module M^* over a Leibniz algebra satisfying the condition:

$$\langle [x, a], y^* \rangle + \langle x, [y^*, a^*] \rangle = 0,$$

where $x \in M$, $y^* \in M^*$, a^* from representation which corresponds to the right module M^* and $\langle x, y^* \rangle$ as usual denotes the value of the linear function y^* at x.

It is clear that $\langle \alpha(a)x, y^* \rangle + \langle x, \alpha(a)y^* \rangle = 0$.

Adding these equalities we obtain:

$$\langle [x, (a - \alpha(a)I)], y^* \rangle + \langle x, [y^*, (a^* + \alpha(a)I)] \rangle = 0.$$

By repeating this procedure, we obtain the equality:

$$\langle [x, (a - \alpha(a)I)^k], y^* \rangle + \langle x, [y^*, (a^* + \alpha(a)I)^k] \rangle = 0.$$

If $k = n$, then $[x, (a - \alpha(a)I)^n] = 0$ for any x, consequently $\langle x, [y^*, (a^* + \alpha(a)I)^n] \rangle = 0$. Therefore $[y^*, (a^* + \alpha(a)I)^n] = 0$ for any $y^* \in M^*$. This shows that M^* is a weight module with the weight $-\alpha$.

Thus, we have proved the following proposition.

Proposition 2.15. *If M is a weight module over a Leibniz algebra with the weight α, then the contradredient module M^* is a weight module with the weight $-\alpha$.*

The next proposition is proved similarly to that in the Lie algebras case (see [106]).

Proposition 2.16. *If M and R are weight modules over a Leibniz algebra with the weight α and β respectively, then $B = M \otimes R$ is a weight module with the weight $\alpha + \beta$.*

A nilpotent Leibniz algebra L of linear transformations is called *a split algebra* if the characteristic roots of each element A of L are contained in the base field.

Let L be a Leibniz algebra, H be a nilpotent subalgebra and let M be a left module over L (and also over H). Suppose that $M = L$ and $\Re(H)$ is a nilpotent split Lie algebra. From Theorem 1.9 we have that $L = L_\alpha \oplus L_\beta \oplus \ldots \oplus L_\delta$, where $\alpha, \beta, \ldots, \delta$ are maps from the subalgebra $\Re(H)$ into F such that if $x_\nu \in L_\nu$, then $(R_h - \nu(R_h)I)^m(x_\nu) = 0$ for some $m = m(\nu)$, where $\nu \in \{\alpha, \beta, \ldots, \delta\}$. The weights $\alpha, \beta, \ldots, \delta$ are called *roots* of the algebra L with respect to the subalgebra H.

Proposition 2.17. *$[L_\alpha, L_\beta] \subseteq L_{\alpha+\beta}$ if $\alpha + \beta$ is a root of the Leibniz algebra L with respect to $\Re(H)$; otherwise $[L_\alpha, L_\beta] = 0$.*

Proof. Elements of $[L_\alpha, L_\beta]$ have the form $\sum_i [x_\alpha^{(i)}, y_\beta^{(i)}]$ where $x_\alpha^{(i)} \in L_\alpha$, $y_\beta^{(i)} \in L_\beta$. From the characteristic property of the tensor product of two spaces it follows that there exists a linear map: $\varphi : L_\alpha \otimes L_\beta \to [L_\alpha, L_\beta]$ such that $\varphi\left(\sum_i x_\alpha^{(i)} \otimes y_\beta^{(i)}\right) = \sum_i \left[x_\alpha^{(i)}, y_\beta^{(i)}\right]$. We show that φ is actually a homomorphism of $\Re(H)$-modules.

Let $R_h \in \Re(H)$. Then by using the Leibniz identity we obtain the following chain of equalities

$$R_h(x_\alpha \otimes y_\beta) = R_h(x_\alpha) \otimes y_\beta + x_\alpha \otimes R_h(y_\beta) = [x_\alpha, h] \otimes y_\beta + x_\alpha \otimes [y_\beta, h],$$

and under φ one has

$$R_h([x_\alpha, y_\beta]) = [[x_\alpha, h], y_\beta] + [x_\alpha, [y_\beta, h]].$$

On the other hand, the image of the element $x_\alpha \otimes y_\beta$ under the homomorphism φ is $[x_\alpha, y_\beta]$. So, we prove that $[L_\alpha, L_\beta]$ is a homomorphic image of the module $L_\alpha \otimes L_\beta$. Moreover, $L_\alpha \otimes L_\beta$ is a weight module with the weight $\alpha + \beta$. But from the definition it is clear that the homomorphic image of the weight module with the weight β is either 0 or a weight module with the weight β. □

Let L be a finite dimensional Leibniz algebra over an algebraically closed field $\mathbb{F}$ and H be a nilpotent subalgebra of L. Let $L = L_\alpha \oplus L_\beta \oplus \ldots \oplus L_\delta$ be a decomposition of the module L into the direct sum of weight submodules with respect to H.

Suppose that H is a Cartan subalgebra. Then it is not difficult to see that $H = L_0$, where L_0 is the root module corresponding to the root 0.

We have also the equality $[L, L] = \sum [L_\alpha, L_\beta]$, where the sum is taken over all roots α, β. From this we obtain $L_0 \cap L^2 = H \cap L^2 = \sum [L_\alpha, L_{-\alpha}]$, where summation is made over all α, such that $-\alpha$ is also a root (in particular $\alpha = 0$).

Definition 2.8. *The form* $K(a, b) = \mathrm{tr}(R_a R_b)$ *for* $a, b \in L$ *is called the Killing form of the Leibniz algebra* L.

A bilinear form $K(a, b)$ on L satisfying the condition:

$$K([a, c], b) + K(a, [b, c]) = 0$$

is called an *invariant form* on L.

The following equalities show that the Killing form is an invariant form on Leibniz algebra. In fact,

$$\begin{aligned} K([a, c], b) + K(a, [b, c]) &= \mathrm{tr}(R_{[a,c]} R_b) + \mathrm{tr}(R_a R_{[b,c]}) \\ &= \mathrm{tr}([R_c, R_a] R_b + R_a [R_c, R_b]) \\ &= \mathrm{tr}((R_c R_a - R_a R_c) R_b + R_a (R_c R_b - R_b R_c)) \\ &= \mathrm{tr}(R_c R_a R_b - R_a R_b R_c) \\ &= \mathrm{tr}[R_c, R_a R_b] = 0. \end{aligned}$$

Note that if $K(a, b)$ is the Killing form then the set

$$L^\perp = \{z \in L \mid K(a, z) = 0 \text{ for any } a \in L\}$$

is an ideal of the algebra L.

Theorem 2.10. *Let* L *be a Leibniz algebra over an algebraically closed field of zero characteristic. Then* L *is solvable if and only if* $\mathrm{tr}(R_a R_a) = 0$ *for any* $a \in L^2$.

Proof. **Necessity.** Taking into account the solvability of $\mathfrak{R}(L)$ and applying Proposition 1.4 we obtain $[\mathfrak{R}(L), \mathfrak{R}(L)] \subseteq \mathcal{N}$. Since $[\mathfrak{R}(L), \mathfrak{R}(L)] = \mathfrak{R}(L^2)$ the operator R_a is nilpotent for any $a \in L^2$, therefore $\mathrm{tr}(R_a R_a) = 0$.

Sufficiency. Let us apply Cartan's criterion for Lie algebras from [106] for the algebra $\mathfrak{R}(L)$ and consider L as $\mathfrak{R}(L)$-module. Then we obtain the solvability of the Lie algebra $\mathfrak{R}(L)$, but this is equivalent to the solvability of the Leibniz algebra L. □

CHAPTER 3

CLASSIFICATION PROBLEM IN LOW DIMENSIONS

3.1 ALGEBRAIC CLASSIFICATION OF LOW-DIMENSIONAL LEIBNIZ ALGEBRAS

The first part of this chapter concerns with the algebraic classification of complex Leibniz algebras of small dimensions. The second part is devoted to the study of varieties of Leibniz algebras. For the Lie algebras case by using the structure of a graded Lie algebra associated to a Leibniz algebra, and based on observations on the cohomologies of Lie and Leibniz algebras, we give a necessary and sufficient condition for a Lie algebra to be Leibniz rigid. For Leibniz algebras case we give necessary conditions (invariance arguments) for existence of degenerations which is helpful in finding a criterion for a Leibniz algebra to be rigid.

3.1.1 Classification of Lie algebras

The classification of finite-dimensional Lie algebras is a fundamental and very difficult problem. It can be split into three parts:

(1) classification of nilpotent Lie algebras;

(2) description of solvable Lie algebras with given nilradical;

(3) description of Lie algebras with given radical.

The third problem reduces to the description of semisimple subalgebras in the algebra of derivations of a given solvable algebra [124]. The classification of semisimple Lie algebras has been known since the works of Cartan and Killing. According to the Cartan-Killing theory the semisimple Lie algebras can be represented as a direct sum of the classical simple Lie algebras from series A_n $(n \geq 1)$, B_n $(n \geq 2)$, C_n $(n \geq 3)$, D_n $(n \geq 4)$ and five exceptional simple Lie algebras G_2, F_2, E_4, E_6, E_7, E_8. The second problem reduces to the description of orbits of certain unipotent linear groups [125]. The first problem is most complicated. Just recall that nilpotent complex Lie algebras are classified only in dimension up to 7. In higher dimensions, there are only partial classifications as subclasses of nilpotent Lie algebras.

There is no non trivial one-dimensional Lie algebra.

On a two-dimensional complex vector space there is only one non trivial Lie algebra structure given by the table of multiplication $[e_1, e_2] = e_1$ in a fixed basis $\{e_1, e_2\}$. The list of representatives of the isomorphism classes of three-dimensional complex Lie algebras can be found in the literature on Lie algebras (for instance see [106]).

3.1.2 Low-dimensional complex Leibniz algebras

There is no non trivial Leibniz algebra in dimension 1. Indeed, let $\dim_{\mathbb{F}}(L) = 1$ and e be a non-zero element in L. If $[e, e] = 0$, then L is an abelian Lie algebra. If $[e, e] \neq 0$, then $[e, e] = \alpha e$ for some nonzero $\alpha \in \mathbb{F}$. But then the Leibniz identities (both right and left) give us a contradiction. Therefore, on a one-dimensional space there is only one (trivial) Leibniz algebra structure.

3.1.2.1 Two-dimensional Leibniz algebras

Two dimensional Leibniz algebras have been classified by Cuvier [59]. Here is the list of all Leibniz algebras structures on two-dimensional vector space over $\mathbb{F}$.

Let $\dim_{\mathbb{F}}(L) = 2$, i.e., $L = \mathbb{F}e_1 + \mathbb{F}e_2$. Then it is easy to verify that there are, up to isomorphism, four Leibniz algebra structures on L :

1. L_1 : abelian algebra
2. L_2 : solvable Lie algebra with the table of multiplication $[e_1, e_2] = -[e_2, e_1] = e_2$.

3. L_3 : nilpotent Leibniz algebra with the table of multiplication $[e_2, e_2] = e_1$.

4. L_4 : solvable Leibniz algebra with the table of multiplication

$$[e_1, e_2] = e_1, [e_2, e_2] = e_1.$$

Note that the algebra L_3 is the left and right Leibniz algebra (obviously), and L_4 is the only right Leibniz algebra. One can check it by a direct verification of identities. In fact, let us test left Leibniz algebra's identity for the triple $\{e_2, e_2, e_2\}$. One has

$$[e_2, [e_2, e_2]] = [e_2, e_1] = 0$$

and

$$[[e_2, e_2], e_2] + [e_2, [e_2, e_2]] = [e_1, e_2] + [e_2, e_1] = e_1 \neq 0.$$

This means that the intersection of the classes of left and right Liebniz algebras strictly contains the class of Lie algebras.

3.1.2.2 *Three-dimensional Leibniz algebras*

The list of isomorphism classes of three-dimensional complex Leibniz algebras has been given in [27, 57]. Here we give a list of isomorphism classes over any field $\mathbb{F}$ of characteristic not two (see [154]). Let L be a three-dimensional algebra over a field $\mathbb{F}$. To give the list we make use of a case by case consideration with respect to the isomorphism invariants $\dim \mathrm{Ann}_r(L)$ and $\dim L^k$.

Case 1 : Let $\dim \mathrm{Ann}_r(L) = 1$ and $\{e_1\}$ be its basis. Since $\mathrm{Ann}_r(L)$ is an ideal of L, the following multiplications may occur:

$$\begin{array}{rclrcl} [e_1, e_2] &=& \alpha_1 e_1, & [e_2, e_2] &=& \alpha_2 e_1, \\ [e_3, e_2] &=& \beta_1 e_1 + \beta_2 e_2 + \beta_3 e_3, & & & \\ [e_3, e_3] &=& \alpha_4 e_1, & [e_1, e_3] &=& \alpha_3 e_1, \\ [e_2, e_3] &=& \gamma_1 e_1 - \beta_2 e_2 - \beta_3 e_3. & & & \end{array}$$

Case 1.1 : Let $\dim L^2 = 2$. Then $(\beta_2, \beta_3) \neq (0, 0)$, otherwise $\dim L^2 = 1$. Since e_2 and e_3 are symmetric, one can set $\beta_2 \neq 0$. Thus, by taking the basis transformation $e_1' = e_1$, $e_2' = \beta_2 e_2 + \beta_3 e_3$, and $e_3' = \frac{1}{\beta_2} e_3$ we obtain $\beta_2 = 1$, $\beta_3 = 0$ and hence the following multiplications occur:

$$\begin{array}{lll}[e_1,e_2] = \alpha_1 e_1, & [e_2,e_2] = \alpha_2 e_1, & \\ [e_1,e_3] = \alpha_3 e_1, & [e_3,e_3] = \alpha_4 e_1, & \quad (3.1)\\ [e_3,e_2] = \beta_1 e_1 + e_2, & [e_2,e_3] = \gamma_1 e_1 - e_2. &\end{array}$$

Applying the Leibniz identity as follows

$$[[e_1,e_3],e_2] = [[e_1,e_2],e_3] + [e_1,[e_3,e_2]]$$
$$[[e_2,e_3],e_2] = [[e_2,e_2],e_3] + [e_2,[e_3,e_2]]$$
$$[[e_3,e_2],e_3] = [[e_3,e_3],e_2] + [e_3,[e_2,e_3]]$$

we obtain the following constraints for the structure constants:

$$\begin{cases}\alpha_1 = 0,\\ \alpha_2(2+\alpha_3) = 0,\\ \beta_1 + \alpha_3\beta_1 + \gamma_1 = 0.\end{cases} \quad (3.2)$$

Case 1.1.1 : Suppose that $\alpha_2 \neq 0$ in (3.2). The application of the basis transformation $e_1' = \alpha_2 e_1,\ e_2' = e_2$ and $e_3' = \frac{\alpha_2\alpha_4-\beta_1^2}{2\alpha_2}e_1 - \frac{\beta_1}{\alpha_2}e_2 + e_3$, yields the following algebra:

$$L_1(\mathbb{F}): \ [e_1,e_3] = -2e_1,\ [e_2,e_2] = e_1,\ [e_2,e_3] = -e_2,\ [e_3,e_2] = e_2.$$

Case 1.1.2 : Assume that $\alpha_2 = 0$, then we have

$$\begin{array}{ll}[e_1,e_3] = \alpha_3 e_1, & [e_3,e_3] = \alpha_4 e_1,\\ [e_3,e_2] = \beta_1 e_1 + e_2, & [e_2,e_3] = -\beta_1(1+\alpha_3)e_1 - e_2.\end{array}$$

It is observed from (3.2) above that, if $(\alpha_3,\alpha_4) = (0,0)$ the algebra obtained is a Lie algebra. Therefore, we may assume that $(\alpha_3,\alpha_4) \neq (0,0)$. By applying the basis transformation $e_1' = e_1,\ e_2' = \beta_1 + e_2$ and $e_3' = e_3$ we get the following table of multiplications:

$$[e_1,e_3] = \alpha_3 e_1,\ [e_3,e_3] = \alpha_4 e_1\ [e_3,e_2] = e_2,\ [e_2,e_3] = -e_2.$$

If $\alpha_3 \neq 0$, then the basis transformation $e_1' = e_1,\ e_2' = e_2$ and $e_3' = e_3 - \frac{\alpha_4}{\alpha_3}e_1$ yields the algebra

$$L_2(\mathbb{F}): [e_1,e_3] = \alpha e_1,\ [e_2,e_3] = -e_2,\ [e_3,e_2] = e_2, \quad \alpha \neq 0.$$

If $\alpha_3 = 0$, then $\alpha_4 \neq 0$ and the basis transformation $e_1' = \alpha_4 e_1,\ e_2' = e_2$ and $e_3' = e_3$ yields the algebra

$$L_3(\mathbb{F}): \ [e_3,e_3] = e_1,\ [e_3,e_2] = e_2,\ [e_2,e_3] = -e_2.$$

Case 1.2: Let $\dim L^2 = 1$. The table of multiplication of L is written as follows:

$$\begin{aligned} [e_1, e_2] &= \alpha_1 e_1, \quad [e_2, e_2] = \alpha_2 e_1, \quad [e_1, e_3] = \alpha_3 e_1, \\ [e_3, e_3] &= \alpha_4 e_1, \quad [e_3, e_2] = \beta_1 e_1, \quad [e_2, e_3] = \gamma_1 e_1. \end{aligned} \tag{3.3}$$

Applying Leibniz identity as follows

$$[[e_2, e_2], e_3] = [[e_2, e_3], e_2] + [e_2, [e_2, e_3]]$$

$$[[e_3, e_2], e_3] = [[e_3, e_3], e_2] + [e_3, [e_2, e_3]]$$

we obtain the following constraints for the structure constants:

$$\begin{cases} \alpha_1 \gamma_1 = \alpha_2 \alpha_3 \\ \alpha_1 \alpha_4 = \alpha_3 \beta_1. \end{cases} \tag{3.4}$$

Case 1.2.1: Assume that $(\alpha_1, \alpha_3) = (0, 0)$. Then due to (3.4) we get the following table of multiplications:

$$\begin{aligned} [e_2, e_2] &= \alpha_2 e_1, \quad [e_3, e_3] = \alpha_4 e_1 \\ [e_3, e_2] &= \beta_1 e_1, \quad [e_2, e_3] = \gamma_1 e_1. \end{aligned} \tag{3.5}$$

If $(\alpha_2, \alpha_4, \beta_1 + \gamma_1) = (0, 0, 0)$, then L is a Lie algebra. Therefore, we assume that $(\alpha_2, \alpha_4, \beta_1 + \gamma_1) \neq (0, 0, 0)$. The basis transformation $e_1' = e_1,\ e_2' = Ae_2 + Be_3$ and $e_3' = e_3$ with $A \neq 0$ reduces (3.5) to the following table of multiplications:

$$\begin{aligned} [e_2, e_2] = [A^2\alpha_2 + B^2\alpha_4 + AB(\beta_1 + \gamma_1)]e_1,\ [e_3, e_3] = (\alpha_4 B + \gamma_1)e_1, \\ [e_3, e_2] = (\alpha_4 B + \beta_1)e_1,\ \ [e_2, e_3] = \gamma_1 e_1. \end{aligned}$$

The condition $(\alpha_2, \alpha_4, \beta_1 + \gamma_1) \neq (0, 0, 0)$ implies that, there exist two numbers $A, B \in \mathbb{F}$ such that:

$$A^2\alpha_2 + B^2\alpha_4 + AB(\beta_1 + \gamma_1) \neq 0.$$

Therefore, we presuppose that $\alpha_2 \neq 0$ in (3.5). The basis transformation $e_1' = \alpha_2 e_1,\ e_2' = e_2$ and $e_3' = e_3 - \frac{\beta_1}{\alpha_2} e_2$ yields $\alpha_2 = 1, \beta_1 = 0$ and hence the following multiplications table is obtained:

$$[e_2, e_2] = e_1,\ [e_3, e_3] = \alpha_4 e_1,\ [e_2, e_3] = \gamma_1 e_1, \tag{3.6}$$

where $(\alpha_4, \gamma_1) \neq (0,0)$, otherwise L is a split algebra. If $\alpha_4 = 0$ in (3.6), then $e_2 - \frac{1}{\gamma_1}e_3 \in \text{Ann}_r(L)$ contradicts $\dim \text{Ann}_r(L) = 1$. Therefore we may assume that $\alpha_4 \neq 0$ which leads to the following cases:

Case 1.2.1.1 : Assume that $\gamma_1 = 0$. This yields the following table of multiplications:

$$[e_2, e_2] = e_1, \quad [e_3, e_3] = \alpha_4 e_1, \quad \text{with } \alpha_4 \neq 0. \tag{3.7}$$

Case 1.2.1.1.A : Suppose that $\mathbb{F}^2 = \mathbb{F}$. Then the equation $x^2 = \alpha_4$ has a solution in $\mathbb{F}$. The basis transformation $e_1' = \alpha_4 e_1$, $e_2' = e_1 + e_3$ and $e_3' = e_1 + \sqrt{\alpha_4}e_2$, applied to (3.7) gives the following algebra:

$$L_4(\mathbb{F}) \ : [e_2, e_2] = e_1, \quad [e_3, e_3] = e_1.$$

Case 1.2.1.1.B : Suppose that $\mathbb{F}^2 \neq \mathbb{F}$. Let the equation $x^2 = \alpha_4$ have no solution in $\mathbb{F}$. We have the following sub cases:

Case 1.2.1.1.B.1 : The equation $x^2 = -\alpha_4$ also has a solution in $\mathbb{F}$. By applying the basis transformation $e_1' = \alpha_4 e_1$, $e_2' = e_1 + e_3$ and $e_3' = e_1 + \sqrt{-\alpha_4}e_2$ to (3.7) we get the following algebra:

$$L_5(\mathbb{F}) \ : [e_2, e_2] = e_1, \quad [e_3, e_3] = -e_1.$$

Case 1.2.1.1.B.2 : Assume the equation $x^2 = -\alpha_4$ has no solution in $\mathbb{F}$. We get the algebra:

$$L_6(\mathbb{F}) \ : [e_2, e_2] = e_1, \quad [e_3, e_3] = \alpha_4 e_1, \qquad \alpha_4 \neq 0.$$

The basis transformation $e_1' = \alpha_4 e_1$, $e_2' = \sqrt{-\alpha_4}e_2$ and $e_3' = \sqrt{-\alpha_4}e_3$ reduce $L_6(\mathbb{F})$ to $L_4(\mathbb{F})$. Thus, the two algebras are isomorphic.

Case 1.2.1.2 : Let us now consider $\gamma_1 \neq 0$ in (3.6). Taking the basis transformation $e_1' = e_1$, $e_2' = e_2$ and $e_3' = \frac{1}{\gamma_1}e_3$, we obtain $\gamma_1 = 1$ and replacing the parameter α_4 by α, we get the following algebras:

$$L_7(\mathbb{F}) \ : [e_2, e_2] = e_1, \quad [e_2, e_3] = e_1, \quad [e_3, e_3] = \alpha e_1, \ \alpha \neq 0.$$

Case 1.2.2 : Let $(\alpha_1, \alpha_3) \neq (0,0)$ in (3.3). We may assume that $\alpha_1 \neq 0$, otherwise we have the following table of multiplications

$$[e_1, e_3] = \alpha_3 e_1, \quad [e_3, e_3] = \alpha_4 e_1, \quad [e_2, e_3] = \gamma_1 e_1. \tag{3.8}$$

But $e_2 \in \text{Ann}_r(L)$ in (3.8). This contradicts the claim that dim $\text{Ann}_r(L) = 1$.

Since $\alpha_1 \neq 0$, then from (3.4) we get $\gamma_1 = \frac{\alpha_2\alpha_3}{\alpha_1}$ and $\alpha_4 = \frac{\alpha_3\beta_1}{\alpha_1}$. Thus we write (3.3) as follows:

$$[e_1, e_2] = \alpha_1 e_1, \ \ [e_2, e_2] = \alpha_2 e_1, \ \ [e_1, e_3] = \alpha_3 e_1, \ \ [e_3, e_3] = \frac{\alpha_3\beta_1}{\alpha_1} e_1,$$
$$[e_3, e_2] = \beta_1 e_1, \ \ [e_2, e_3] = \frac{\alpha_2\alpha_3}{\alpha_1} e_1.$$

Similarly, we have $e_3 - \frac{\alpha_3}{\alpha_1} e_2 \in \operatorname{Ann}_r(L)$. This contradicts the claim that $\dim \operatorname{Ann}_r(L) = 1$.

Case 2 : Suppose that $\dim \operatorname{Ann}_r(L) = 2$ and $\{e_1, e_2\}$ is its basis. Let $\{e_1, e_2, e_3\}$ be the basis of L, then the following table of multiplications occur:

$$\begin{array}{rcl} [e_3, e_3] & = & \alpha_1 e_1 + \alpha_2 e_2, \quad [e_1, e_3] = \beta_1 e_1 + \beta_2 e_2, \\ [e_2, e_3] & = & \gamma_1 e_1 + \gamma_2 e_2. \end{array} \tag{3.9}$$

Case 2.1 : Let $\dim L^2 = 1$. This reduces (3.9) to the following table of multiplications:

$$[e_3, e_3] = \alpha_1 e_1, \ \ [e_1, e_3] = \beta_1 e_1, \ \ [e_2, e_3] = \gamma_1 e_1. \tag{3.10}$$

Evidently, if $\gamma_1 = 0$, then (3.10) will be a split algebra. Applying the basis transformation $e_1' = \gamma_1 e_1, \ e_2' = e_2$ and $e_3' = e_3 - \frac{\alpha_1}{\gamma_1} e_2$ we get $\alpha_1 = 0, \gamma_1 = 1$ and hence the following table of multiplications:

$$[e_1, e_3] = \beta_1 e_1, \ \ [e_2, e_3] = e_1. \tag{3.11}$$

Case 2.1.1 : Assume that $\beta_1 = 0$. Then from (3.11) we get the following algebra:

$$L_8(\mathbb{F}) \ : [e_2, e_3] = e_1.$$

Case 2.1.2 : Assume that $\beta_1 \neq 0$. The basis transformation $e_1' = e_1, \ e_2' = e_2 - e_1$ and $e_3' = e_3$ yields the following split algebra:

$$[e_1, e_3] = e_1.$$

Case 2.2 : Let $\dim L^2 = 2$. This gives the following table of multiplications:

$$\begin{array}{rcl} [e_3, e_3] & = & \alpha_1 e_1 + \alpha_2 e_2, \\ [e_1, e_3] & = & \beta_1 e_1 + \beta_2 e_2, \\ [e_2, e_3] & = & \gamma_1 e_1 + \gamma_2 e_2, \end{array} \tag{3.12}$$

where the rank of the matrix $\begin{pmatrix} \alpha_1 & \alpha_2 \\ \beta_1 & \beta_2 \\ \gamma_1 & \gamma_2 \end{pmatrix}$ is 2.

Case 2.2.1 : Suppose that $\det(R_{e_3|\mathrm{Ann}_r(L)}) \neq 0$, it is equivalently defined as

$$\beta_1\gamma_2 - \gamma_1\beta_2 \neq 0. \tag{3.13}$$

Case 2.2.1.1 : Suppose that $\beta_2 \neq 0$. Since $R_{e_3|\mathrm{Ann}_r(L)}$ is non-degenerate, it yields $\gamma_1 \neq 0$. The basis transformation

$$e_1' = e_1,\ e_2' = e_2 \text{ and } e_3' = \left(\frac{\alpha_1\gamma_2}{\gamma_1} - \alpha_2\right)e_1 - \frac{\alpha_1}{\gamma_1}e_2 + e_3$$

reduces the table (3.12) to

$$[e_1, e_3] = e_2,\ [e_2, e_3] = \gamma_1 e_1 + \gamma_2 e_2,\ \gamma_1 \neq 0. \tag{3.14}$$

If $\gamma_2 = 0$ one has

$$[e_1, e_3] = e_2, \quad [e_2, e_3] = \gamma_1 e_1, \quad \gamma_1 \neq 0. \tag{3.15}$$

Case 2.2.1.1.A : Suppose that $\mathbb{F}^2 = \mathbb{F}$. Assume that the equation $x^2 = \gamma_1$ has a solution in $\mathbb{F}$. The basis transformation $e_1' = \sqrt{\gamma_1}e_1,\ e_2' = e_2$ and $e_3' = \frac{1}{\sqrt{\gamma_1}}e_3$ in (3.15) yields the following algebra:

$$L_9(\mathbb{F})\ : [e_1, e_3] = e_2, \quad [e_2, e_3] = e_1.$$

Case 2.2.1.1.B : Suppose that $\mathbb{F}^2 \neq \mathbb{F}$. Let the equation $x^2 = \gamma_1$ have no solution in $\mathbb{F}$. Then we consider the following sub cases:

Case 2.2.1.1.B.1 : Let $x^2 = -\gamma_1$ have a solution in $\mathbb{F}$. Applying the basis transformation $e_1' = \sqrt{-\gamma_1}e_1,\ e_2' = e_2$ and $e_3' = \frac{1}{\sqrt{-\gamma_1}}e_3$ to (3.15) we get

$$L_{10}(\mathbb{F})\ : [e_1, e_3] = e_2, \quad [e_2, e_3] = -e_1.$$

Case 2.2.1.1.B.2 : Let $x^2 = -\gamma_1$ have no solution in $\mathbb{F}$. We get the following algebra

$$L_{11}(\mathbb{F}) : [e_1, e_3] = e_2, \quad [e_2, e_3] = \gamma_1 e_1, \quad \gamma_1 \neq 0.$$

Algebras in $L_{11}(\mathbb{F})$ can be reduced to $L_9(\mathbb{F})$ via basis transformations $e_1' = \gamma_1 e_1,\ e_2' = \gamma_1 e_2$ and $e_3' = e_3$. Thus, the algebras $L_9(\mathbb{F})$ and $L_{11}(\mathbb{F})$ are isomorphic.

Suppose that $\gamma_1 \neq 0$, then (3.14) yields the following algebra

$$L_{12}(\mathbb{F}) : [e_1, e_3] = e_2, \ [e_2, e_3] = \gamma_1 e_1 + e_2, \ \gamma_1 \neq 0.$$

Case 2.2.1.2 : Let $\beta_2 = 0$. Setting the basis transformation $e'_1 = e_1$ and $e'_2 = e_2 - e_1, \ e'_3 = \frac{1}{\beta_1} e_3$, the table of multiplication (3.12) is reduced to

$$[e_3, e_3] = \alpha_1 e_1 + \alpha_2 e_2, \ [e_1, e_3] = e_1, \ [e_2, e_3] = \gamma_1 e_1 + \gamma_2 e_2.$$

Due to (3.13) we have $\gamma_2 \neq 0$. Applying the basis transformation $e'_1 = e_1, \ e'_2 = e_2$ and $e'_3 = \left(\frac{\alpha_1 \gamma_1}{\gamma_2} - \alpha_1\right) e_1 - \frac{\alpha_2}{\gamma_2} e_2 + e_3$, we get $\alpha_1 = \alpha_2 = 0$ and hence the table of multiplications above is reduced to:

$$[e_1, e_3] = e_1, \ [e_2, e_3] = \gamma_1 e_1 + \gamma_2 e_2, \ \gamma_2 \neq 0. \tag{3.16}$$

If $(\gamma_1, \gamma_2) \neq (0, 1)$, then there exist numbers $A, B \in \mathbb{F}$ such that

$$AB(\gamma_2 - 1) - B^2 \gamma_1 \neq 0.$$

The composition of the basis transformation

$$e'_1 = Ae_1 + Be_2, \ e'_2 = e_2, \ e'_3 = e_3, \ (A \neq 0)$$

and

$$e''_1 = e'_1, \ e''_2 = \left(1 + \frac{B}{A}\gamma_1\right) e'_1 + \left[B\gamma_2 - \left(B + \frac{B^2}{A}\gamma_1\right)\right] e'_2, \ e''_3 = e'_3,$$

brings (3.16) to **Case 2.2.1.1** already considered. Furthermore, $(\gamma_1, \gamma_2) = (0, 1)$ yields the following algebra

$$L_{13}(\mathbb{F}) \ : [e_1, e_3] = e_1, \ [e_2, e_3] = e_2.$$

Case 2.2.2 : Let $\det(R_{e_3|\mathrm{Ann}_r(L)}) = 0$. In this case, we have the following general table of multiplications:

$$\begin{array}{rcl} [e_3, e_3] & = & \alpha_1 e_1 + \alpha_2 e_2, \\ [e_1, e_3] & = & t_1(\beta_1 e_1 + \beta_2 e_2), \\ [e_2, e_3] & = & t_2(\gamma_1 e_1 + \gamma_2 e_2) \end{array} \tag{3.17}$$

Since in (3.17), the vectors e_1 and e_2 are "symmetric", we may assume that $t_1 \neq 0$. The basis transformation $e'_1 = e_1,\ e'_2 = \frac{t_2}{t_1}e_1 - e_2$ and $e'_3 = e_3$ reduces (3.17) to the following table of multiplications:

$$[e_3, e_3] = \alpha_1 e_1 + \alpha_2 e_2, \ [e_1, e_3] = \beta_1 e_1 + \beta_2 e_2. \tag{3.18}$$

Applying the basis transformation $e'_1 = \alpha_1 e_1 + \alpha_2 e_2,\ e'_2 = e_2$ and $e'_3 = e_3$ to the table above we get the following table of multiplications:

$$[e_3, e_3] = e_1, \ [e_1, e_3] = \beta_1 e_1 + \beta_2 e_2. \tag{3.19}$$

The basis transformation $e'_1 = e_1,\ e'_2 = \beta_2 e_2$ and $e'_3 = e_3$ reduces (3.19) to the following table of multiplications:

$$[e_3, e_3] = e_1, \ [e_1, e_3] = \beta_1 e_1 + e_2. \tag{3.20}$$

We consider the following cases in (3.20).

Case 2.2.2.1: If $\beta_1 = 0$, we get the following algebra:

$$L_{14}(\mathbb{F}) \ : [e_1, e_3] = e_2, \ [e_3, e_3] = e_1. \tag{3.21}$$

Case 2.2.2.2 : Let $\beta_1 \neq 0$. Taking the basis transformation $e'_1 = \frac{1}{\beta_1^2}e_1,\ e'_2 = \frac{1}{\beta_1^3}e_2$, and $e'_3 = \frac{1}{\beta_1}e_3$, we obtain the following algebra:

$$L_{15}(\mathbb{F}) \ : [e_1, e_3] = e_1 + e_2, \ [e_3, e_3] = e_1.$$

3.2 APPLICATION

We now apply the algorithm illustrated above to $\mathbb{R}$. In order to make our classification clear herein and after we use the following notation $L_i^j(\mathbb{F})$ when we list the algebras, where i represents number of algebra from above obtained list $L_1(\mathbb{F}) - L_{15}(\mathbb{F})$ and j stands for the number of the algebra in respective theorem.

Theorem 3.1. *Up to isomorphism, there exist three one parametric families and eleven explicit representatives of three dimensional non-Lie Leibniz algebras over the real field:*

$$L_1^1(\mathbb{R}) : [e_1, e_3] = -2e_1, \ [e_2, e_2] = e_1, \ [e_2, e_3] = -e_2,$$

$$[e_3, e_2] = e_2;$$

$L_2^2(\mathbb{R}) : [e_1, e_3] = \alpha e_1, [e_2, e_3] = -e_2, [e_3, e_2] = e_2, \quad \alpha \in \mathbb{R} \setminus \{0\};$

$L_3^3(\mathbb{R}) : [e_3, e_3] = e_1, \ [e_3, e_2] = e_2, \ [e_2, e_3] = -e_2;$

$L_4^4(\mathbb{R}) : [e_2, e_2] = e_1, \ [e_3, e_3] = e_1;$

$L_5^5(\mathbb{R}) : [e_2, e_2] = e_1, \ [e_3, e_3] = -e_1;$

$L_7^6(\mathbb{R}) : [e_2, e_2] = e_1, \ [e_2, e_3] = e_1, \ [e_3, e_3] = \alpha e_1, \quad \alpha \in \mathbb{R} \setminus \{0\};$

$L_8^7(\mathbb{R}) : [e_2, e_3] = e_1;$

$L_9^8(\mathbb{R}) : [e_1, e_3] = e_2, \ [e_2, e_3] = e_1;$

$L_{10}^9(\mathbb{R}) : [e_1, e_3] = e_2, \ [e_2, e_3] = -e_1;$

$L_{12}^{10}(\mathbb{R}) : [e_1, e_3] = e_2, \ [e_2, e_3] = \alpha e_1 + e_2, \quad \alpha \in \mathbb{R} \setminus \{0\};$

$L_{13}^{11}(\mathbb{R}) : [e_1, e_3] = e_1, \ [e_2, e_3] = e_2;$

$L_{14}^{12}(\mathbb{R}) : [e_1, e_3] = e_2, \ [e_3, e_3] = e_1;$

$L_{15}^{13}(\mathbb{R}) : [e_1, e_3] = e_1 + e_2, \ [e_3, e_3] = e_1.$

The classification of three-dimensional complex Lie algebras can be found in [106]. Combining Leibniz and Lie algebras cases, we give in Table 3.1 the list of isomorphism classes of all three-dimensional complex Leibniz algebras.

Algebra	Multiplication Table	Classification
$L_1(\alpha)$ $\alpha \neq 0$, $\alpha \in \mathbb{C}$	$[e_1, e_3] = \alpha e_1$, $[e_2, e_3] = e_1 + e_2$, $[e_3, e_3] = e_1$	Solvable Leibniz
L_2	$[e_3, e_3] = e_1$, $[e_2, e_3] = e_1 + e_2$	Solvable Leibniz
L_3	$[e_1, e_2] = e_3$, $[e_1, e_3] = -2e_3$, $[e_2, e_1] = -e_3$, $[e_2, e_3] = 2e_3$, $[e_3, e_1] = 2e_3$, $[e_3, e_2] = -2e_3$	Simple Lie
$L_4(\alpha)$	$[e_1, e_3] = \alpha e_1$, $[e_2, e_3] = -e_2$, $[e_3, e_2] = e_2$, $[e_3, e_3] = e_1$	Solvable Leibniz
L_5	$[e_1, e_3] = e_1$, $[e_2, e_3] = e_1$, $[e_3, e_3] = e_1$	Solvable Leibniz
L_6	$[e_1, e_3] = e_2$, $[e_3, e_3] = e_1$	Nilpotent Leibniz
L_7	$[e_1, e_2] = e_1$, $[e_1, e_3] = e_1$, $[e_3, e_2] = e_1$, $[e_3, e_3] = e_1$	Solvable Leibniz
L_8	$[e_1, e_1] = e_2$, $[e_2, e_1] = e_2$	Solvable Leibniz
$L_9(\alpha)$, $\alpha \neq 0, 1$, $\alpha \leftrightarrow \alpha^{-1}$	$[e_1, e_2] = e_2$, $[e_1, e_3] = \alpha e_3$, $[e_2, e_1] = -e_2$, $[e_3, e_1] = -\alpha e_3$	Solvable Lie
L_{10}	$[e_1, e_2] = e_2$, $[e_2, e_1] = -e_2$	Solvable Lie
L_{11}	$[e_1, e_2] = e_2$, $[e_1, e_3] = e_2 + e_3$, $[e_2, e_1] = -e_2$, $[e_3, e_1] = -e_2 - e_3$	Solvable Lie
$L_{12}(\alpha)$ $\alpha \in \mathbb{C}$	$[e_2, e_2] = e_1$, $[e_2, e_3] = e_1$, $[e_3, e_3] = \alpha e_1$	Nilpotent Lie
L_{13}	$[e_2, e_2] = e_1$, $[e_2, e_3] = e_1$, $[e_3, e_2] = e_1$	Composable, nilpotent Leibniz
L_{14}	$[e_1, e_3] = e_1$, $[e_2, e_3] = e_2$, $[e_3, e_3] = e_1$	Solvable Leibniz
L_{15}	$[e_1, e_1] = e_2$	Composable, nilpotent Leibniz
L_{16}	$[e_1, e_2] = e_2$, $[e_1, e_3] = e_3$, $[e_2, e_1] = -e_2$, $[e_3, e_1] = -e_3$	Solvable Lie
L_{17}	$[e_1, e_2] = e_3$, $[e_2, e_1] = -e_3$	Nilpotent Lie
L_{18}	-	Abelian

Table 3.1 Isomorphism classes of three-dimensional complex Leibniz algebras $LB_3(\mathbb{C})$

3.3 LOW-DIMENSIONAL NILPOTENT LEIBNIZ ALGEBRAS

Further results deal with the nilpotent Leibniz algebras and their subclasses. There are results on classification of p-filiform, quasi-filiform Leibniz algebras. They can be found, for instance, in [46] and [48].

The definition of the nilpotency has been given in Section 2.2. Let us specify some subclasses of LN_n.

Definition 3.1. *A Leibniz algebra L is said to be a k-filiform, if* $\dim L^2 = n - k - 1$, *and* $\dim L^i = n - k - (i - 1)$, *where* $3 \le i \le n$.

It has been proved (see [24]) that 0-filiform (null-filiform) Leibniz algebras in each fixed dimension are mutually isomorphic, i.e., there is a unique null-filiform Leibniz algebra in any fixed dimension. This fact is only inherent in Leibniz algebras, i.e., there is no a null-filiform Lie algebra.

The class of one-filiform (cited in this book as "filiform" as in [179]) algebras, denoted by Lb_n, is most developed part of LN_n, which is focused in this book. There are classification results on k-filiform Leibniz algebras as well, but the technique applied [179] is slightly different than one considered here.

We begin with assembling a few simple observations on associative and Leibniz algebras.

Let L be a non-unital associative algebra and A be an associative algebra, obtained from L by the external adjoining a unit $A = L \oplus \mathbb{C}\mathbf{1}$. Then one has

Proposition 3.1. *Let L be a finite dimensional nilpotent associative algebra. Then the algebra* $A = L \oplus \mathbb{C}\mathbf{1}$ *does not contain non trivial idempotent.*

Proof. Let $\mathbf{1} + a$ be an idempotent element of $A = L \oplus \mathbb{C}\mathbf{1}$, where a is an element of L with the index of nilpotency m. Obviously, $\mathbf{1} + a = (\mathbf{1} + a)^n$ for any natural $n \ge 2$. On the other hand, one has $(\mathbf{1} + a)^n = \mathbf{1} + \sum_{k=1}^{n} \binom{n}{k} a^k$, this gives $a = na + \sum_{k=2}^{n} \binom{n}{k} a^k$, and multiplying both sides of this equality by a^{m-2} we get $(n-1)a^{(m-1)} = 0$, i.e., $a^{(m-1)} = 0$, while the index of nilpotency of a is m. This contradiction shows that A has no non trivial idempotent. □

Note that if L is a Leibniz algebra with $L^3 = 0$, then it is associative. We make use of the following corollary of Proposition 3.1.

Corollary 3.1. *Let L be a finite dimensional Leibniz algebra with $L^3 = 0$. Then the algebra $A = L \oplus \mathbb{C}\mathbf{1}$ has no non-trivial idempotent.*

For the further references we state the following simple fact.

Proposition 3.2. *Let L_1 and L_2 be finite-dimensional associative algebras without unit. Then*

$$L_1 \oplus \mathbb{C}\mathbf{1} \cong L_2 \oplus \mathbb{C}\mathbf{1} \text{ if only if } L_1 \cong L_2.$$

Another result which we make use in this chapter is Mazzola's classification result of five-dimensional unital associative algebras [127]. There Mazzola listed, up to isomorphism, all unital associative algebra structures on five-dimensional vector space. However, the list is too long to be given here (there are 59 isomorphism classes there). We suppose that the reader is familiar with the list.

For a given n-dimensional nilpotent Leibniz algebra L we define the following isomorphism invariant

$$\chi(L) = (\dim L^1, \dim L^2, \ldots, \dim L^{n-1}, \dim L^n).$$

Evidently,

$$\dim L^1 > \dim L^2 > \cdots > \dim L^n.$$

Proposition 3.3. *If for an n-dimensional nilpotent Leibniz algebra L the first two components of the invariant $\chi(L)$ are equal to n and $n-1$, respectively, then L is a null-filiform Leibniz algebra.*

Proof. From the hypotheses of the proposition we can easily conclude that L is one-generated algebra. Let $x \in L \setminus L^2$ be a generator of L. Then the following set of vectors $\{x, [x, x], \ldots, \underbrace{[[[x, x], x], \ldots, x]}_{n \text{ times}}\}$ forms a basis of L, with $\dim L^i = n - i + 1$, where $2 \leq i \leq n + 1$. □

First we begin with the following description of null-filiform Leibniz algebras.

Theorem 3.2. *Up to isomorphism, there is only one n-dimensional null-filiform Leibniz algebra. Namely, in every such Leibniz algebra L there exists a basis* $\{x_1, x_2, \dots, x_n\}$ *with respect to which L has the following table of multiplication.*

$$[x_i, x_1] = x_{i+1}, \quad [x_i, x_j] = 0, \ 1 \le i \le n-1, \ j \ge 2. \tag{3.22}$$

Proof. Let L be a null-filiform Leibniz algebra of dimension n and let $\{e_1, e_2, \dots, e_n\}$ be a basis of L such that $e_i \in L^i \setminus L^{i+1}$ with $1 \le i \le n$ (such a basis can be chosen always). Since $e_2 \in L^2$, for some elements a_{2p}, b_{2p} of L we have

$$e_2 = \sum \alpha_p [a_{2p}, b_{2p}] = \sum \alpha_{ij}^2 [e_i, e_j] = \alpha_{11}^2 [e_1, e_1] + (*),$$

where $(*) \in L^3$; i.e., $e_2 = \alpha_{11}^2 [e_1, e_1] + (*)$. Notice that $\alpha_{11}^2 [e_1, e_1] \neq 0$ (otherwise $e_2 \in L^3$). Similarly,

$$e_3 = \sum \alpha_{ijk}^3 [[e_i, e_j], e_k] = \alpha_{111}^3 [[e_1, e_1], e_1] + (**),$$

where $(**) \in L^4$; i.e., $e_3 = \alpha_{111}^3 [[e_1, e_1], e_1] + (**)$. Notice that $\alpha_{111}^3 [[e_1, e_1], e_1] \neq 0$ (otherwise $e_3 \in L^4$).

Continuing likewise, we find the set of non zero elements of L

$$x_1 := e_1, \ x_2 := [e_1, e_1], \ x_3 := [[e_1, e_1], e_1], \ \dots, \ x_n := [[[e_1, e_1], e_1], \dots, e_1].$$

It is easy to check that these elements are linearly independent. Hence, they form a basis of L. Thus, $[x_i, x_1] = x_{i+1}$ for $1 \le i \le n-1$; moreover, $[x_i, x_j] = 0$ for $j \ge 2$. Indeed, if $j = 2$ then

$$[x_i, x_2] = [x_i, [x_1, x_1]] = [[x_i, x_1], x_1] - [[x_i, x_1], x_1] = 0.$$

Assume that it is true for $j > 2$. The validity for $j + 1$ follows from the inductive hypothesis and the equality

$$[x_i, x_{j+1}] = [x_i, [x_j, x_1]] = [[x_i, x_j], x_1] - [[x_i, x_1], x_j] = 0.$$

□

Henceforth we denote the algebra with multiplication (3.22) by NF_n.

3.3.1 Four-dimensional nilpotent complex Leibniz algebras

In this section we shall expose, up to isomorphism, the list of all four-dimensional nilpotent complex Leibniz algebras. Here first we use four-dimensional case of Theorem 4.3 (will be given in Chapter 4), where the description of filiform non Lie Leibniz algebras is reduced to three families of algebras and a verification of the isomorphisms inside each of the families. To find other isomorphic algebras one uses the classification of unital associative algebras of dimension five [127]. From the list in [127], according to some constraints we pick up Leibniz algebras. Recall that the rings of commuting and non-commuting variables $x_1, x_2, \ldots, x_n$ over a field $\mathbb{F}$ are denoted by $\mathbb{F}[x_1, x_2, \ldots, x_n]$ and $\mathbb{F}\langle x_1, x_2, \ldots, x_n \rangle$, respectively.

First we give the following auxiliary result.

Proposition 3.4. *Any four-dimensional nilpotent complex Leibniz algebra L belongs to one of the following types of algebras*

(i) *null-filiform Leibniz algebras, that is* $\chi(L) = (4, 3, 2, 1)$;

(ii) *filiform Leibniz algebras, that is* $\chi(L) = (4, 2, 1, 0)$;

(iii) *associative algebras, with* $\chi(L) := (4, 2, 0, 0)$ *or* $(4, 1, 0, 0)$;

(iv) *abelian, that is* $\chi(L) = (4, 0, 0, 0)$.

Proof. Let us consider all possible cases for $\chi(L)$. The following can occur.

a) Let $\chi(L) = (4, 3, 2, 1)$. Then in view of Proposition 3.3 the algebra L is null-filiform.

b) Let $\chi(L) = (4, 2, 1, 0)$. Then L is a filiform.

It is clear that in the following three cases the algebra L is associative.

c) $\chi(L) = (4, 2, 0, 0)$.

d) $\chi(L) = (4, 1, 0, 0)$.

e) Let $\chi(L) = (4, 0, 0, 0)$. Then we get an abelian algebra. □

Note that Proposition 3.4 implies that the cases $\chi(L) = (4, 3, 2, 0)$, $(4, 3, 1, 0)$ and $(4, 3, 0, 0)$ are impossible.

From now on as a matter of convenience we assume that the omitted products of basis vectors are zero, we also do not consider the abelian case.

Let us consider a Leibniz algebra L with a basis $\{e_1, e_2, e_3, e_4\}$.

Theorem 3.3. *Up to isomorphism all nilpotent Leibniz algebra structures on four-dimensional vector space are given by the following representatives.*

$\mathcal{R}_1 : [e_1, e_1] = e_2, [e_2, e_1] = e_3, [e_3, e_1] = e_4.$
$\mathcal{R}_2 : [e_1, e_1] = e_3, [e_1, e_2] = e_4, [e_2, e_1] = e_3, [e_3, e_1] = e_4.$
$\mathcal{R}_3 : [e_1, e_1] = e_3, [e_2, e_1] = e_3, [e_3, e_1] = e_4;$
$\mathcal{R}_4(\alpha) : [e_1, e_1] = e_3, [e_1, e_2] = \alpha e_4, [e_2, e_1] = e_3, [e_2, e_2] = e_4, [e_3, e_1] = e_4, \ \alpha \in \{0, 1\}.$
$\mathcal{R}_5 : [e_1, e_1] = e_3, [e_1, e_2] = e_4, [e_3, e_1] = e_4.$
$\mathcal{R}_6 : [e_1, e_1] = e_3, [e_2, e_2] = e_4, [e_3, e_1] = e_4.$
$\mathcal{R}_7 : [e_1, e_1] = e_4, [e_2, e_1] = e_3, [e_3, e_1] = e_4, [e_1, e_2] = -e_3, [e_1, e_3] = -e_4.$
$\mathcal{R}_8 : [e_1, e_1] = e_4, [e_2, e_1] = e_3, [e_3, e_1] = e_4, [e_1, e_2] = -e_3 + e_4, [e_1, e_3] = -e_4.$
$\mathcal{R}_9 : [e_1, e_1] = e_4, [e_2, e_1] = e_3, [e_2, e_2] = e_4, [e_3, e_1] = e_4, [e_1, e_2] = -e_3 + 2e_4, [e_1, e_3] = -e_4.$
$\mathcal{R}_{10} : [e_1, e_1] = e_4, [e_2, e_1] = e_3, [e_2, e_2] = e_4, [e_3, e_1] = e_4, [e_1, e_2] = -e_3, [e_1, e_3] = -e_4.$
$\mathcal{R}_{11} : [e_1, e_1] = e_4, [e_1, e_2] = e_3, [e_2, e_1] = -e_3, [e_2, e_2] = -2e_3 + e_4,$
$\mathcal{R}_{12} : [e_1, e_2] = e_3, [e_2, e_1] = e_4, [e_2, e_2] = -e_3,$
$\mathcal{R}_{13}(\alpha) : [e_1, e_1] = e_3, [e_1, e_2] = e_4, [e_2, e_1] = -\alpha e_3, [e_2, e_2] = -e_4, \quad \alpha \in \mathbb{C},$
$\mathcal{R}_{14}(\alpha) : [e_1, e_1] = e_4, [e_1, e_2] = \alpha e_4, [e_2, e_1] = -\alpha e_4, [e_2, e_2] = e_4, [e_3, e_3] = e_4, \ \alpha \in \mathbb{C},$
$\mathcal{R}_{15} : [e_1, e_2] = e_4, [e_1, e_3] = e_4, [e_2, e_1] = -e_4, [e_2, e_2] = e_4, [e_3, e_1] = e_4,$
$\mathcal{R}_{16} : [e_1, e_1] = e_4, [e_1, e_2] = e_4, [e_2, e_1] = -e_4, [e_3, e_3] = e_4,$
$\mathcal{R}_{17} : [e_1, e_2] = e_3, [e_2, e_1] = e_4,$
$\mathcal{R}_{18} : [e_1, e_2] = e_3, [e_2, e_1] = -e_3, [e_2, e_2] = e_4,$
$\mathcal{R}_{19} : [e_2, e_1] = e_4, [e_2, e_2] = e_3,$
$\mathcal{R}_{20}(\alpha) : [e_1, e_2] = e_4, [e_2, e_1] = (1 + \alpha)/(1 - \alpha)e_4, [e_2, e_2] = e_3, \ \alpha \in \mathbb{C} \setminus \{1\},$
$\mathcal{R}_{21} : [e_1, e_2] = e_4, [e_2, e_1] = -e_4, [e_3, e_3] = e_4,$
$\mathcal{R}_{22} : [e_1, e_1] = e_2,$
$\mathcal{R}_{23} : [e_1, e_1] = e_2, [e_3, e_3] = e_4,$
$\mathcal{R}_{24} : [e_1, e_2] = e_3, [e_2, e_1] = -e_3,$
$\mathcal{R}_{25} : [e_1, e_1] = e_3, [e_1, e_2] = e_3, [e_2, e_2] = \beta e_3,$

$\mathcal{R}_{26} : [e_1, e_1] = e_3, [e_1, e_2] = e_3, [e_2, e_1] = e_3,$
$\mathcal{R}_{27} : [e_1, e_2] = e_3, [e_2, e_1] = -e_3, [e_1, e_3] = e_4, [e_3, e_1] = -e_4,$
$\mathcal{R}_{28} : [e_1, e_1] = e_2, [e_2, e_1] = e_3.$

Proof. According to Proposition 3.4 any four-dimensional nilpotent Leibniz algebra is either null-filiform or filiform or associative. Let us consider each of these classes separately.

Let L be null-filiform. Then in view of Theorem 3.2 there is only one null-filiform Leibniz algebra and it can be given by the table of multiplication

$$[e_1, e_1] = e_2, [e_2, e_1] = e_3, [e_3, e_1] = e_4.$$

This is $\mathcal{R}_1$ in the list.

Let now L be a filiform Leibniz algebra and $\{e_1, e_2, e_3, e_4\}$ be a basis such that $e_1, e_2 \in L \setminus L^2$, $e_3 \in L^2 \setminus L^3$ and $e_4 \in L^3$. We choose the basis $\{e_1, e_2, e_3, e_4\}$ so that $[e_2, e_1] = e_3, [e_3, e_1] = e_4$ and $[e_4, e_1] = 0$.

Let

$$\begin{cases} [e_1, e_1] = \alpha_1 e_3 + \alpha_2 e_4, & [e_2, e_2] = \beta_1 e_3 + \beta_2 e_4, \\ [e_1, e_2] = \gamma_1 e_3 + \gamma_2 e_4, & [e_3, e_2] = \delta_1 e_4, \\ [e_1, e_3] = \delta_2 e_4, & [e_2, e_3] = \delta_3 e_4. \end{cases}$$

Applying the following equalities

$$0 = [e_1, [e_1, e_1]] = [e_1, \alpha_1 e_3 + \alpha_2 e_4] = \alpha_1 \delta_2 e_4,$$

$$0 = [e_2, [e_1, e_1]] = [e_2, \alpha_1 e_3 + \alpha_2 e_4] = \alpha_1 \delta_3 e_4,$$

we derive $\alpha_1 \delta_2 = \alpha_1 \delta_3 = 0$.

Case i). Let $\alpha_1 \neq 0$. Then $\delta_2 = \delta_3 = 0$ and taking the basis transformation as follows:

$$e_i' = \alpha_1^i e_i, \quad 1 \leq i \leq 4,$$

one can assume that $\alpha_1 = 1$.

Putting $e_1' = e_1 - \alpha_2 e_3$, we derive $\alpha_2 = 0$.

From the equalities

$$0 = [e_1, [e_2, e_1]] = [[e_1, e_2], e_1] - [[e_1, e_1], e_2] = \gamma_1 e_4 - \delta_1 e_4,$$

$$0 = [e_2, [e_2, e_1]] = [[e_2, e_2], e_1] - [[e_2, e_1], e_2] = \beta_1 e_4 - \delta_1 e_4,$$

we get $\beta_1 = \gamma_1 = \delta_1$.

So, the table of multiplication of L has the following form:

$$\begin{cases} [e_1, e_1] = e_3, & [e_2, e_2] = \beta_1 e_3 + \beta_2 e_4, \\ [e_2, e_1] = e_3, & [e_3, e_1] = e_4, \\ [e_1, e_2] = \beta_1 e_3 + \gamma_2 e_4, & [e_3, e_2] = \beta_1 e_4. \end{cases}$$

Taking $e_2' = e_2 - \beta_1 e_1$ we transform the table of multiplication to the following form:

$$\begin{cases} [e_2, e_1] = (1 - \beta_1) e_3, & [e_3, e_1] = e_4, \\ [e_1, e_1] = e_3, & [e_2, e_2] = \beta_2 e_4, \\ [e_1, e_2] = \gamma_2 e_4. \end{cases}$$

If $\beta_1 \neq 1$, then taking the basis transformation

$$e_1' = (1 - \beta_1) e_1, \ e_2' = e_2, \ e_3' = (1 - \beta_1)^2 e_3, \ e_4' = (1 - \beta_1)^3 e_4$$

we obtain the family of algebras

$F_1(\alpha, \beta) : [e_1, e_1] = e_3, [e_1, e_2] = \alpha e_4,$
$[e_2, e_1] = e_3, [e_2, e_2] = \beta e_4, [e_3, e_1] = e_4.$

If $\beta_1 = 1$, then we obtain the family of algebras

$F_2(\alpha, \beta) : [e_1, e_1] = e_3, [e_1, e_2] = \alpha e_4, [e_2, e_2] = \beta e_4, [e_3, e_1] = e_4.$

Case ii). Let $\alpha_1 = 0$. Then taking the base change

$$e_2' = e_2 - \delta_1 e_1, \quad e_3' = e_3 - \alpha_2 \delta_1 e_4,$$

we derive $\delta_1 = 0$.

From the equalities

$$\gamma_1 \delta_2 e_3 = [e_1, \gamma_1 e_3 + \gamma_2 e_4] = [e_1, [e_1, e_2]]$$
$$= [[e_1, e_1], e_2] - [[e_1, e_2], e_1] = -\gamma_1 e_4,$$

$$\gamma_1 \delta_3 e_3 = [e_2, \gamma_1 e_3 + \gamma_2 e_4] = [e_2, [e_1, e_2]]$$
$$= [[e_2, e_1], e_2] - [[e_2, e_2], e_1] = -\beta_1 e_4,$$

we get

$$\gamma_1(\delta_2 + 1) = 0, \quad \beta_1 = -\gamma_1 \delta_3.$$

Further, the equalities

$$\delta_2 e_4 = [e_1, e_3] = [e_1, [e_2, e_1]] = [[e_1, e_2], e_1] - [[e_1, e_1], e_2] = \gamma_1 e_4,$$

$$\delta_3 e_4 = [e_2, e_3] = [e_2, [e_2, e_1]] = [[e_2, e_2], e_1] - [[e_2, e_1], e_2] = \beta_1 e_4,$$

$$0 = [e_1, [e_2, e_2]] = [e_1, \beta_1 e_3 + \beta_2 e_4] = \delta_2 \beta_1 e_4,$$

imply

$$\delta_2 \beta_1 = 0, \quad \gamma_1 = \delta_2, \quad \beta_1 = \delta_3.$$

Therefore, we get $\beta_1 = 0$ and the table of multiplication of L has the following form

$$\begin{cases} [e_2, e_1] = e_3, & [e_3, e_1] = e_4, \\ [e_1, e_1] = \alpha_2 e_4, & [e_2, e_2] = \beta_2 e_4, \\ [e_1, e_2] = \gamma_1 e_3 + \gamma_2 e_4, & [e_1, e_3] = \gamma_1 e_4 \end{cases}$$

with the restriction on γ_1 as follows

$$\gamma_1(\gamma_1 + 1) = 0.$$

If $\gamma_1 = 0$, then the base change

$$e_1' = e_1 + e_2 - (\gamma_2 + \alpha_2) e_3, \ e_3' = e_3 + \beta_2 e_4$$

gives the family of algebras $F_1(\alpha, \beta)$.

If $\gamma_1 = -1$, then we obtain the family of algebras

$$F_3(\alpha, \beta, \gamma) : [e_1, e_1] = \alpha e_4, [e_2, e_2] = \beta e_4, [e_2, e_1] = e_3, [e_3, e_1] = e_4,$$

$$[e_1, e_2] = -e_3 + \gamma e_4, [e_1, e_3] = -e_4,$$

where in the latter case at least one of α, β, γ is not zero.

Let us consider isomorphisms inside each of the families $F_1(\alpha, \beta)$ $F_1(\alpha, \beta)$ and $F_3(\alpha, \beta, \gamma)$.

We start with $F_1(\alpha, \beta)$. The following cases may occur:

Case 1.1: $\beta = 0$ and $\alpha \neq 0$. Then the following base change reduces $F_1(\alpha, \beta)$ to $F_1(1, 0)$:

$$e_1' = \alpha e_1, e_2' = \alpha e_2, e_3' = \alpha^2 e_3, e_4' = \alpha^3 e_4.$$

In this case we get $\mathcal{R}_2$.

Case 1.2: $\beta = 0$ and $\alpha = 0$. Then $F_1(0, 0)$ is $\mathcal{R}_3$.

Case 2: $\beta \neq 0$. By changing the basis $\{e_1, e_2, e_3, e_4\}$ as follows

$$e_1' = \beta e_1, e_2' = \beta e_2, e_3' = \beta^2 e_3, e_4' = \beta^3 e_4$$

the algebra $F_1(\alpha, \beta)$ can be reduced to $F_1(\alpha, 1)$.

Thus we get the table of multiplication for the family of algebras $F_1(\alpha, 1)$ with parameter α as follows

$$[e_1, e_1] = e_3, [e_1, e_2] = \alpha e_4, [e_2, e_1] = e_3, [e_2, e_2] = e_4, [e_3, e_1] = e_4.$$

Now we specify isomorphisms within $F_1(\alpha, 1)$. It is not difficult to see that the following change of the basis

$$e_1' = A_1e_1 + A_2e_2, \quad e_2' = (A_1 + A_2)e_2 + A_2(\alpha - 1)e_3,$$

$$e_3' = A_1(A_1 + A_2)e_3 + A_2(A_1\alpha + A_2)e_4, \quad e_4' = A_1^2(A_1 + A_2)e_4$$

where $A_1 \neq 0, \pm A_2$ transforms algebras from $F_1(\alpha, 1)$ either to

$$\mathcal{R}_4(1) : [e_1, e_1] = e_3, [e_2, e_1] = e_3, [e_2, e_2] = e_4, [e_3, e_1] = e_4, [e_1, e_2] = e_4$$

or to

$$\mathcal{R}_4(0) : [e_1, e_1] = e_3, [e_2, e_1] = e_3, [e_2, e_2] = e_4, [e_3, e_1] = e_4.$$

Indeed, for α, α', A_1, A_2 we obtain the following relations $\alpha' = \frac{A_1\alpha+A_2}{A_1^2}$ and $A_2 = A_1^2 - A_1$. Clearly at $\alpha = 1$ we get $\alpha' = 1$. But if $\alpha \neq 1$ then setting $A_1 = 1 - \alpha$ we obtain $\alpha' = 0$.

We show that the exposed algebras are not isomorphic. Note that the dimensions of maximal abelian subalgebras of the algebras $\mathcal{R}_2$ and $\mathcal{R}_4(\alpha)$ are different and therefore, $\mathcal{R}_2$ can not be isomorphic to the algebra $\mathcal{R}_4(\alpha)$ for any $\alpha \in \{0, 1\}$. The algebra $\mathcal{R}_3$ is not isomorphic to the algebras $\mathcal{R}_2$ and $\mathcal{R}_4(\alpha)$ for any $\alpha \in \{0, 1\}$, by dimension reasons of the right annihilators.

Let us now consider the family of algebras $F_2(\alpha,\beta)$. We treat again a few cases. Let $\beta = 0$ and $\alpha \neq 0$. The transformation

$$e_1' = e_1, \quad e_2' = \alpha^{-1}e_2, \quad e_3' = e_3, \quad e_4' = e_4$$

gives $F_2(1, 0)$ that is $\mathcal{R}_5$.

Suppose that $\beta \neq 0$. Then it is easy to see that the transformation

$$e_1' = \beta e_1, \; e_2' = \beta e_2, \; e_3' = \beta^2 e_3, \; e_4' = \beta^3 e_4$$

leads to the type

$$F_2(\alpha, 1) : [e_1, e_1] = e_3, \; [e_1, e_2] = \alpha e_4, \; [e_2, e_2] = e_4, \; [e_3, e_1] = e_4.$$

Note that the possible base change inside of $F_2(\alpha, 1)$ is as follows

$$e_1' = A_1e_1 + A_2e_2, \quad e_2' = B_2e_2 - A_1A_2B_2^{-1}e_3,$$
$$e_3' = A_1^2e_3 + A_2(A_1\alpha + A_2)e_4, \quad e_4' = A_1^3e_4,$$

where $A_1B_2 \neq 0$. This yields $\alpha' = \frac{B_2(A_1\alpha+A_2)}{A_1^3}$ and $B_2^2 = A_1^3$. Now we put $A_2 = -A_1\alpha$ to obtain $\alpha' = 0$. Therefore the algebras $F_2(\alpha, 1)$ are isomorphic to the algebra $\mathcal{R}_6$ with the following table of multiplications:

$$[e_1, e_1] = e_3, [e_2, e_2] = e_4, [e_3, e_1] = e_4.$$

We note that the algebra $F_2(0, 0)$ is split.

It just remains to show that the obtained algebras are not isomorphic to each other. Indeed, the algebra $\mathcal{R}_5$ is not isomorphic to the algebra $\mathcal{R}_6$, by dimension reasons of the left annihilators.

Let us consider the last family of filiform Leibniz algebras: $F_3(\alpha,\beta,\gamma)$.

An algebra of the family $F_3(\alpha,\beta,\gamma)$ is non Lie if and only if at least one of α,β,γ is not zero. Since we know already the classification of four-dimensional filiform Lie algebras (the non split filiform Lie algebra of dimension four has the table of multiplication $[e_2, e_1] = e_3, [e_3, e_1] = e_4, [e_1, e_2] = -e_3, [e_1, e_3] = -e_4$), we shall consider only non Lie algebras of the family $F_3(\alpha,\beta,\gamma)$, that is, at least one of α,β,γ is not zero. Without loss of generality, one can assume that $\alpha \neq 0$. Taking the transformation

$$e_1' = e_1, \ e_2' = \alpha e_2, \ e_3' = \alpha e_3, \ e_4' = \alpha e_4$$

we obtain $\alpha = 1$.

To treat the family $F_3(1,\beta,\gamma)$ we consider the general base change

$$e_1' = A_1e_1 + A_2e_2 + A_3e_3, \ e_2' = B_1e_1 + B_2e_2 + B_3e_3.$$

Then express the new basis $\{e_1', e_2', e_3', e_4'\}$ of the algebras from $F_3'(1,\beta',\gamma')$ with respect to the old basis $\{e_1, e_2, e_3, e_4\}$ and comparing the coefficients we obtain the following relations

$$A_1^2 + A_1A_2\gamma + A_2^2\beta = A_1^2B_2, \ B_1 = 0, \ A_1^2B_2 \neq 0,$$
$$\beta' = \frac{B_2\beta}{A_1^2}, \ \gamma' = \frac{A_1\gamma + 2A_2\beta}{A_1^2}.$$

Note that

$$\gamma'^2 - 4\beta' = \frac{1}{A_1^2}(\gamma^2 - 4\beta).$$

Consider the case $\beta = 0$. Then $\beta' = 0$.

So, in this case if $\gamma = 0$, then $\gamma' = 0$ and we get $\mathcal{R}_7$. But if $\gamma \neq 0$, then setting $A_1 = \gamma, B_2 = 1$ and $A_2 = 0$ we obtain $\gamma' = 1$, and we get $\mathcal{R}_8$.

Consider the case $\beta \neq 0$. Then putting $B_2 = \frac{A_1^2}{\beta}$ we get $\beta' = 1$.

If $\gamma^2 - 4\beta = 0$, then taking $A_2 = \frac{-2A_1\gamma+4A_1^2}{\gamma^2}$ and as A_1 any non zero complex number we obtain $\gamma' = 2$. Thus, in this case we get $\mathcal{R}_9$.

If $\gamma^2 - 4\beta \neq 0$, then putting $A_1 = \sqrt{\frac{4\beta-\gamma^2}{4}}$ and $A_2 = -\frac{\gamma}{2\beta}\sqrt{\frac{4\beta-\gamma^2}{4}}$ we obtain $\gamma' = 0$ and it is $\mathcal{R}_{10}$.

Now we suppose that the algebra L has the type iii) in Proposition 3.4, that L is a Leibniz algebra with $\chi(L) := (4,2,0,0)$ or $(4,1,0,0)$, in particular, L is associative. Then all the results on associative algebras are applicable to this case and we deal with associative algebras.

We consider the associative algebra $A = L \oplus \mathbb{C}\mathbf{1}$. It is five-dimensional and unital. Now by using properties of L we remove from Mazzola's list [127] inappropriate algebras. The first condition concerns the number of central idempotents. In view of this condition A is not isomorphic to the algebras with numbers 1–15, 25, 26, 38, 39, 55. Then in view of Corollary 3.1 and due to the fact that the image of an idempotent element under isomorphism is an idempotent element we obtain that A is not isomorphic to each of the algebras 16–22, 27, 28, 29, 40, 41, 46, 47, 52, 58 in the list of Mazzola. Moreover, the condition $L^3 = 0$ implies that A is not isomorphic to the algebras 23, 24, 33, 34, 44. Now it remains to pick up the remaining algebras.

N 30. Let $A = \mathbb{C}\langle x, y\rangle/(xy + yx, xy - yx + y^2 - x^2) + (x, y)^3$. Choose the following basis of A: $e_0 = \mathbf{1}, e_1 = x, e_2 = y, e_3 = xy, e_4 = x^2$. Then the subalgebra $L = \text{Span}\{e_1, e_2, e_3, e_4\}$ has the composition law:

$$[e_1, e_1] = e_4, [e_1, e_2] = e_3, [e_2, e_1] = -e_3, [e_2, e_2] = -2e_3 + e_4.$$

This is the algebra $\mathcal{R}_{11}$.

N 31. Let $A = \mathbb{C}\langle x, y\rangle/(x^2, xy + y^2) + (x, y)^3$. Choose the basis of A

as $e_0 = \mathbf{1}, e_1 = x, e_2 = y, e_3 = xy, e_4 = yx$. Then the subalgebra $L = \text{Span}\{e_1, e_2, e_3, e_4\}$ has the following composition law:

$$[e_1, e_2] = e_3, [e_2, e_1] = e_4, [e_2, e_2] = -e_3.$$

This coincides with $\mathcal{R}_{12}$.

N 32. Let $A = \mathbb{C}\langle x, y\rangle / (xy + y^2, \alpha x^2 + yx) + (x, y)^3$. As a basis of A we take the vectors $e_0 = \mathbf{1}, e_1 = x, e_2 = y, e_3 = x^2, e_4 = xy$ of A. Consider the subalgebra $L = \text{Span}\{e_1, e_2, e_3, e_4\}$. It has the following table of multiplication:

$$[e_1, e_1] = e_3, [e_1, e_2] = e_4, [e_2, e_1] = -\alpha e_3, [e_2, e_2] = -e_4.$$

For distinct values of the parameter α these algebras are not isomorphic. In this case L is $\mathcal{R}_{13}(\alpha)$.

N 35. Let $A = \mathbb{C}\langle x, y, z\rangle / (xz, yz, zx, zy, x^2 - y^2, x^2 - z^2, xy + yx, \alpha x^2 + yx)$, where $\alpha \neq 0$. Choose the following basis for A : $e_0 = \mathbf{1}, e_1 = x, e_2 = y, e_3 = z, e_4 = x^2$. Then the subalgebra $L = \text{Span}\{e_1, e_2, e_3, e_4\}$ has the following composition law:

$$[e_1, e_1] = e_4, [e_1, e_2] = \alpha e_4, [e_2, e_1] = -\alpha e_4, [e_2, e_2] = e_4,$$

$[e_3, e_3] = e_4$ and it is $\mathcal{R}_{14}(\alpha)$.

We note that the algebras $\mathcal{R}_{14}(\alpha_1)$ and $\mathcal{R}_{14}(\alpha_2)$ $(\alpha_1 \neq \alpha_2)$ are not isomorphic except of the case, where $\alpha_2 = -\alpha_1$, in the latter case they are isomorphic [127].

N 36. Let $A = \mathbb{C}\langle x, y, z\rangle / (x^2, yz, zy, z^2, xy - xz, xy + yx, yx + y^2, yx + zx)$. As the required basis here we choose $e_0 = \mathbf{1}, e_1 = x, e_2 = y, e_3 = z, e_4 = xy$. Then the subalgebra $L = \text{Span}\{e_1, e_2, e_3, e_4\}$ has the following law of multiplication

$$[e_1, e_2] = e_4, [e_1, e_3] = e_4, [e_2, e_1] = -e_4, [e_2, e_2] = e_4, [e_3, e_1] = e_4$$

and it is isomorphic to $\mathcal{R}_{15}$ from the theorem.

N 37. Let $A = \mathbb{C}\langle x, y, z\rangle / (xz, y^2, yz, zx, zy, x^2 - z^2, x^2 - xy, x^2 + yx)$. As a basis we take $e_0 = \mathbf{1}, e_1 = x, e_2 = y, e_3 = z, e_4 = xy$. Then the subalgebra $L = \text{Span}\{e_1, e_2, e_3, e_4\}$ is isomorphic to the algebra:

$$\mathcal{R}_{16} : [e_1, e_1] = e_4, [e_1, e_2] = e_4, [e_2, e_1] = -e_4, [e_3, e_3] = e_4.$$

N 42. Let $A = \mathbb{C}\langle x, y\rangle/(x^2, y^2)+(x, y)^3$. As a basis we take $e_0 = \mathbf{1}, e_1 = x, e_2 = y, e_3 = xy, e_4 = yx$. Then the table of multiplication of the subalgebra $L = \text{Span}\{e_1, e_2, e_3, e_4\}$ is the same as:

$$\mathcal{R}_{17} : [e_1, e_2] = e_3, [e_2, e_1] = e_4.$$

N 43. Let $A = \mathbb{C}[x, y]/(y^3, xy, x^3)$ and choose the basis

$$e_0 = \mathbf{1}, e_1 = x, e_2 = y, e_3 = x^2, e_4 = y^2$$

of A. Then the subalgebra $L = \text{Span}\{e_1, e_2, e_3, e_4\}$ with the table of multiplication:

$$[e_1, e_1] = e_3, [e_2, e_2] = e_4$$

is split and it is isomorphic to the algebra $\mathcal{R}_{23}$.

N 48. Let $A = \mathbb{C}\langle x, y\rangle/(x^2, xy + yx) + (x, y)^3$. As a basis we choose $e_0 = \mathbf{1}, e_1 = x, e_2 = y, e_3 = xy, e_4 = y^2$. Then the subalgebra $L = \text{Span}\{e_1, e_2, e_3, e_4\}$ coincides with the algebra:

$$\mathcal{R}_{18} : [e_1, e_2] = e_3, [e_2, e_1] = -e_3, [e_2, e_2] = e_4.$$

N 49 ($\alpha = 1$). Let $A = \mathbb{C}\langle x, y\rangle/(x^2, xy) + (x, y)^3$. As a basis we take $e_0 = \mathbf{1}, e_1 = x, e_2 = y, e_3 = y^2, e_4 = yx$. Then the subalgebra $L = \text{Span}\{e_1, e_2, e_3, e_4\}$ is isomorphic to:

$$\mathcal{R}_{19} : [e_2, e_1] = e_4, [e_2, e_2] = e_3.$$

N 49 ($\alpha \neq 1$). Let us now consider N49($\alpha \neq 1$) i.e.,

$$A = \mathbb{C}\langle x, y\rangle / \left(x^2, (1 + \alpha)xy + (1 - \alpha)yx\right) + (x, y)^3.$$

As a basis of A we can choose the vectors

$$e_0 = \mathbf{1}, e_1 = x, e_2 = y, e_3 = y^2, e_4 = xy$$

and the subalgebra $L = \text{Span}\{e_1, e_2, e_3, e_4\}$ is isomorphic to

$$\mathcal{R}_{20}(\alpha) : [e_1, e_2] = e_4, [e_2, e_1] = \frac{(1 + \alpha)}{(1 - \alpha)}e_4, [e_2, e_2] = e_3.$$

It is easy to see that for different values of α we obtain non isomorphic algebras.

N 50. Let $A = \mathbb{C}\langle x, y, z\rangle / (x^2, xz, y^2, yz, zx, zy, xy+yx, yx+z^2)$. As a basis of A we choose the vectors $e_0 = \mathbf{1}, e_1 = x, e_2 = y, e_3 = z, e_4 = xy$. Then the subalgebra $L = \mathrm{Span}\{e_1, e_2, e_3, e_4\}$ is isomorphic to the algebra:

$$\mathcal{R}_{21} : [e_1, e_2] = e_4, [e_2, e_1] = -e_4, [e_3, e_3] = e_4.$$

N 51. Let $A = \mathbb{C}\langle x, y, z\rangle / (xz, yz, zx, zy, x^2 - y^2, x^2 - z^2, xy, yx)$. As a basis we choose the set of vectors $e_0 = \mathbf{1}, e_1 = x, e_2 = y, e_3 = z, e_4 = x^2$. Then the subalgebra $L = \mathrm{Span}\{e_1, e_2, e_3, e_4\}$ has the following table of multiplication:

$$[e_1, e_1] = e_4, [e_2, e_2] = e_4, [e_3, e_3] = e_4.$$

But this algebra can be included into the family of algebras $\mathcal{R}_{14}(\alpha)$ with $\alpha = 0$.

Due to Proposition 3.2 the obtained algebras are not pairwise isomorphic.

By adding Lie algebras and split Leibniz algebras, i.e., Leibniz algebras which are direct sums of proper ideals we get the algebras $\mathcal{R}_{22} - \mathcal{R}_{28}$. □

Note 3.1. *All the other algebras of Mazzola's list with* $L^3 = \{0\}$ *are either Lie algebras or split Leibniz algebras.*

Summarizing the classification result of the above theorem, the result on classifications of complex nilpotent Lie algebras with dimensions at most four and complex nilpotent Leibniz algebras of dimensions at most three, we obtain the complete classification of complex nilpotent Leibniz algebras of the dimension at most four.

3.4 FOUR-DIMENSIONAL SOLVABLE LEIBNIZ ALGEBRAS

Here we shall consider only non nilpotent and non split complex solvable Leibniz algebras. This section along with [50], [54] and [55] completes the classification of four-dimensional solvable Leibniz algebras. We divide this section into two subsections. In the first subsection we study four-dimensional solvable Leibniz algebras with a two-dimensional nilradical, while the second part is devoted to the study of solvable Leibniz algebras with a three-dimensional nilradical.

Let L be a solvable Leibniz algebra. Then it can be decomposed as a direct sum of vector spaces $L = N + Q$, where N is the nilradical and Q is the complementary vector space. Since the square of a solvable algebra is a nilpotent ideal (due to Proposition 2.2) we obtain that L^2 is nilpotent, i.e., $L^2 \subseteq N$ and consequently, $Q^2 \subseteq N$.

Definition 3.2. *Let $d_1, d_2, \ldots, d_n$ be derivations of a Leibniz algebra L. The derivations $d_1, d_2, \ldots, d_n$ are said to be nil-independent if any non trivial linear combination $a_1d_1 + a_2d_2 + \cdots + a_nd_n$ is not nilpotent, where $a_1, a_2, \ldots, a_n \in \mathbb{C}$.*

In other words, if for $a_1, a_2, \ldots, a_n \in \mathbb{C}$ there exists a natural number k such that $(a_1d_1 + a_2d_2 + \cdots + a_nd_n)^k = 0$ then $a_1 = a_2 = \ldots = a_n = 0$.

Lemma 3.1. *Suppose that for $x \in Q$ the operator $R_x|_N$ is nilpotent. Then the subspace $V = \mathrm{Span}\{x + N\}$ is a nilpotent ideal of the algebra L.*

Proof. Since $L^2 \subseteq N$, the subspace V is an ideal of L. We claim that it is nilpotent as well. If $a \in N$, then $R_a|_N$ is a nilpotent operator. Let us suppose that $k \in \mathbb{N}$ is such that $(R_a|_N)^k = 0$, then $(R_a|_V)^{k+1} = 0$. Hence $R_a|_V$ is nilpotent. Since V is an ideal of the solvable Leibniz algebra L, then the space of inner derivations Inner(V) of V is a solvable Lie subalgebra of End(V), and thus by Lie theorem for Lie algebras there exists a basis such that $R_a|_V$ and $R_x|_V$ are upper triangular; moreover, $R_a|_V$ is nilpotent, which means that $R_a|_V$ has zero diagonal elements. On the other hand, by the assumption, $R_x|_N$ is nilpotent, then with the similar argument as the previous one, there exists $s \in N$ such that $(R_x|_N)^s = 0$, then $(R_x|_V)^{s+1} = 0$. Summarizing, we obtain that $R_a|_V$ and $R_x|_V$ are nilpotent and upper triangular, hence $R_a|_V + R_x|_V$ also is nilpotent. Thus, by Engel's theorem for Leibniz algebras, V is a nilpotent ideal. □

In [54], Casas et al. proved the following theorem, which is applied in this section.

Theorem 3.4. *Let L be a solvable Leibniz algebra and N be its nilradical. Then the dimension of the complementary vector space to N is not greater than the maximal number of nil-independent derivations of N.*

Proof. We claim that for $x \in Q$, the operator $R_x|_N$ is a non nilpotent outer derivation of N. Indeed, if there exists $x \in Q$ such that the operator $R_x|_N$ is nilpotent, then the subspace $V = \operatorname{Span}\{x + N\}$ is a nilpotent ideal of the algebra L by Lemma 3.1 contradicting the maximality of N. Let $\{x_1, \ldots, x_m\}$ be a basis of Q. Then the operators $R_{x_1}|_N, R_{x_2}|_N, \ldots, R_{x_m}|_N$ are nil-independent, since if for some scalars $\alpha_1, \ldots, \alpha_m \in \mathbb{C}$ we have $\left(\sum_{i=1}^{m} \alpha_i R_{x_i}|_N\right)^k = 0$, then $R^k_{y|_N} = 0$, where $y = \sum_{i=1}^{m} \alpha_i x_i$. Hence $y = 0$, and so $\alpha_i = 0$ for all $i = 1, \ldots, m$. Therefore, we see that the dimension of Q is bounded by the maximal number of nil-independent derivations of the nilradical N. Moreover, similar to the case of Lie algebras, for a solvable Leibniz algebra L we also have the inequality $\dim N \geq \frac{\dim L}{2}$. □

According to the theorem above the dimension of the nilradical of four-dimensional solvable Leibniz algebra can be equal to two or three.

The classifications of two and three-dimensional nilpotent complex Leibniz algebras have been given in Section 3.1.2. We give the results in the following two theorems.

Theorem 3.5. *Let L be a two-dimensional nilpotent Leibniz algebra. Then L either is $\mathbb{C}^2$ (abelian) or isomorphic to*

$$\mu_1 : [e_1, e_1] = e_2.$$

Theorem 3.6. *Let L be a three-dimensional nilpotent Leibniz algebra. Then L is isomorphic to one of the following pairwise non-isomorphic algebras:*

$$\begin{aligned}
\lambda_1 &:= \mathbb{C}^3 - \textit{abelian},\\
\lambda_2 &:= \mu_1 \oplus \mathbb{C},\\
\lambda_3 &:= [e_1, e_2] = e_3, \ [e_2, e_1] = -e_3,\\
\lambda_4(\alpha) &:= [e_1, e_1] = e_3, \ [e_2, e_2] = \alpha e_3, \ [e_1, e_2] = e_3,\\
\lambda_5 &:= [e_2, e_1] = e_3, \ [e_1, e_2] = e_3,\\
\lambda_6 &:= [e_1, e_1] = e_2, \ [e_2, e_1] = e_3.
\end{aligned}$$

Further for a convenience we shall consider another form of the family of $\lambda_4(\alpha)$. In fact, we shall transform the family of algebras $\lambda_4(\alpha)$

to the form in which nil-independent derivations of new algebras λ_4', $\lambda_4'(\beta)$ have diagonal forms.

Namely, by choosing an appropriate basis we represent the parametric family $\lambda_4(\alpha)$ as two non isomorphic algebras:

$$\lambda_4'(\beta) := \begin{cases} [e_2, e_1] = e_3, \\ [e_1, e_2] = \beta e_3, \ \text{with}\, \beta = \frac{\sqrt{1-4\alpha}-1}{\sqrt{1-4\alpha}+1}, \alpha \neq \frac{1}{4}, \end{cases}$$

$$\lambda_4' := \begin{cases} [e_1, e_1] = e_3, \\ [e_2, e_1] = e_3, \\ [e_1, e_2] = -e_3. \end{cases}$$

Later we shall use a description of the derivation algebras of three-dimensional nilpotent complex Leibniz algebras given above. The required description is given in a matrix form as follows.

Proposition 3.5. *There exists a basis such that the derivations of the algebras λ_1, λ_2, λ_3, λ_4', $\lambda_4'(\beta)$, λ_5 and λ_6 are given as follows*

$$\mathrm{Der}(\lambda_1) = M_3(\mathbb{C});$$

$$\mathrm{Der}(\lambda_2) = \left\{ A \in M_3(\mathbb{C}) | \ A = \begin{pmatrix} a_1 & a_2 & a_3 \\ 0 & b_2 & b_3 \\ 0 & 0 & 2a_1 \end{pmatrix} \right\};$$

$$\mathrm{Der}(\lambda_3) = \left\{ A \in M_3(\mathbb{C}) | \ A = \begin{pmatrix} a_1 & a_2 & a_3 \\ b_1 & b_2 & b_3 \\ 0 & 0 & a_1 + b_2 \end{pmatrix} \right\};$$

$$\mathrm{Der}(\lambda_4') = \left\{ A \in M_3(\mathbb{C}) | \ A = \begin{pmatrix} a_1 & a_2 & a_3 \\ 0 & 2a_1 & b_3 \\ 0 & 0 & 3a_1 \end{pmatrix} \right\};$$

$$\mathrm{Der}(\lambda_4'(\beta)) = \left\{ A \in M_3(\mathbb{C}) | \ A = \begin{pmatrix} a_1 & 0 & a_3 \\ 0 & b_2 & b_3 \\ 0 & 0 & a_1 + b_2 \end{pmatrix} \right\};$$

$$\mathrm{Der}(\lambda_5) = \left\{ A \in M_3(\mathbb{C}) | \ A = \begin{pmatrix} a_1 & 0 & a_3 \\ 0 & b_2 & b_3 \\ 0 & 0 & a_1 + b_2 \end{pmatrix} \right\};$$

$$\mathrm{Der}(\lambda_6) = \left\{ A \in M_3(\mathbb{C}) | \ A = \begin{pmatrix} a_1 & 0 & a_3 \\ 0 & 2a_1 & a_2 \\ 0 & 0 & 3a_1 \end{pmatrix} \right\},$$

where $a_i, b_j, c_k \in \mathbb{C}$ with $i, j, k = 1, 2, 3$.

Proof. The proof is carried out by straightforward checking the derivation property and using the table of multiplications of the algebras. □

3.4.1 Four-dimensional solvable Leibniz algebras with two-dimensional nilradical

Let us now consider Leibniz algebras whose nilradical is two-dimensional. Here is the classification result of these algebras.

Theorem 3.7. *Let L be a four-dimensional solvable Leibniz algebra with a two-dimensional nilradical. Then, it is isomorphic to one of the following pairwise non-isomorphic algebras with the basis $\{e_1, e_2, x, y\}$:*

$$R_1 := \begin{cases} [e_1, x] = e_1, \\ [e_2, y] = e_2, \end{cases} \quad R_2 := \begin{cases} [e_1, x] = e_1, \\ [e_2, y] = e_2, \\ [x, e_1] = -e_1, \\ [y, e_2] = -e_2, \end{cases} \quad R_3 := \begin{cases} [e_1, x] = e_1, \\ [e_2, y] = e_2, \\ [y, e_2] = -e_2. \end{cases}$$

Note that the basis $\{e_1, e_2, x, y\}$ is the extension of the basis $\{e_1, e_2\}$ of the corresponding nilradicals.

Proof. Recall that the classification of two-dimensional nilpotent Leibniz algebras is given by Theorem 3.5. Let L be a four-dimensional solvable Leibniz algebra with two-dimensional nilradical μ. Since the number of nil-independent derivations of μ_1 is equal to 1, we conclude that μ must be an abelian algebra. Therefore, considering the basis $\{x, y, e_1, e_2\}$, we write the table of L as follows

$$\begin{cases} [e_1, x] = a_1e_1 + a_2e_2, & [e_2, x] = a_3e_1 + a_4e_2, \\ [e_1, y] = b_1e_1 + b_2e_2, & [e_2, y] = b_3e_1 + b_4e_2. \end{cases}$$

Since R_x and R_y are nil-independent derivations of μ, we conclude that $a_1 \neq 0$. Without loss of generality, we can assume that $a_1 = 1$, and taking the basis transformation $y' = y - b_1x$ we conclude that $b_1 = 0$.

It is easy to prove that the matrix of R_x may have one of the following forms:

$$\begin{pmatrix} 1 & 0 \\ 0 & a_4 \end{pmatrix} \text{ or } \begin{pmatrix} 1 & 1 \\ 0 & 1 \end{pmatrix}.$$

Therefore, we consider the following cases:

Case 1. If $\begin{pmatrix} 1 & a_2 \\ a_3 & a_4 \end{pmatrix} \simeq \begin{pmatrix} 1 & 0 \\ 0 & a_4 \end{pmatrix}$, then we can write:

$$\begin{cases} [e_1, x] = e_1, & [e_2, x] = a_4e_2, \\ [e_1, y] = b_2e_2, & [e_2, y] = b_3e_1 + b_4e_2. \end{cases}$$

Applying the Leibniz identity to $[e_1,[x,y]]$ and $[e_2,[x,y]]$, we get the restrictions $b_2(1-a_4)=0$ and $b_3(1-a_4)=0$. Thus, we have to consider the following subcases:

Case 1.1. Let $a_4 = 1$. It is clear that $R_y :\simeq \begin{pmatrix} 0 & b_2 \\ b_3 & b_4 \end{pmatrix}$ is congruent to one of the following matrices:

$$\begin{pmatrix} 1 & 0 \\ 0 & b_4 \end{pmatrix} \quad \text{or} \quad \begin{pmatrix} 1 & 1 \\ 0 & 1 \end{pmatrix}.$$

It is easy to see that $R_y :\simeq \begin{pmatrix} 1 & 0 \\ 0 & b_4 \end{pmatrix}$, with $b_4 \neq 1$, otherwise $R_x - R_y$ would be nilpotent, which is impossible. Taking the basis transformation $x' = \frac{b_4}{b_4-1}x - \frac{1}{b_4-1}y$ and $y' = \frac{1}{b_4-1}y - \frac{1}{b_4-1}x$, we obtain the following products

$$[e_1,x]=e_1, \quad [e_2,x]=0, \quad [e_1,y]=0, \quad [e_2,y]=e_2.$$

Case 1.2. If $a_4 \neq 1$, then we get $b_2 = b_3 = 0$. Since R_x and R_y are nil-independent derivations of μ, we obtain $b_4 \neq 0$. Considering the base change $x' = x - \frac{a_4}{b_4}y$, $y' = \frac{1}{b_4}y$, we again get

$$[e_1,x]=e_1, \quad [e_2,x]=0, \quad [e_1,y]=0, \quad [e_2,y]=e_2.$$

Case 2. If $R_x :\simeq \begin{pmatrix} 1 & 1 \\ 0 & 1 \end{pmatrix}$, applying the Leibniz identity to $[e_1,[x,y]]$, we get $b_3 = b_4 = 0$, giving rise to a contradiction with the assumption that R_y is not nilpotent.

Therefore,

$$[e_1,x]=e_1, \quad [e_2,x]=0, \quad [e_1,y]=0, \quad [e_2,y]=e_2.$$

Let us consider the remaining products in L. Suppose that

$$\begin{cases} [x,e_1]=\alpha_1e_1+\alpha_2e_2, & [x,e_2]=\alpha_3e_1+\alpha_4e_2 \\ [y,e_1]=\beta_1e_1+\beta_2e_2, & [y,e_2]=\beta_3e_1+\beta_4e_2, \\ [x,x]=c_1e_1+c_2e_2, & [x,y]=c_3e_1+c_4e_2, \\ [y,x]=d_1e_1+d_2e_2, & [y,y]=d_3e_1+d_4e_2. \end{cases}$$

By using the basis change $x' = x - c_1e_1 - c_4e_2, y' = y - d_1e_1 - d_4e_2$, we get

$$c_1 = c_4 = d_1 = d_4 = 0.$$

Applying the Leibniz identity we obtain

$$\alpha_2 = \alpha_3 = \alpha_4 = \beta_1 = \beta_2 = \beta_3 = c_2 = c_3 = d_2 = d_3 = 0,$$

and

$$\alpha_1^2 + \alpha_1 = 0, \quad \beta_3^2 + \beta_3 = 0.$$

As a result we have the following cases:

Case 2.1. If $\alpha_1 = \beta_3 = 0$, we have R_1.

Case 2.2. If $\alpha_1 = \beta_3 = -1$, we get R_2.

Case 2.3. If $(\alpha_1, \beta_3) = (0, -1)$ or $(\alpha_1, \beta_3) = (-1, 0)$, we obtain R_3.

□

3.4.2 Four-dimensional solvable Leibniz algebras with three-dimensional nilradical

In this section we give a complete list of four-dimensional solvable complex Leibniz algebras with three-dimensional nilradical. We make use of the list of algebras given in Theorem 3.6.

Note that for a four-dimensional solvable Leibniz algebra L with three-dimensional nilradical N, there exists a basis $\{e_1, e_2, e_3, x\}$ such that the right multiplication operator R_x is non-nilpotent derivation on N, where $\{e_1, e_2, e_3\}$ is a basis of N.

Proposition 3.6. *Let L be a four-dimensional solvable Leibniz algebra, whose nilradical is three-dimensional abelian algebra. Then L is isomorphic to one of the following pairwise non isomorphic algebras*

$$L_1(\mu_2, \mu_3) := \begin{cases} [e_1, x] = e_1, \\ [e_2, x] = \mu_2 e_2, \\ [e_3, x] = \mu_3 e_3, \\ [x, e_1] = -e_1, \\ [x, e_2] = -\mu_2 e_2, \\ [x, e_3] = -\mu_3 e_3, \end{cases} \qquad L_2(\mu_2, \mu_3) := \begin{cases} [e_1, x] = e_1, \\ [e_2, x] = \mu_2 e_2, \\ [e_3, x] = \mu_3 e_3, \\ [x, e_1] = -e_1, \\ [x, e_2] = -\mu_2 e_2, \\ \textit{where } \mu_3 \neq 0, \end{cases}$$

$$L_3(\mu_2,\mu_3) := \begin{cases} [e_1,x] = e_1, \\ [e_2,x] = \mu_2 e_2, \\ [e_3,x] = \mu_3 e_3, \\ [x,e_1] = -e_1, \\ \text{where } \mu_2,\mu_3 \neq 0, \end{cases} \qquad L_4(\mu_2,\mu_3) := \begin{cases} [e_1,x] = e_1, \\ [e_2,x] = \mu_2 e_2, \\ [e_3,x] = \mu_3 e_3, \end{cases}$$

$$L_5(\mu_2) := \begin{cases} [e_1,x] = e_1, \\ [e_2,x] = \mu_2 e_2, \\ [x,e_1] = -e_1, \\ [x,e_2] = -\mu_2 e_2, \\ [x,x] = e_3. \end{cases} \qquad L_6(\mu_2) := \begin{cases} [e_1,x] = e_1, \\ [e_2,x] = \mu_2 e_2, \\ [x,e_1] = -e_1, \\ [x,x] = e_3, \\ \text{where } \mu_2 \neq 0, \end{cases}$$

$$L_7(\mu_2) := \begin{cases} [e_1,x] = e_1, \\ [e_2,x] = \mu_2 e_2, \\ [x,x] = e_3. \end{cases} \qquad L_8 := \begin{cases} [e_1,x] = e_1, \\ [x,e_1] = -e_1, \\ [x,e_2] = e_3, \end{cases}$$

$$L_9 := \begin{cases} [e_1,x] = e_1, \\ [x,e_2] = e_3. \end{cases} \qquad L_{10}(\mu_3) := \begin{cases} [e_1,x] = e_1 + e_2, \\ [e_2,x] = e_2, \\ [e_3,x] = \mu_3 e_3, \\ [x,e_1] = -e_1 - e_2, \\ [x,e_2] = -e_2, \end{cases}$$

$$L_{11} := \begin{cases} [e_1,x] = e_1 + e_2, \\ [e_2,x] = e_2, \\ [x,e_1] = -e_1 - e_2, \\ [x,e_2] = -e_2, \\ [x,x] = e_3, \end{cases} \qquad L_{12}(\mu_3 \neq 0) := \begin{cases} [e_1,x] = e_1 + e_2, \\ [e_2,x] = e_2, \\ [e_3,x] = \mu_3 e_3, \\ [x,e_1] = -e_1 - e_2, \\ [x,e_2] = -e_2, \\ [x,e_3] = -\mu_3 e_3, \end{cases}$$

$$L_{13}(\mu_3) := \begin{cases} [e_1,x] = e_1 + e_2, \\ [e_2,x] = e_2, \\ [e_3,x] = \mu_3 e_3, \end{cases} \qquad L_{14} := \begin{cases} [e_1,x] = e_1 + e_2, \\ [e_2,x] = e_2, \\ [x,x] = e_3, \end{cases}$$

$$L_{15}(\mu_3 \neq 0) := \begin{cases} [e_1,x] = e_1 + e_2, \\ [e_2,x] = e_2, \\ [e_3,x] = \mu_3 e_3, \\ [x,e_3] = -\mu_3 e_3, \end{cases} \qquad L_{16}(\alpha) := \begin{cases} [e_1,x] = e_2, \\ [e_3,x] = e_3, \\ [x,e_1] = \alpha e_2, \\ [x,e_3] = -e_3, \end{cases}$$

$$L_{17} := \begin{cases} [e_1, x] = e_2, \\ [e_3, x] = e_3, \\ [x, e_1] = -e_2, \\ [x, e_3] = -e_3, \\ [x, x] = e_2. \end{cases} \qquad L_{18} := \begin{cases} [e_1, x] = e_2, \\ [e_3, x] = e_3, \\ [x, e_3] = -e_3, \\ [x, x] = e_1. \end{cases}$$

$$L_{19}(\alpha) := \begin{cases} [e_1, x] = e_2, \\ [e_3, x] = e_3, \\ [x, e_1] = \alpha e_2, \end{cases} \qquad L_{20} := \begin{cases} [e_1, x] = e_2, \\ [e_3, x] = e_3, \\ [x, e_1] = -e_2, \\ [x, x] = e_2. \end{cases}$$

$$L_{21} := \begin{cases} [e_1, x] = e_2, \\ [e_3, x] = e_3, \\ [x, x] = e_1. \end{cases} \qquad L_{22} := \begin{cases} [e_1, x] = e_1 + e_2, \\ [e_2, x] = e_2 + e_3, \\ [e_3, x] = e_3, \end{cases}$$

$$L_{23} := \begin{cases} [e_1, x] = e_1 + e_2, \\ [e_2, x] = e_2 + e_3, \\ [e_3, x] = e_3, \\ [x, e_1] = e_1 - e_2, \\ [x, e_2] = e_2 - e_3, \\ [x, e_3] = e_3, \end{cases}$$

where $\alpha, \mu_i \in \mathbb{C}$.

Proof. Let $\{x, e_1, e_2, e_3\}$ be the basis of L extended from a basis $\{e_1, e_2, e_3\}$ of N. It is easy to see that the matrix of R_x may have one of the following forms:

$$\begin{pmatrix} \mu_1 & 0 & 0 \\ 0 & \mu_2 & 0 \\ 0 & 0 & \mu_3 \end{pmatrix}, \begin{pmatrix} \mu_1 & 1 & 0 \\ 0 & \mu_1 & 0 \\ 0 & 0 & \mu_3 \end{pmatrix}, \begin{pmatrix} \mu_1 & 1 & 0 \\ 0 & \mu_1 & 1 \\ 0 & 0 & \mu_1 \end{pmatrix}.$$

Let $R_x :\simeq \begin{pmatrix} \mu_1 & 0 & 0 \\ 0 & \mu_2 & 0 \\ 0 & 0 & \mu_3 \end{pmatrix}$. Then the multiplication in L can be written as follows:

$$\begin{cases} [e_1, x] = \mu_1 e_1, & [x, e_1] = \alpha_1 e_1 + \alpha_2 e_2 + \alpha_3 e_3, \\ [e_2, x] = \mu_2 e_2, & [x, e_2] = \beta_1 e_1 + \beta_2 e_2 + \beta_3 e_3, \\ [e_3, x] = \mu_3 e_3, & [x, e_3] = \gamma_1 e_1 + \gamma_2 e_2 + \gamma_3 e_3, \\ [x, x] = \delta_1 e_1 + \delta_2 e_2 + \delta_3 e_3. \end{cases}$$

Applying the Leibniz identity to $[x,[e_1,x]]$, $[x,[e_2,x]]$ and $[x,[e_3,x]]$, we obtain the following restrictions for the structure constants

$$\begin{array}{ll} \alpha_2(\mu_1-\mu_2)=0, & \alpha_3(\mu_1-\mu_3)=0, \\ \beta_1(\mu_1-\mu_2)=0, & \beta_3(\mu_2-\mu_3)=0, \\ \gamma_1(\mu_1-\mu_3)=0, & \gamma_2(\mu_2-\mu_3)=0. \end{array} \tag{3.23}$$

Applying the Leibniz identity to $[x,[x,e_1]]$, $[x,[x,e_2]]$ and $[x,[x,e_3]]$, we get

$$\begin{array}{l} (\mu_1+\alpha_1)\alpha_1+\alpha_2\beta_1+\alpha_3\gamma_1=0, \\ (\mu_1+\alpha_1)\alpha_2+\alpha_2\beta_2+\alpha_3\gamma_2=0, \\ (\mu_1+\alpha_1)\alpha_3+\alpha_2\beta_3+\alpha_3\gamma_3=0, \\ \beta_1\alpha_1+(\beta_2+\mu_2)\beta_1+\beta_3\gamma_1=0, \\ \beta_1\alpha_2+(\beta_2+\mu_2)\beta_2+\beta_3\gamma_2=0, \\ \beta_1\alpha_3+(\beta_2+\mu_2)\beta_3+\beta_3\gamma_3=0, \\ \gamma_1\alpha_1+\gamma_2\beta_1+(\gamma_3+\mu_3)\gamma_1=0, \\ \gamma_1\alpha_2+\gamma_2\beta_2+(\gamma_3+\mu_3)\gamma_2=0, \\ \gamma_1\alpha_3+\gamma_2\beta_3+(\gamma_3+\mu_3)\gamma_3=0, \end{array} \tag{3.24}$$

and the Leibniz identity applied to $[x,[x,x]]$ gives

$$\begin{array}{l} \alpha_1\delta_1+\beta_1\delta_2+\gamma_1\delta_3=0,\ \alpha_2\delta_1+\beta_2\delta_2+\gamma_2\delta_3=0, \\ \alpha_3\delta_1+\beta_3\delta_2+\gamma_3\delta_3=0. \end{array} \tag{3.25}$$

To specify the structure constants, we distinguish the following cases:

Case 1. Let $\mu_1\neq\mu_2$, $\mu_1\neq\mu_3$, $\mu_2\neq\mu_3$. Thanks to the restrictions (3.23) and (3.24) we get

$$\alpha_2=\alpha_3=\beta_1=\beta_3=\gamma_1=\gamma_2=0,$$

$$\begin{array}{lll} (\mu_1+\alpha_1)\alpha_1=0, & (\beta_2+\mu_2)\beta_2=0, & (\gamma_3+\mu_3)\gamma_3=0, \\ \alpha_1\delta_1=0, & \beta_2\delta_2=0, & \gamma_3\delta_3=0. \end{array} \tag{3.26}$$

Case 1.1. Let $\mu_1\mu_2\mu_3\neq 0$. Then, without loss of generality, we can assume that $\mu_1=1$. Therefore, we have to consider the following cases:

- If $\alpha_1\neq 0, \beta_2\neq 0, \gamma_3\neq 0$, then according to the restriction (3.26), we have

$$\alpha_1=-1,\quad \beta_2=-\mu_2,\quad \gamma_3=-\mu_3,\quad \delta_1=\delta_2=\delta_3=0.$$

Hence we obtain $L_1(\mu_2,\mu_3)$ for $\mu_2\neq 1$, $\mu_3\neq 1$, $\mu_2\neq\mu_3$ and $\mu_2\mu_3\neq 0$.

- If one of the parameters $\alpha_1,\beta_2,\gamma_3$ equals to zero and the other two do not, then without loss of generality we may assume that $\alpha_1 \neq 0, \beta_2 \neq 0$ and $\gamma_3 = 0$. Due to the restriction (3.26), we come to

$$\alpha_1 = -1, \quad \beta_2 = -\mu_2, \qquad \delta_1 = \delta_2 = 0.$$

  It suffices to make the transformation $x' = x - \frac{\delta_3}{\mu_3}e_3$, to get $\delta_3 = 0$. As a result we obtain $L_2(\mu_2,\mu_3)$ for $\mu_2 \neq 1,\ \mu_3 \neq 1,\ \mu_2 \neq \mu_3$ and $\mu_2\mu_3 \neq 0$.
- If two of the parameters $\alpha_1,\beta_2,\gamma_3$ are equal to zero and another one is not zero, then, without loss of generality, we can suppose that $\alpha_1 \neq 0,\ \beta_2 = 0$ and $\gamma_3 = 0$, . Due to the restriction (3.26), we get

$$\alpha_1 = -1, \qquad \delta_1 = 0.$$

  Considering the base change $x' = x - \frac{\delta_2}{\mu_2}e_2 - \frac{\delta_3}{\mu_3}e_3$, we get $\delta_2 = \delta_3 = 0$. This gives $L_3(\mu_2,\mu_3)$ for $\mu_2 \neq 1,\ \mu_3 \neq 1,\ \mu_2 \neq \mu_3$ and $\mu_2\mu_3 \neq 0$.
- If $\alpha_1 = 0, \beta_2 = 0$ and $\gamma_3 = 0$, then applying the base change $x' = x - \frac{\delta_1}{\mu_2}e_1 - \frac{\delta_2}{\mu_2}e_2 - \frac{\delta_3}{\mu_3}e_3$ we get $\delta_1 = \delta_2 = \delta_3 = 0$. Hence we obtain $L_4(\mu_2,\mu_3)$ with $\mu_2 \neq 1,\ \mu_3 \neq 1,\ \mu_2 \neq \mu_3$ and $\mu_2\mu_3 \neq 0$.

Case 1.2. Let one of the parameters μ_1,μ_2,μ_3 be equal to zero. Then, without loss of generality, we can assume that $\mu_1 = 1$, $\mu_3 = 0$. Moreover, according to the restriction (3.26), we have $\gamma_3 = 0$.

- Let $\alpha_1 \neq 0,\ \beta_2 \neq 0$. Then, by the restriction (3.26), we get

$$\alpha_1 = -1, \quad \beta_2 = -\mu_2, \qquad \delta_1 = \delta_2 = 0.$$

  Therefore the following cases occur:
  - If $\delta_3 = 0$, we obtain $L_1(\mu_2, 0)$ with $\mu_2 \notin \{0, 1\}$.
  - If $\delta_3 \neq 0$, by taking the basis transformation $e_3' = \delta_3 e_3$, we get $L_5(\mu_2)$ with $\mu_2 \notin \{0, 1\}$.
- Let only one of the parameters α_1,β_2 equal zero. Then, without loss of generality, we can assume that $\alpha_1 \neq 0$

and $\beta_2 = 0$. Moreover, due to the restriction (3.26) we conclude that

$$\alpha_1 = -1, \quad \delta_1 = 0.$$

The base change $x' = x - \frac{\delta_2}{\mu_2}e_2$ gives $\delta_2 = 0$. Again the following cases may occur:

- If $\delta_3 = 0$, we derive the algebra which is isomorphic to $L_2(0, \mu_3)$ with $\mu_3 \notin \{0, 1\}$.
- If $\delta_3 \neq 0$, by taking the basis transformation $e_3' = \delta_3 e_3$, we obtain $L_6(\mu_2)$ with $\mu_2 \notin \{0, 1\}$.

- Let $\alpha_1 = 0, \beta_2 = 0$. We apply the basis transformation $x' = x - \frac{\delta_1}{\mu_1}e_2 - \frac{\delta_2}{\mu_2}e_2$, to obtain $\delta_1 = \delta_2 = 0$. It is sufficient here to consider the following cases:
  - If $\delta_3 = 0$, then we get $L_4(\mu_2, 0)$ with $\mu_2 \notin \{0, 1\}$.
  - If $\delta_3 \neq 0$, then taking the basis transformation $e_3' = \delta_3 e_3$, we get $L_7(\mu_2)$ with $\mu_2 \notin \{0, 1\}$.

Case 2. Let two of the structure constants μ_1, μ_2, μ_3 be equal. Then, without loss of generality, we can assume that $\mu_1 = \mu_2$. We have to distinguish the following cases:

Case 2.1. Let $\mu_1 = \mu_2 \neq 0$. Then it is clear that we can assume that $\mu_1 = \mu_2 = 1$ and $\mu_3 \neq 1$. Moreover, due to (3.23), (3.24) and (3.25), we come to the following restrictions:

$$\alpha_3 = \beta_3 = \gamma_1 = \gamma_2 = 0,$$

$$\begin{aligned}
&(1 + \alpha_1)\alpha_1 + \alpha_2\beta_1 = 0, \\
&(1 + \alpha_1)\alpha_2 + \alpha_2\beta_2 = 0, \\
&\beta_1\alpha_1 + (1 + \beta_2)\beta_1 = 0, \\
&\beta_1\alpha_2 + (1 + \beta_2)\beta_2 = 0, \\
&(\gamma_3 + \mu_3)\gamma_3 = 0, \\
&\alpha_1\delta_1 + \beta_1\delta_2 = 0, \\
&\alpha_2\delta_1 + \beta_2\delta_2 = 0, \\
&\gamma_3\delta_3 = 0.
\end{aligned} \tag{3.27}$$

Note that for $y \in \{e_1, e_2\}$, we have $[y, x] = y$. Thus, by any change of the basis $\{e_1, e_2\}$ the products $[e_1, x] = e_1$ and $[e_2, x] = e_2$ are unchanged.

The matrix $\begin{pmatrix} \alpha_1 & \alpha_2 \\ \beta_1 & \beta_2 \end{pmatrix}$ may have the following two Jordan forms $\begin{pmatrix} \alpha_1 & 0 \\ 0 & \beta_2 \end{pmatrix}$ and $\begin{pmatrix} \alpha_1 & 1 \\ 0 & \alpha_1 \end{pmatrix}$.

Case 2.1.1. If $\begin{pmatrix} \alpha_1 & \alpha_2 \\ \beta_1 & \beta_2 \end{pmatrix} :\simeq \begin{pmatrix} \alpha_1 & 0 \\ 0 & \beta_2 \end{pmatrix}$, i.e., $\alpha_2 = \beta_1 = 0$, then thanks to (3.27), we have the following restrictions:

$$(1+\alpha_1)\alpha_1, \quad (1+\beta_2)\beta_2 = 0, \quad (\gamma_3+\mu_3)\gamma_3 = 0,$$

$$\alpha_1\delta_1 = 0, \quad \beta_2\delta_2 = 0, \quad \gamma_3\delta_3 = 0.$$

It is clear that this case is similar to that of Case 1, with $\mu_1 = \mu_2 = 1$. Hence, similarly to Case 1 we get the following algebras:

- If $\mu_3 \neq 0$,
  - We get $L_1(1,\mu_3)$ with $\mu_3 \neq 1$, by considering $\alpha_1 \neq 0, \beta_2 \neq 0$ and $\gamma_3 \neq 0$.
  - We obtain $L_2(1,\mu_3)$ with $\mu_3 \neq 1$, by considering $\alpha_1 \neq 0, \beta_2 \neq 0$ and $\gamma_3 = 0$.
  - We derive $L_2(\mu_2, 1)$ with $\mu_2 \neq 1$, by considering $\alpha_1 \neq 0, \beta_2 = 0$ and $\gamma_3 \neq 0$.
  - We get $L_3(1,\mu_3)$ with $\mu_3 \neq 1$, by considering $\alpha_1 \neq 0, \beta_2 = 0$ and $\gamma_3 = 0$.
  - And finally we obtain $L_4(1,\mu_3)$ with $\mu_3 \neq 1$, by considering $\alpha_1 = 0, \beta_2 = 0$ and $\gamma_3 = 0$.
- If $\mu_3 = 0$, we obtain
  - $L_5(1)$ by considering $\alpha_1 \neq 0$ and $\beta_2 \neq 0$.
  - $L_6(1)$ by considering $\alpha_1 \neq 0$ and $\beta_2 = 0$.
  - $L_7(1)$ by considering $\alpha_1 = 0$ and $\beta_2 = 0$.

Case 2.1.2. If $\begin{pmatrix} \alpha_1 & \alpha_2 \\ \beta_1 & \beta_2 \end{pmatrix} :\simeq \begin{pmatrix} \alpha_1 & 1 \\ 0 & \alpha_1 \end{pmatrix}$, i.e., $\alpha_2 = 1$, $\beta_1 = 0$, $\beta_2 = \alpha_2$. Due to the restriction (3.27) we obtain the system of equations $(1+\alpha_1)\alpha_1 = 0, \quad 1+2\alpha_1 = 0$, which has no solution. Therefore in this case such algebra does not exist.

Case 2.2. Let $\mu_1 = \mu_2 = 0$. Then, without loss of generality, we can take $\mu_3 = 1$. A simple basis transformations shows that $\mu_1 = 1, \mu_2 = \mu_3 = 0$.

Due to the restrictions (3.23,) (3.24) and (3.25), we get $\alpha_2 = \alpha_3 = \beta_1 = \gamma_1 = 0$ and

$$\begin{aligned} &(1+\alpha_1)\alpha_1 = 0, && \beta_2^2 + \beta_3\gamma_2 = 0, && \beta_2\beta_3 + \beta_3\gamma_3 = 0,\\ & && \gamma_2\beta_2 + \gamma_3\gamma_2 = 0, && \gamma_2\beta_3 + \gamma_3^2 = 0,\\ &\alpha_1\delta_1 = 0, && \beta_2\delta_2 + \gamma_2\delta_3 = 0, && \beta_3\delta_2 + \gamma_3\delta_3 = 0. \end{aligned} \tag{3.28}$$

Similar to Case 2.1, we have to consider the following two subcases:

Case 2.2.1. Let $\begin{pmatrix} \beta_2 & \beta_3 \\ \gamma_2 & \gamma_3 \end{pmatrix} :\simeq \begin{pmatrix} \beta_2 & 0 \\ 0 & \gamma_3 \end{pmatrix}$, i.e., $\beta_3 = \gamma_2 = 0$. By (3.28), we obtain the following restrictions $\beta_2 = \gamma_3 = 0$, $(1+\alpha_1)\alpha_1 = 0$ and $\alpha_1\delta_1 = 0$.

- If $\alpha_1 = -1$, then $\delta_1 = 0$.
  If $(\delta_2, \delta_3) = (0, 0)$, we obtain $L_1(0, 0)$.
  If $(\delta_2, \delta_3) \neq (0, 0)$, we make the base change $e_3' = \delta_2 e_2 + \delta_3 e_3$, to obtain the algebra $L_5(0)$.
- If $\alpha_1 = 0$, then the base change $x' = x - \delta_1 e_1$, gives $[x, x] = \delta_2 e_2 + \delta_3 e_3$.
  If $(\delta_2, \delta_3) = (0, 0)$, we get $L_4(0, 0)$.
  If $(\delta_2, \delta_3) \neq (0, 0)$, it suffices to make the basis transformation $e_3' = \delta_2 e_2 + \delta_3 e_3$, to obtain $L_7(0)$.

Case 2.2.2. If $\begin{pmatrix} \beta_2 & \beta_3 \\ \gamma_2 & \gamma_3 \end{pmatrix} :\simeq \begin{pmatrix} \beta_2 & 1 \\ 0 & \beta_2 \end{pmatrix}$, i.e., $\beta_3 = 1$, $\gamma_2 = 0$ and $\gamma_3 = \beta_2$. Due to (3.28), we have the following restrictions $\beta_2 = 0$, $(1+\alpha_1)\alpha_1 = 0$, $\alpha_1\delta_1 = 0$ and $\delta_2 = 0$.

- If $\alpha_1 = -1$, hence $\delta_1 = 0$ and the structural constant δ_3 is determined by taking the change $x' = x - \delta_3 e_2$. Thus, we obtain the algebra L_8.
- If $\alpha_1 = 0$, then making the change of basis $x' = x - \delta_1 e_1 - \delta_3 e_2$, the algebra L_9 is obtained.

Case 3. Let $\mu_1 = \mu_2 = \mu_3 = 1$, then we consider the following subcases.

Case 3.1. Let $\begin{pmatrix} \alpha_1 & \alpha_2 & \alpha_3 \\ \beta_1 & \beta_2 & \beta_3 \\ \gamma_1 & \gamma_2 & \gamma_3 \end{pmatrix} :\simeq \begin{pmatrix} \alpha_1 & 0 & 0 \\ 0 & \beta_2 & 0 \\ 0 & 0 & \gamma_3 \end{pmatrix}$, i.e., $\alpha_2 = \alpha_3 = \beta_1 = \beta_3 = \gamma_1 = \gamma_2 = 0$.

In view of the restrictions (3.24) and (3.25), we have $(1+\alpha_1)\alpha_1 = 0$, $(1+\beta_2)\beta_2 = 0$, $(1+\gamma_3)\gamma_3 = 0$, $\alpha_1\delta_1 = 0$, $\beta_2\delta_2 = 0$ and $\gamma_3\delta_3 = 0$. It is clear that this case is similar to Case 1, with $\mu_1 = \mu_2 = \mu_3 = 1$. Hence, by treating similarly, we get the following algebras:

- $L_1(1,1)$ by considering $\alpha_1 \neq 0, \beta_2 \neq 0$ and $\gamma_3 \neq 0$.
- $L_2(1,1)$ by considering $\alpha_1 \neq 0, \beta_2 \neq 0$ and $\gamma_3 = 0$.
- $L_3(1,1)$ by considering $\alpha_1 \neq 0, \beta_2 = 0$ and $\gamma_3 = 0$.
- $L_4(1,1)$ by considering $\alpha_1 = 0, \beta_2 = 0$ and $\gamma_3 = 0$.

Case 3.2. Let $\begin{pmatrix} \alpha_1 & \alpha_2 & \alpha_3 \\ \beta_1 & \beta_2 & \beta_3 \\ \gamma_1 & \gamma_2 & \gamma_3 \end{pmatrix} :\simeq \begin{pmatrix} \alpha_1 & 1 & 0 \\ 0 & \alpha_1 & 0 \\ 0 & 0 & \gamma_3 \end{pmatrix}$, i.e., $\alpha_2 = 1$, $\alpha_3 = \beta_1 = \beta_3 = \gamma_1 = \gamma_2 = 0$, and $\beta_2 = \alpha_1$. Due to the restrictions (3.24) and (3.25), we obtain the following system of equations:

$$(1+\alpha_1)\alpha_1 = 0, \quad 1 + 2\alpha_1 = 0,$$

which has no solution. Therefore this case is impossible.

Case 3.3. Let $\begin{pmatrix} \alpha_1 & \alpha_2 & \alpha_3 \\ \beta_1 & \beta_2 & \beta_3 \\ \gamma_1 & \gamma_2 & \gamma_3 \end{pmatrix} :\simeq \begin{pmatrix} \alpha_1 & 1 & 0 \\ 0 & \alpha_1 & 1 \\ 0 & 0 & \alpha_1 \end{pmatrix}$, i.e., $\alpha_2 = \beta_3 = 1$, $\alpha_3 = \beta_1 = \gamma_1 = \gamma_2 = 0$, $\beta_2 = \gamma_3 = \alpha_1$. Due to the restrictions (3.24) and (3.25), we have the following system of equations:

$$(1+\alpha_1)\alpha_1 = 0, \quad 1 + 2\alpha_1 = 0,$$

which has no solution. Similarly, if $R_x :\simeq \begin{pmatrix} \mu_1 & 1 & 0 \\ 0 & \mu_1 & 0 \\ 0 & 0 & \mu_3 \end{pmatrix}$, we obtain the algebras $L_{10}(\mu_3) - L_{21}$. On the other hand, if $R_x :\simeq \begin{pmatrix} \mu_1 & 1 & 0 \\ 0 & \mu_1 & 1 \\ 0 & 0 & \mu_1 \end{pmatrix}$, the algebras L_{22}, L_{23} are obtained.

□

Proposition 3.7. *Let L be a four-dimensional solvable Leibniz algebra, whose nilradical is isomorphic to λ_2. Then, L is isomorphic to one* *of the following pairwise non isomorphic algebras:*

$$L_{24}(\gamma) := \begin{cases} [e_1, e_2] = e_3, \\ [e_1, x] = e_1, \\ [e_2, x] = \gamma e_2, \\ [e_3, x] = 2e_3, \\ [x, e_1] = -e_1, \\ [x, e_2] = -\gamma e_2. \end{cases} \qquad L_{25}(\delta) := \begin{cases} [e_1, e_1] = e_3, \\ [e_1, x] = e_1, \\ [e_2, x] = \delta e_2, \\ [e_3, x] = 2e_3 \\ [x, e_1] = -e_1. \end{cases}$$

$$L_{26} := \begin{cases} [e_1, e_1] = e_3, \\ [e_1, x] = e_1, \\ [e_3, x] = 2e_3, \\ [x, e_1] = -e_1, \\ [x, x] = e_2. \end{cases} \qquad L_{27} := \begin{cases} [e_1, e_1] = e_3, \\ [e_1, x] = e_1, \\ [e_2, x] = 2e_2 + e_3, \\ [e_3, x] = 2e_3, \\ [x, e_1] = -e_1. \end{cases}$$

$$L_{28} := \begin{cases} [e_1, e_1] = e_3, \\ [e_1, x] = e_1 + e_2, \\ [e_2, x] = e_2, \\ [e_3, x] = 2e_3, \\ [x, e_1] = -e_1 - e_2, \\ [x, e_2] = -e_2. \end{cases} \qquad L_{29} := \begin{cases} [e_1, e_1] = e_3, \\ [e_2, x] = e_2, \\ [x, e_2] = -e_2. \end{cases}$$

$$L_{30}(\lambda) := \begin{cases} [e_1, e_1] = e_3, \\ [e_2, x] = e_2, \\ [x, e_1] = e_3, \\ [x, e_2] = -e_2, \\ [x, x] = \lambda e_3. \end{cases} \qquad L_{31} := \begin{cases} [e_1, e_1] = e_3, \\ [e_2, x] = e_2, \\ [x, e_2] = -e_2, \\ [x, x] = -2e_3. \end{cases}$$

$$L_{32} := \begin{cases} [e_1, e_1] = e_3, \\ [e_2, x] = e_2. \end{cases} \qquad L_{33}(\mu) := \begin{cases} [e_1, e_1] = e_3, \\ [e_2, x] = e_2, \\ [x, e_1] = e_3, \\ [x, x] = \mu e_3. \end{cases}$$

$$L_{34} := \begin{cases} [e_1, e_1] = e_3, \\ [e_2, x] = e_2, \\ [x, e_1] = e_3, \\ [e_1, x] = e_3. \end{cases}$$

where $\gamma, \lambda, \mu \in \mathbb{C}$ and $\delta \in \mathbb{C} \setminus \{0\}$.

Proof. The proof is similar to that of the previous propositions. □

Below we present the classification of four-dimensional solvable Leibniz algebras whose nilradical is λ_3.

Proposition 3.8. *Let L be a four-dimensional solvable Leibniz algebra, whose nilradical is isomorphic to λ_3. Then, L is isomorphic to one of the following pairwise non-isomorphic algebras:*

$$L_{35}(\gamma) := \begin{cases} [e_1, e_2] = e_3, \\ [e_2, e_1] = -e_3, \\ [e_1, x] = e_1, \\ [e_2, x] = \gamma e_2, \\ [e_3, x] = (1+\gamma)e_3, \\ [x, e_1] = -e_1, \\ [x, e_2] = -\gamma e_2, \\ [x, e_3] = -(1+\gamma)e_3. \end{cases} \qquad L_{36} := \begin{cases} [e_1, e_2] = e_3, \\ [e_2, e_1] = -e_3, \\ [e_1, x] = e_1, \\ [e_2, x] = -e_2, \\ [x, e_1] = -e_1, \\ [x, e_2] = e_2, \\ [x, x] = e_3. \end{cases}$$

$$L_{37} := \begin{cases} [e_1, e_2] = e_3, \\ [e_2, e_1] = -e_3, \\ [e_1, x] = e_1 + e_2, \\ [e_2, x] = e_2, \\ [e_3, x] = 2e_3, \\ [x, e_1] = -e_1 - e_2, \\ [x, e_2] = -e_2, \\ [x, e_3] = -2e_3. \end{cases}$$

Proof. According to Proposition 3.5 and the multiplication law of λ_3, the multiplication of L can be written as follows:

$$\begin{aligned}
[e_1, e_2] &= e_3, \\
[e_2, e_1] &= -e_3, \\
[e_1, x] &= a_1 e_1 + a_2 e_2 + a_3 e_3, \\
[e_2, x] &= b_1 e_1 + b_2 e_2 + b_3 e_3, \\
[e_3, x] &= (a_1 + b_2) e_3.
\end{aligned}$$

Since $e_1, e_2 \notin \mathrm{Ann}_r(L)$, applying properties of the right annihilator we write:

$$\begin{aligned}
[x, e_1] &= -a_1 e_1 - a_2 e_2 + \alpha_3 e_3, \\
[x, e_2] &= -b_1 e_1 - b_2 e_2 + \beta_3 e_3, \\
[x, e_3] &= -(a_1 + b_2) e_3, \\
[x, x] &= \gamma_3 e_3.
\end{aligned}$$

Let us distinguish the following cases:

Case 1. If $(a_1, b_2) \neq (0, 0)$, without loss of generality, we may assume that $a_1 \neq 0$. Also note that $b_1 = 0$. Otherwise, we could consider two cases: $a_2 = 0$ or $a_2 \neq 0$. In the first case, considering the base change $e_1' = e_2$, $e_2' = e_1$, $e_3' = -e_3$, we get $b_1 = 0$. On the other hand, if $a_2 \neq 0$, applying the basis transformation $e_2' = e_2 + \frac{-(b_2-a_1)+\sqrt{(b_2-a_1)^2+4a_2b_1}}{2a_2} e_1$, we come to $b_1 = 0$. It suffices to make the basis transformation $x' = \frac{1}{a_1} x$ to obtain $a_1 = 1$.

Let us consider the following cases:

Case 1.1. If $b_2 \notin \{0, 1\}$, then applying the basis transformation:

$$e_1' = e_1 - \frac{a_2}{b_2 - 1} e_2 - \frac{a_3(b_2 - 1) - a_2 b_3}{b_2(b_2 - 1)} e_3, \quad e_2' = e_2 - b_3 e_3,$$

we obtain $a_2 = a_3 = b_3 = 0$.

The Leibniz identity applied to $[x, [e_1, x]]$, $[x, [e_2, x]]$ and $[x, [x, x]]$ gives

$$\alpha_3 = 0, \quad \beta_3 = 0, \quad \gamma_3(1 + b_2) = 0.$$

Therefore there are the following options:

- If $\gamma_3 = 0$, we get $L_{35}(b_2)$, where $b_2 \notin \{0, 1\}$.
- If $\gamma_3 \neq 0$, then $b_2 = -1$. We make the base change

$$e'_1 = \gamma_3 e_1, \quad e'_3 = \gamma_3 e_3,$$

to obtain the algebra L_{36}.

Case 1.2. If $b_2 \in \{0, 1\}$, then $e_3 \notin \mathrm{Ann}_r(L)$ and in view of $[x, x], [e_i, x] + [x, e_i] \in \mathrm{Ann}_r(L)$ for $1 \leq i \leq 2$, we get $\gamma_3 = 0$, $\alpha_3 = -a_3$ and $\beta_3 = -b_3$.

- If $b_2 = 0$, the base change

$$e'_1 = e_1 + a_2 e_2, \quad e'_2 = e_2 - b_3 e_3, \quad x' = x - (a_3 + a_2 b_3) e_2$$

gives $a_2 = a_3 = b_3 = 0$. The result is $L_{35}(0)$.

- If $b_2 = 1$, and we make the following change of basis:

$$e'_1 = e_1 - (a_3 + \alpha_2 b_3) e_3, \quad e'_2 = e_2 - b_3 e_3,$$

we come to $a_3 = b_3 = 0$.
Finally if $a_2 = 0$, we obtain the algebra $L_{35}(1)$. Otherwise, if $a_2 \neq 0$, we assert $a_2 = 1$ by considering the basis transformation $e'_2 = a_2 e_2$, $e'_3 = a_2 e_3$. Therefore, we get L_{37}.

Case 2. If $(a_1, b_2) = (0, 0)$, due to the non nilpotency of R_x, we have $a_2 b_1 \neq 0$. Taking the base change $e'_1 = e_1 + e_2$, we obtain

$$[e'_1, x] = [e_1 + e_2, x] = b_1(e_1 + e_2) + (a_2 - b_1)e_2 + (a_3 + b_3)e_3$$
$$= b_1 e'_1 + a'_2 e_2 + a'_3 e_3.$$

Since $b_1 \neq 0$, we come to Case 1.

□

Proposition 3.9. *Let L be a four-dimensional solvable Leibniz algebra, whose nilradical is isomorphic to $\lambda_4(\alpha)$. Then, L is isomorphic to one of the following pairwise non isomorphic algebras:*

$$L_{38}(\gamma) := \begin{cases} [e_2, e_1] = e_3, \\ [e_1, x] = e_1, \\ [e_2, x] = \gamma e_2, \ \gamma \in \mathbb{C}, \\ [e_3, x] = (1 + \gamma) e_3, \\ [x, e_1] = -e_1. \end{cases} \qquad L_{39} := \begin{cases} [e_2, e_1] = e_3, \\ [e_1, x] = e_1, \\ [e_2, x] = -e_2, \\ [x, e_1] = -e_1, \\ [x, x] = e_3. \end{cases}$$

$$L_{40} := \begin{cases} [e_2, e_1] = e_3, \\ [e_1, x] = e_1 + e_3, \\ [e_3, x] = e_3, \\ [x, e_1] = -e_1, \\ [x, x] = -e_2. \end{cases} \qquad L_{41} := \begin{cases} [e_2, e_1] = e_3, \\ [e_2, x] = e_2, \\ [e_3, x] = e_3. \end{cases}$$

$$L_{42}(\beta) := \begin{cases} [e_2, e_1] = e_3, \\ [e_1, e_2] = \beta e_3, \\ [e_1, x] = e_1, \\ [e_2, x] = \beta e_2, \\ [e_3, x] = (\beta + 1)e_3, \\ [x, e_1] = -e_1, \\ [x, e_2] = -\beta e_2, \end{cases}$$

where $\beta = \frac{\sqrt{1-4\alpha}-1}{\sqrt{1-4\alpha}+1}$, *with* $\alpha \notin \left\{0, \frac{1}{4}\right\}$

Proof. We have the following options:

1. If $\mathcal{N}$ is isomorphic to $\lambda_4'(0)$ then applying the same approach as in Proposition 3.8, we obtain the algebras $L_{38}(\gamma) - L_{41}$.

2. If $\mathcal{N}$ is isomorphic to one of $\lambda_4'(\alpha)$ $(\alpha \neq 0)$ then we get the family $L_{42}(\beta)$.

□

The proofs of the following two propositions are similar to those of the above propositions.

Proposition 3.10. *Let L be a four-dimensional solvable Leibniz algebra, whose nilradical is isomorphic to* λ_5. *Then L is isomorphic to the following algebra*

$$L_{43} := \begin{cases} [e_1, e_2] = e_3, & [e_1, x] = e_1, \\ [e_2, e_1] = e_3, & [e_2, x] = e_2, \\ [x, e_1] = -e_1, & [e_3, x] = 2e_3, \\ [x, e_2] = -e_2. \end{cases}$$

Proposition 3.11. *Let L be a four-dimensional solvable Leibniz algebra, whose nilradical is isomorphic to* λ_6. *Then L is isomorphic to the following algebra*

$$L_{44} := \begin{cases} [e_1, e_1] = e_2, & [e_1, x] = e_1, \\ [e_2, e_1] = e_3, & [e_2, x] = 2e_1, \\ [x, e_1] = -e_1, & [e_3, x] = 3e_1. \end{cases}$$

Remark. To verify that the above algebras are non isomorphic, we use a computer program, implemented in *Mathematica*. The computer code is available from the authors.

3.5 RIGIDITY OF LIE AND LEIBNIZ ALGEBRAS

In 1951 I.E.Segal [168] introduced the notion of contractions of Lie algebras on physical grounds: if two physical theories (like relativistic and classical mechanics) are related by a limiting process, then the associated invariance groups (like the Poincaré and Galilean groups) should also be related by some limiting process. If the velocity of light is assumed to go to infinity, relativistic mechanics "transforms" into classical mechanics. This also induces a singular transition from the Poincaré algebra to the Galilean one. Another example is a limiting process from quantum mechanics to classical mechanics under $\hbar \longrightarrow 0$, that corresponds to the contraction of the Heisenberg algebras to the abelian ones of the same dimensions [62].

There are two approaches to the contraction problems of algebras. The first of them is based on physical considerations that are mainly oriented to applications of contractions. Contractions were used to establish connections between various kinematical groups and to shed a light on their physical meaning. In this way relationships between the conformal and Schrödinger groups were elucidated and various Lie algebras including a relativistic position operator were interrelated. Under dynamical group description of interacting systems, contractions corresponding to the coupling constant going to zero give non interacting systems. Application of contractions allows us to derive interesting results in the special function theory as well. The second consideration is purely algebraical, dealing with abstract algebraic structures. We will deal with this case and focus mainly on algebraic aspects of the contractions.

Let V be a vector space of dimension n over an algebraically closed field $\mathbb{F}$ (char$\mathbb{F}$=0). The set of bilinear maps $V \times V \to V$ forms a vector space $\mathrm{Hom}(V \otimes V, V)$ of dimension n^3, which can be considered together with its natural structure of an affine algebraic variety over $\mathbb{F}$. It is denoted by $\mathrm{Alg}_n(\mathbb{F})$. An n-dimensional algebra A over $\mathbb{F}$ may be regarded as an element λ of $\mathrm{Hom}(V \otimes V, V)$ via the bilinear mapping $\lambda : A \otimes A \to A$ defining a binary algebraic operation on A. In this section, we have occasion to write its composition law λ instead of the algebra A for the sake of simplicity.

In fact, in terms of λ the Leibniz identity is as follows

$$\lambda((x, \lambda(y, z)) = \lambda(\lambda(x, y), z) - \lambda(\lambda(x, z), y).$$

Recall, that the set of all n-dimensional Leibniz algebras over a field $\mathbb{F}$ is denoted by $LB_n(\mathbb{F})$. The set $LB_n(\mathbb{F})$ can be included in the above mentioned n^3-dimensional affine space as follows: let $\{e_1, e_2, \dots, e_n\}$ be a basis of the Leibniz algebra L. Then the table of multiplication of L is represented by a point (γ_{ij}^k) of this affine space as follows:

$$\lambda(e_i, e_j) = \sum_{k=1}^{n} \gamma_{ij}^k e_k.$$

Thus, the algebra L corresponds to the point $(\gamma_{ij}^k) \in \mathbb{F}^{n^3}$. The Leibniz identity gives polynomial relations among γ_{ij}^k. Hence we regard $LB_n(\mathbb{F})$ as a subvariety of $\mathbb{F}^{n^3}$.

The linear reductive group $GL_n(\mathbb{F})$ acts on $\mathrm{Alg}_n(\mathbb{F})$ by

$$(g * \lambda)(x, y) = g(\lambda(g^{-1}(x), g^{-1}(y))) \quad \text{[transport of structure]}.$$

Two algebras λ_1 and λ_2 are isomorphic if and only if they belong to the same orbit under this action.

For given two algebras λ and μ we say that λ *degenerates to* μ, if μ lies in the Zariski closure of the orbit of λ. We denote this by $\lambda \to \mu$.

In this case entire orbit $\mathrm{Orb}(\mu)$ lies in the closure of $\mathrm{Orb}(\lambda)$. We denote this, as has been mentioned above, by $\lambda \to \mu$, i.e., $\mu \in \overline{\mathrm{Orb}(\lambda)}$. Degeneration is transitive, that is if $\lambda \to \mu$ and $\mu \to \nu$ then $\lambda \to \nu$.

From now on we shall consider $LB_n(\mathbb{F})$ as a subvariety of $\mathbb{F}^{n^3}$.

It is easy to see that the scalar matrices of $GL_n(\mathbb{F})$ act on $LB_n(\mathbb{F})$ scalarly. Therefore, the orbits Orb (L) are cones with the deleted vertex $\{0\}$ that corresponds to the abelian complex algebra (denoted by $\mathbb{F}^n$). Thus, $\mathbb{F}^n$ belongs to $\overline{\text{Orb}\,(L)}$ for all $L \in LB_n(\mathbb{F})$. In particular, among the orbits Orb (L) only one is closed and that is the orbit of $\mathbb{F}^n$.

A Leibniz algebra λ is said to be *degenerate to a Leibniz algebra* μ, if μ is represented by a structure which lies in the Zariski closure of the $GL_n(\mathbb{F})$-orbit of the structure which represents λ.

There are algebras the orbits of which are open in $LB_n(\mathbb{F})$. These algebras are called rigid. In that case the corresponding algebra does not admit any non trivial deformation. The orbits of rigid algebras give irreducible components of the variety $LB_n(\mathbb{F})$. Hence to find the rigid orbits of variety of algebras is of great interest. By the Noetherian consideration they are of finite number. However, in general, not every irreducible component is generated by rigid algebras. Flanigan [78] has shown that there exists a component in Alg_3 which consists of union of infinitely many orbits of non isomorphic algebras having the same dimension (so-called *rigid families*). If L is a rigid algebra in $LB_n(\mathbb{F})$ then there exists an irreducible component $\mathfrak{C}$ of $LB_n(\mathbb{F})$ such that $\text{Orb}(L) \cap \mathfrak{C}$ is non empty open subset of $\mathfrak{C}$. The closure of $\text{Orb}(L)$ is contained in $\mathfrak{C}$. Then the dimension of the irreducible component $\mathfrak{C}$ of $LB_n(\mathbb{F})$ is given by

$$\dim \mathfrak{C} = n^2 - \dim \text{Aut}L.$$

Here is a proposition from [53] in Lie algebras case counting the number of irreducible components and the number of open orbits.

Proposition 3.12. *Let $r(n)$ and $s(n)$ be the number of irreducible components and the number of open orbits, respectively, in the variety $L_n(\mathbb{C})$ of n-dimensional complex Lie algebras. Then one has*

$$(r(1), r(2), \ldots, r(7)) = (1, 1, 2, 4, 7, 17, 49)$$

and

$$(s(1), s(2), \ldots, s(7)) = (1, 1, 1, 2, 3, 6, 14).$$

Let us recall a few useful facts from the algebraic groups theory, concerning the degenerations. The first of them concerns constructive subsets of algebraic varieties over $\mathbb{C}$, the closures of which relative to Euclidean and Zariski topologies coincide. Since $GL_n(\mathbb{C})$-orbits are constructive sets, the usual Euclidean topology on $\mathbb{C}^{n^3}$ leads to the same degenerations as does the Zariski topology. Now we may express the concept of degeneration in a slightly different way, that is the following condition will imply that $\lambda \to \mu$:

$$\exists g_t \in GL_n(\mathbb{C}(t)) \text{ such that } \lim_{t\to 0} g_t * \lambda = \mu,$$

where $\mathbb{C}(t)$ is the field of fractions of the polynomial ring $\mathbb{C}[t]$.

The second fact concerns the closure of $GL_n(\mathbb{C})$-orbits stating that the boundary of each orbit is a union of finitely many orbits with dimensions strictly less than the dimension of the given orbit. It follows that each irreducible component of the variety, on which algebraic group acts, contains only one open orbit that has a maximal dimension. It is obvious that in the content of variety of algebras the representatives of this kind orbits are rigid.

It is an interesting and at the same time difficult problem to determine the number of irreducible components of an algebraic variety. The dimension of the algebraic variety can be found by using degeneration approach. In this case no need to find all the degenerations, just to find the rigid algebras and rigid families. The closure of the orbit of a rigid algebra gives a component of the variety. In order to find the dimension of the variety it is sufficient to find rigid algebras and rigid families having a maximal orbit dimension. Here are some cases where the degenerations of classes of algebras have been studied. For associative algebras $\mathrm{Alg}_n(\mathbb{C}) : \mathrm{Alg}_4(\mathbb{C})$ see [81], $\mathrm{Alg}_5(\mathbb{C})$ see [128] and [121], [127]; for nilpotent associative algebras case (see [122]); for the class of Jordan algebras see [107], [108], for nilpotent Lie algebras $NL_n(\mathbb{C})$: at $n \leq 5$ it can be found in [41], [99], [100] and NL_6 was described by Seeley [166], NL_7 and NL_8 were investigated by Goze, Ancochea-Bermúdez [95]; the variety of filiform Lie algebras were investigated by Goze, Khakimdjanov Yu. [97]; for nilpotent Leibniz algebras in dimensions less than five, the geometric classification can be found in [13]. A slightly different approach to the geometric classification problem of algebras can be found in [70] [72], [73], [74], [75] and [76].

Let A be a complex n-dimensional algebra (underlying vector space is denoted by V) with the binary operation $\lambda : V \otimes V \to V$. Consider a continuous function $g_t : (0, 1] \to GL(V)$. In other words, g_t is a non singular linear operator on V for all $t \in (0, 1]$. Define a parameterized family λ_t of new algebra structures on V via the old binary operation λ

as follows:

$$\lambda_t(x,y) = (g_t * \lambda)(x,y) = g_t^{-1}\lambda(g_t(x), g_t(y)), \ x,y \in V.$$

Definition 3.3. *If the limit* $\lim\limits_{t\to+0} \lambda_t(x,y) = \lambda_0(x,y)$ *exists for all* $x,y \in V$, *then the algebraic structure* λ_0 *defined by this way on* V *is said to be a contraction of the algebra* A.

Note. Obviously, the contractions can be considered in the basis level. Namely, let $\{e_1, e_2, \ldots, e_n\}$ be a basis of an n-dimensional algebra A. If the limit $\lim\limits_{t\to+0} \lambda_t(e_i, e_j) = \lambda_0(e_i, e_j)$ exists then the algebra (V, λ_0) is a contraction of A.

Definition 3.4. *A contraction from an algebra* A *to algebra* A_0 *is said to be* ***trivial*** *if* A_0 *is abelian, and to be* ***improper*** *if* A_0 *is isomorphic to* A.

Note that both trivial and improper contractions always exist. Here is an example of the trivial and the improper contractions.

Example 3.1. *Let* $A = (V, \lambda)$ *be a complex n-dimensional algebra. If we take* $g_t = diag(t, t, \ldots, t)$ *then* $g_t * \lambda$ *is abelian and for* $g_t = diag(1, 1, \ldots, 1)$ *we get* $g_t * \lambda = A$.

We equally use the notions of rigidity with respect to degeneration and deformation as the same, since the rigidity with respect to degeneration follows from the rigidity with respect to deformation, but the converse does not always hold. In the case of Lie algebras over a field characteristic zero, these two definitions of rigidity are equivalent [70],[122], [182]. By using the results of [70] this can be extended to Leibniz algebras case.

Algebraic deformation theory was introduced, for associative algebras, in 1963 by Gerstenhaber [84], and it was extended to Lie algebras by Nijenhuis and Richardson [133], [134]. Following Gerstenhaber, Nijenhuis and Richardson, many authors have published papers on some aspect of the deformations of given type of algebraic structures (associative, commutative, Lie, Jordan, etc.).

One of the fundamental concepts in deformation theory is the notion of rigidity. Loosely speaking, an algebra A is rigid if an arbitrary "infinitesimal" deformation of A produces an algebra A' isomorphic to A. Focusing on Lie algebras, a geometric formulation of rigidity is given as follows.

Let V be a complex vector space of dimension n and L_n denote the set of all Lie algebra structures on V. The set L_n has a natural structure of an affine variety. There is an action of the group GL_n, of all automorphisms of V, on L_n. Under this action each orbit represents an isomorphism class of Lie algebra structures. A Lie algebra L is rigid if its orbit under the action of GL_n is Zariski open in L_n. A theorem of Nijenhuis and Richardson [133] shows that the vanishing of the second cohomology of a Lie algebra, with coefficients in the adjoint representation, is a sufficient condition for a Lie algebra to be rigid. Lie algebras are part of a bigger class of Leibniz algebras. Balavoine [28] published a paper on the deformations of Leibniz algebras. In particular, using the theory of Fox [79], he obtained a result similar to that of Nijenhuis and Richardson, i.e., the vanishing of the second Leibniz cohomology is a sufficient condition for the rigidity of a Leibniz algebra. Since a Lie algebra is also a Leibniz algebra, we need to consider the rigidity of Lie algebras both as Lie algebras and as Leibniz algebras.

3.6 LEIBNIZ COHOMOLOGY COMPUTATIONS

This section is dedicated to the cohomology of Leibniz algebras, built on the computational methods of [69]; a large part of the material is due to [141].

3.6.1 Leibniz cohomology

Let $\mathfrak{g}$ be a Lie algebra over a field $\mathbb{F}$, and M a left $\mathfrak{g}$-module. Cohomology theory associates to $\mathfrak{g}$ two complexes, namely the Chevalley-Eilenberg complex

$$C^*(\mathfrak{g}, M) := (\mathrm{Hom}(\Lambda^*\mathfrak{g}, M), d),$$

and the Leibniz or Loday complex

$$CL^*(\mathfrak{g}, M) := (\mathrm{Hom}(\otimes^*\mathfrak{g}, M), d),$$

for the Leibniz cohomology of $\mathfrak{g}$ with values in the symmetric (or antisymmetric) Leibniz $\mathfrak{g}$-bimodule M. The coboundary operator on the complex $C^*(\mathfrak{g}, M)$ is the standard Chevalley-Eilenberg coboundary operator, see e.g. [104]. The coboundary operator on $CL^*(\mathfrak{g}, M)$ is the

Leibniz or Loday coboundary operator $d : CL^n(\mathfrak{g}, M) \to CL^{n+1}(\mathfrak{g}, M)$ defined by

$$\begin{aligned}(df)(x_1, \ldots, x_{n+1}) &= x_1 \cdot f(x_2, \ldots, x_{n+1}) \\ &+ \textstyle\sum_{i=1}^{n+1}(-1)^i f(x_1, \ldots, \hat{x}_i, \ldots, x_{n+1}) \cdot x_i \\ &+ \textstyle\sum_{1\le i<j\le n+1}(-1)^{j+1} f(x_1, \ldots, x_{i-1}, [x_i, x_j], x_{i+1}, \ldots, \hat{x}_j, \ldots, x_{n+1})\end{aligned}$$

for any $f \in CL^n(\mathfrak{g}, M)$ and all elements $x_1, \ldots, x_{n+1} \in \mathfrak{g}$. Leibniz cohomology is more generally defined for any Leibniz algebra $\mathfrak{h}$ and any Leibniz $\mathfrak{h}$-bimodule M with the same coboundary operator d, see e.g. [69] for further details.

With values in the symmetric $\mathfrak{g}$-bimodule M, the natural epimorphism $\bigotimes^* \mathfrak{g} \to \Lambda^* \mathfrak{g}$ induces a monomorphism of complexes

$$\varphi : C^*(\mathfrak{g}, M) \hookrightarrow CL^*(\mathfrak{g}, M),$$

which is an isomorphism in degree 0 and 1. There is then a short exact sequence of complexes inducing a long exact sequence in cohomology which mediates between Chevalley-Eilenberg and Leibniz cohomology.

It is shown in Lemma 1.5 in [69] that Leibniz cohomology of a Leibniz algebra $\mathfrak{h}$ with values in an antisymmetric $\mathfrak{h}$-bimodule M^a reduces to lower degree cohomology with values in a symmetric $\mathfrak{h}$-bimodule:

$$HL^p(\mathfrak{h}, M^a) \cong HL^{p-1}(\mathfrak{h}, \mathrm{Hom}(\mathfrak{h}, M)^s) \tag{3.29}$$

for all $p \geq 1$.

As usual in cohomology, for a short exact sequence of Leibniz $\mathfrak{h}$-bimodules

$$0 \to M' \to M \to M'' \to 0,$$

there is a long exact sequence in cohomology

$$\ldots \to HL^n(\mathfrak{h}, M') \to HL^n(\mathfrak{h}, M) \to HL^n(\mathfrak{h}, M'') \to HL^{n+1}(\mathfrak{h}, M') \to \ldots,$$

for all $n \geq 0$ and starting with a monomorphism

$$HL^0(\mathfrak{h}, M') \to HL^0(\mathfrak{h}, M).$$

3.6.2 Leibniz cohomology of rigid Lie algebras

In this subsection, fix the base field $\mathbb{F}$ to be the field $\mathbb{C}$ of complex numbers.

Observe that in general $H^n(\mathfrak{g}, M) = 0$ for a certain n does not necessarily imply that $HL^n(\mathfrak{g}, M) = 0$. For example, the trivial Lie algebra $\mathfrak{g} = \mathbb{F}$ has $H^2(\mathbb{F}, \mathbb{F}) = 0$, but $HL^2(\mathbb{F}, \mathbb{F}) = \mathbb{F} \neq 0$, (see [71]). For another counter-example, a short computation with the Hochschild-Serre spectral sequence [104] shows that we have also $H^2(\mathfrak{g}, \mathfrak{g}) = 0$ for the direct sum $\mathfrak{g} = \mathfrak{sl}_2(\mathbb{C}) \oplus \mathbb{C}$, while $HL^2(\mathfrak{g}, \mathfrak{g}) \cong \mathbb{C}$ (see [71]).

We will consider in this subsection the question for which finite-dimensional Lie algebras $\mathfrak{g} \neq \mathbb{F}$ the hypothesis $H^2(\mathfrak{g}, \mathfrak{g}) = 0$ implies that $HL^2(\mathfrak{g}, \mathfrak{g}) = 0$. This assertion is true for nilpotent Lie algebras of dimension ≥ 2, since Theorem 2 in [60] shows that for a non-trivial nilpotent Lie algebra $\dim H^2(\mathfrak{g}, \mathfrak{g}) \geq 2$ as its center is non-trivial. The assertion is also true for (non-nilpotent) solvable Lie algebras. In order to show this, let us elaborate a little on an article of Carles [51].

Carles [51] investigates Lie algebras $\mathfrak{g}$ possessing a one-codimensional ideal. For these, Carles shows that the dimension $\dim \mathrm{Der}(\mathfrak{g})$ of the Lie algebra of derivations $\mathrm{Der}(\mathfrak{g})$ of $\mathfrak{g}$ is greater or equal to the dimension $\dim \mathfrak{g}$ of $\mathfrak{g}$. As a corollary, this is true for Lie algebras $\mathfrak{g}$ with $[\mathfrak{g}, \mathfrak{g}] \neq \mathfrak{g}$, because in this case $\mathfrak{g}$ admits an ideal of codimension one. In a later section, Carles investigates Lie algebras $\mathfrak{g}$ which satisfy $\dim \mathrm{Der}(\mathfrak{g}) = \dim \mathfrak{g}$. He shows that any such a Lie algebra is algebraic and therefore admits a decomposition $S \oplus U \oplus \mathcal{N}$ where S is a Levi's subalgebra, $\mathcal{N}$ is the nilradical and U is a subalgebra consisting of elements (called ad-semisimple) with $[S + U, U] = 0$ and such that the exterior torus (denoted by $\mathrm{ad}(U)$) is algebraic. In addition, Carles shows that if $\mathrm{codim}\, [\mathfrak{g}, \mathfrak{g}] > 1$, then $\mathfrak{g}$ is complete, i.e. $H^0(\mathfrak{g}, \mathfrak{g}) = H^1(\mathfrak{g}, \mathfrak{g}) = \{0\}$. In [51] (Lemma 5.1), Carles shows that if $[\mathfrak{g}, \mathfrak{g}] \neq \mathfrak{g}$, $\dim \mathrm{Der}(\mathfrak{g}) \leq \dim \mathfrak{g} + \dim H^2(\mathfrak{g}, \mathfrak{g})$.

Let us draw conclusions from these results with respect to the above question. Suppose that $\mathfrak{g}$ is a finite-dimensional solvable Lie algebra with $H^2(\mathfrak{g}, \mathfrak{g}) = 0$ and $\mathrm{codim}\, [\mathfrak{g}, \mathfrak{g}] > 1$. By [51], we have on the one hand $\dim \mathrm{Der}(\mathfrak{g}) \geq \dim \mathfrak{g}$ (by Corollary 2.20, because $\mathfrak{g}$ is solvable it implies $[\mathfrak{g}, \mathfrak{g}] \neq \mathfrak{g}$ and thus $\mathfrak{g}$ admits an ideal of codimension one). On the other hand, we have $\dim \mathrm{Der}(\mathfrak{g}) \leq \dim \mathfrak{g}$ (by Lemma 5.1, because $H^2(\mathfrak{g}, \mathfrak{g}) = 0$). Thus we conclude that $\dim \mathrm{Der}(\mathfrak{g}) = \dim \mathfrak{g}$, which implies by Proposition 3.1 in [51] that $\mathfrak{g}$ is algebraic and, thanks to $\mathrm{codim}\, [\mathfrak{g}, \mathfrak{g}] > 1$, $\mathfrak{g}$ is complete, i.e., $H^0(\mathfrak{g}, \mathfrak{g}) = H^1(\mathfrak{g}, \mathfrak{g}) = \{0\}$. This implies that $\mathit{Center}(\mathfrak{g}) = \{0\}$ and therefore by Theorem 3.8 below that $HL^2(\mathfrak{g}, \mathfrak{g}) = \{0\}$.

Let us cite a part of Theorem 2 from [71]:

Theorem 3.8. *Let* $\mathfrak{g}$ *be a finite-dimensional complex Lie algebra. Then* $H^2(\mathfrak{g}, \mathfrak{g})$ *is a direct factor of* $HL^2(\mathfrak{g}, \mathfrak{g})$. *Furthermore, the supplementary subspace vanishes in the case when the center of* $\mathfrak{g}$ *is zero.*

Proof. A proof of this result (as well as further discussion and extensions) is available in [69]. □

Observe that this implies in particular that Richardson's example $M \dot{+} \mathfrak{g}$ (see [159]) has non-trivial $HL^2(M \dot{+} \mathfrak{g}, M \dot{+} \mathfrak{g})$.

Corollary 3.2. *A finite dimensional solvable non-nilpotent Lie algebra* $\mathfrak{g}$ *with* $H^2(\mathfrak{g}, \mathfrak{g}) = 0$ *is the semidirect product of its nilradical* $\mathcal{N}$ *and an exterior torus of derivations* Q, *i.e.,* $\mathfrak{g} = \mathcal{N} \dot{+} Q$. *Furthermore, if* $\dim Q > 1$, *then* $HL^2(\mathfrak{g}, \mathfrak{g}) = \{0\}$.

Proof. By the above discussion of the results of Carles in [51], it follows that any solvable Lie algebra $\mathfrak{g}$ with $H^2(\mathfrak{g}, \mathfrak{g}) = 0$ is algebraic, i.e., it is isomorphic to the Lie algebra of an algebraic group. As the algebraicity implies (by Proposition 1.5 in [51]) the decomposability of the algebra, it follows that for solvable Lie algebras $\mathfrak{g}$ with $H^2(\mathfrak{g}, \mathfrak{g}) = 0$, we have a decomposition $\mathfrak{g} = \mathcal{N} \dot{+} Q$, where $\mathcal{N}$ is the nilradical of $\mathfrak{g}$ and Q is an exterior torus of derivations in the sense of Malcev; that is, Q is an abelian subalgebra of $\mathfrak{g}$ such that $\mathrm{ad}(x)$ is semisimple for all $x \in Q$.

If $\dim Q > 1$, it follows from the above discussion before Theorem 3.8 that $HL^2(\mathfrak{g}, \mathfrak{g}) = \{0\}$. □

3.6.3 Non-triviality of Leibniz cohomology

Let $f : \mathfrak{h} \to \mathfrak{q}$ be the quotient morphism which sends a Leibniz algebra $\mathfrak{h}$ onto its quotient by some two-sided ideal I, and let M be a Leibniz $\mathfrak{q}$-bimodule. Then M is also a Leibniz $\mathfrak{h}$-bimodule via f. There is a monomorphism of cochain complexes

$$f^* : CL^*(\mathfrak{q}, M) \to CL^*(\mathfrak{h}, M),$$

and a quotient complex, called the *relative complex*. The cohomology spaces are denoted accordingly $HL^*(\mathfrak{h}; \mathfrak{q}, M)$. The corresponding short exact sequence of complexes induces a long exact sequence in cohomology:

Proposition 3.13. *The map f induces a long exact sequence*

$$0 \to HL^1(\mathfrak{q}, M) \to HL^1(\mathfrak{h}, M) \to HL^1(\mathfrak{h}; \mathfrak{q}, M) \to HL^2(\mathfrak{q}, M) \to \dots$$

Proof. This is Proposition 3.1 in [69]. □

Let us consider the semidirect product Lie algebra $\widehat{\mathfrak{g}} = M \dot{+} \mathfrak{g}$ where $\mathfrak{g}$ is a semisimple Lie algebra (over $\mathbb{C}$) and M is a non-trivial finite-dimensional irreducible $\mathfrak{g}$-module. From $\widehat{\mathfrak{g}}$, we construct as our main object the Leibniz algebra $\mathfrak{h}$ which is the hemisemidirect product $\mathfrak{h} := I \dot{+} \widehat{\mathfrak{g}}$ of $\widehat{\mathfrak{g}}$ with ideal of squares I which is another non-trivial finite-dimensional irreducible $\mathfrak{g}$-module. We will need now the above observation that the hemisemidirect product is a split extension.

Proposition 3.14. *Let $\mathfrak{g}$ be a semisimple Lie algebra with non-trivial finite-dimensional irreducible $\mathfrak{g}$-modules M and I. Then the Leibniz algebra $\mathfrak{h} = I \dot{+} (M \dot{+} \mathfrak{g})$ satisfies*

$$H^2(\widehat{\mathfrak{g}}, \widehat{\mathfrak{g}}) \hookrightarrow HL^2(\mathfrak{h}, \mathfrak{h}),$$

where $\widehat{\mathfrak{g}} = M \dot{+} \mathfrak{g}$.

Proof. By Theorem 3.8, we have that $H^2(\widehat{\mathfrak{g}}, \widehat{\mathfrak{g}})$ is a direct factor of $HL^2(\widehat{\mathfrak{g}}, \widehat{\mathfrak{g}})$. On the other hand, the long exact sequence for $f : \mathfrak{h} \to \widehat{\mathfrak{g}}$ of Proposition 3.13 splits, because $\mathfrak{h}$ is the hemisemidirect product of $\widehat{\mathfrak{g}}$ and I. Therefore all connecting homomorphisms are zero and we have a monomorphism for all $n \geq 1$

$$HL^n(\widehat{\mathfrak{g}}, \widehat{\mathfrak{g}}) \hookrightarrow HL^n(\mathfrak{h}, \widehat{\mathfrak{g}}).$$

Now consider the long exact sequence for the Leibniz cohomology of $\mathfrak{h}$ associated with values in the short exact sequence

$$0 \to I \to \mathfrak{h} \to \widehat{\mathfrak{g}} \to 0.$$

Recall the construction of the connecting homomorphism ∂ : $HL^2(\mathfrak{h}, \widehat{\mathfrak{g}}) \to HL^3(\mathfrak{h}, I^a)$ in the long exact sequence. We claim that the subspace $HL^2(\widehat{\mathfrak{g}}, \widehat{\mathfrak{g}}) \subset HL^2(\mathfrak{h}, \widehat{\mathfrak{g}})$ is in the kernel of ∂. This is clear, because lifting a 2-cocycle $c \in CL^2(\widehat{\mathfrak{g}}, \widehat{\mathfrak{g}})$ to a cochain in $CL^2(\mathfrak{h}, \mathfrak{h})$, it will remain a cocycle and thus the preimage of its coboundary in $CL^3(\mathfrak{h}, I^a)$ is zero. Therefore we have an epimorphism

$$HL^2(\mathfrak{h}, \mathfrak{h}) \twoheadrightarrow HL^2(\widehat{\mathfrak{g}}, \widehat{\mathfrak{g}}).$$

This ends the proof of the proposition, because all cohomology spaces are $\mathbb{C}$-vector spaces and thus the above epimorphism splits. □

3.7 RIGID LEIBNIZ ALGEBRA WITH NON-TRIVIAL HL^2

In this section, we finally obtain an analogue of Richardson's theorem [159] for Leibniz algebras. Namely, we give an example of a finite-dimensional Leibniz algebra $\mathfrak{h}$ over a field $\mathbb{F}$ which is geometrically rigid, but $\mathfrak{h}$ has $HL^2(\mathfrak{h}, \mathfrak{h}) \neq 0$. It is showed that $\mathfrak{h}$ is geometrically rigid, but satisfies $HL^2(\mathfrak{h}, \mathfrak{h}) \neq 0$.

Let V be a finite dimensional complex vector space. Let LB be the algebraic variety of all Leibniz algebra structures on V, defined in $\mathrm{Hom}(V^{\otimes 2}, V)$ by the quadratic equations corresponding to the left Leibniz identity, i.e., for $\mu \in \mathrm{Hom}(V^{\otimes 2}, V)$, we require

$$\mu \circ \mu(x, y, z) := \mu(x, \mu(y, z)) - \mu(\mu(x, y), z) + \mu(\mu(x, z), y) = 0$$

for all $x, y, z \in V$.

Recall that two Leibniz algebra structures μ and μ' give rise to isomorphic Leibniz algebras (V, μ) and (V, μ') in case there exists $g \in GL(V)$ such that $g \cdot \mu = \mu'$, where the action of $GL(V)$ on LB is defined by

$$g \cdot \mu(x, y) := g(\mu(g^{-1}(x), g^{-1}(y)))$$

for all $x, y \in V$.

Let $\mathfrak{h} = (V, \mu)$ be a Leibniz algebra on V, and consider a Leibniz subalgebra $\mathfrak{s}$ of $\mathfrak{h}$. We will denote the complex subspace of V corresponding to $\mathfrak{s}$ by S. Let W be a supplementary subspace of S in V. For an element $\phi \in CL^2(\mathfrak{h}, V)$, we denote by $r(\phi)$ the restriction to the union of $S \otimes S$, $S \otimes W$ and $W \otimes S$. Let

$$N_1 := \{(g, m) \in GL(V) \times LB \mid r(g \cdot m) = r(m)\},$$

denote by $\mathrm{proj}_{LB} : GL(V) \times LB \to LB$ the projection map and let $p_1 : N_1 \to LB$ be the restriction of proj_{LB} to N_1.

Definition 3.5. *The subalgebra $\mathfrak{s}$ is called a stable subalgebra of $\mathfrak{h}$ if p_1 maps every neighborhood of $(1, \mu) \in N_1$ onto a neighborhood of μ in LB.*

Remark. (a) This is not the original definition of stable subalgebra in [159], but the strong version of stability which permits Page and Richardson to show the strengthened stability theorem at the end of their paper.

(b) The definition implies that if $\mathfrak{h}_1 = (V, \mu_1)$ is a Leibniz algebra sufficiently near to $\mathfrak{h}$, then $\mathfrak{h}_1$ is isomorphic to a Leibniz algebra $\mathfrak{h}_2 = (V, \mu_2)$ with the following property: for all $s \in \mathfrak{s}$ and all $x \in V$, we have $\mu_2(x, s) = \mu(x, s)$ and $\mu_2(s, x) = \mu(s, x)$.

The following theorem was proved in [141].

Theorem 3.9 (Stability Theorem). *Let $\mathfrak{h} = (V, \mu)$ be a finite-dimensional complex Leibniz algebra, and let $\mathfrak{s}$ be a subalgebra of $\mathfrak{h}$ such that $E^2(\mathfrak{h}, \mathfrak{s}, V) = 0$. Then $\mathfrak{s}$ is a stable subalgebra of $\mathfrak{h}$.*

Let $\mathfrak{g}$ be a semisimple Lie algebra and M be an irreducible (left) $\mathfrak{g}$-module (of dimension ≥ 2). We put $\widehat{\mathfrak{g}} = \mathfrak{g} \dot{+} M$ semidirect product of $\mathfrak{g}$ and M (with $[M, M] = 0$).

Theorem 3.10 (Richardson's Theorem). *Let $\widehat{\mathfrak{g}} = \mathfrak{g} \dot{+} M$ as above. Then $\widehat{\mathfrak{g}}$ is not rigid if and only if there exists a semisimple Lie algebra $\widehat{\mathfrak{g}'}$ which satisfies the following conditions:*

(a) there exists a semisimple subalgebra $\mathfrak{g}'$ of $\widehat{\mathfrak{g}'}$ which is isomorphic to $\mathfrak{g}$,

(b) if we identify $\mathfrak{g}'$ with $\mathfrak{g}$, then $\widehat{\mathfrak{g}'} / \mathfrak{g}'$ is isomorphic to M as a $\mathfrak{g}$-module.

Proof. This is Theorem 2.1 in [159]. □

Richardson shows in [159] that for $\mathfrak{g} = \mathfrak{sl}_2(\mathbb{C})$ and for $M = M_k$ the standard irreducible $\mathfrak{sl}_2(\mathbb{C})$-module of dimension $k + 1$ and highest weight k, the Lie algebra $\widehat{\mathfrak{g}}$ is not rigid if and only if $k = 2, 4, 6, 10$. The semisimple Lie algebras $\widehat{\mathfrak{g}'}$ are in this case the standard semisimple Lie algebras of dimension 6, 8, 10 and 14. It turns out that $\widehat{\mathfrak{g}}$ has necessarily rank 2 (see [159]), and these are all rank 2 semi-simple Lie algebras ($A_1 \times A_1$, A_2, B_2 and G_2).

We will extend Richardson's theorem to Leibniz algebras in the following sense. First of all, we will restrict to simple Lie algebras $\mathfrak{g}$. Let I be another irreducible (right) $\mathfrak{g}$-module (of dimension ≥ 2). We also set $\mathfrak{h} = I \dot{+} \widehat{\mathfrak{g}}$ the hemisemidirect product with the $\mathfrak{g}$-module I (in particular $[\mathfrak{g}, I] = 0$, $[I, \mathfrak{g}] = I$ and $[M, I] = 0$). Observe that $\mathfrak{h}$ is a non-Lie Leibniz algebra with ideal of squares I and with quotient Lie algebra $\widehat{\mathfrak{g}}$.

So, we have $\mathfrak{h} = \mathfrak{g} \oplus M \oplus I$ as vector spaces. In all the following, we fix the complex vector space $\mathfrak{g} \oplus M \oplus I$ and we will be interested in the different Leibniz algebra structures on $\mathfrak{g} \oplus M \oplus I$.

Theorem 3.11. *Let $\mathfrak{h} = (V, \mu)$ be a finite-dimensional complex Leibniz algebra, and let $\mathfrak{g}$ be a subalgebra of $\mathfrak{h}$ such that $E^2(\mathfrak{h}, \mathfrak{g}, V) = 0$.*

Then the Leibniz algebra $\mathfrak{h}$ is not rigid if and only if there exists a Leibniz algebra $\mathfrak{h}'$ which satisfies the following conditions:

a) There exists a semisimple Lie subalgebra $\mathfrak{s}'$ of $\mathfrak{h}'$ with a semisimple Lie subalgebra $\mathfrak{g}' \subset \mathfrak{s}'$ which is isomorphic to $\mathfrak{g}$, i.e. $\mathfrak{g} \cong \mathfrak{g}'$;

b) if we identify the subalgebra $\mathfrak{g}'$ with $\mathfrak{g}$ by this isomorphism, then $\mathfrak{s}'/\mathfrak{g}$ is isomorphic to M as a $\mathfrak{g}$-module and $\mathfrak{h}'/\mathfrak{s}'$ is isomorphic to I as a Leibniz antisymmetric $\mathfrak{g}$-module.

Proof. Let $V := \mathfrak{g} \oplus M \oplus I$. We will consider Leibniz algebras $\mathfrak{h}'$ on this fixed vector space V, i.e. $\mathfrak{h} = (V, \mu)$ and $\mathfrak{h}' = (V, \mu')$.

"$\Leftarrow$" Suppose that there exists a Leibniz algebra $\mathfrak{h}'$ which satisfies the conditions a) and b).

We may assume that

$$\mu(x, x') = \mu'(x, x'), \ \mu(x, m) = \mu'(x, m),$$

$$\mu(m, x) = \mu'(m, x), \ \mu(i, x) = \mu'(i, x)$$

for all $x, x' \in \mathfrak{g}$, all $m \in M$ and all $i \in I$.

Putting $g_t(x) = x, \ g_t(m) = tm, \ g_t(i) = ti$, we have that $g_t \in \mathrm{GL}(V)$ for all $t \neq 0$ and

$$\lim_{t \to 0} g_t \cdot \mu' = \mu.$$

Therefore, L is not rigid.

"$\Rightarrow$" Let LB be the set of all Leibniz algebras defined on the vector space V. We are considering the Leibniz subalgebra $\mathfrak{s} := \mathfrak{g}$ of $\mathfrak{h}$. It satisfies the cohomological condition in order to apply the stability theorem. From Theorem 3.9, we therefore get the existence of a neighborhood U of μ in LB such that if $\mu_1 \in U$, the Leibniz algebra $L_1 = (V, \mu_1)$ is isomorphic to a Leibniz algebra $L' = (V, \mu')$ which satisfies the following conditions:

1) $\mu(x, x') = \mu'(x, x')$,

2) $\mu(x, m) = \mu'(x, m), \ \mu(m, x) = \mu'(m, x)$,

3) $\mu(x, i) = \mu'(x, i) = 0,\ \mu(i, x) = \mu'(i, x)$,

for all $x, x' \in \mathfrak{g}$, all $m \in M$ and all $i \in I$.

Since $\mathfrak{h}$ is a non-Lie algebra, the Leibniz algebra $\mathfrak{h}'$ is also a non-Lie algebra. Therefore, the ideal of squares of the algebra $\mathfrak{h}'$ is also non zero. We denote it by I'.

From conditions 1) and 2) we conclude that $I' \cap (\mathfrak{g} + M) = \{0\}$, and thus $J := I' \cap I \neq \{0\}$.

The condition 3) implies that J is left module over $\mathfrak{g}$. Since I is an irreducible left $\mathfrak{g}$-module, we have $J = I$ and thus $I' = I$ as a vector spaces.

We conclude that the restriction to $\mathfrak{g} \oplus M$ places us in the Lie situation, i.e. in exactly the same situation as in Richardson's theorem. By Theorem 3.10, we thus obtain that there exists a semisimple Lie algebra $\mathfrak{s}'$ with a semisimple subalgebra $\mathfrak{g}'$, isomorphic to $\mathfrak{g}$ and (when $\mathfrak{g}$ is identified with $\mathfrak{g}'$) an isomorphism of $\mathfrak{g}$-modules $\mathfrak{s}'/\mathfrak{g} \cong M$. Furthermore, the conditions 1)-3) imply that *a*) and *b*) are satisfied. □

Corollary 3.3. *The Leibniz algebra* $\mathfrak{h} := I_l \dot{+} (M_k \dot{+} \mathfrak{sl}_2(\mathbb{C}))$ *for two standard irreducible left* $\mathfrak{sl}_2(\mathbb{C})$*-modules* M_k *and* I_l *of highest weights* $k = 2n$ *and* l *respectively with odd integer* $n > 5$ *and odd* $l > 2$ *is rigid and satisfies* $HL^2(\mathfrak{h}, \mathfrak{h}) \neq 0$.

Proof. As discussed earlier, Richardson shows in [159] that for $\mathfrak{g} = \mathfrak{sl}_2(\mathbb{C})$ and for M_k the standard irreducible $\mathfrak{sl}_2(\mathbb{C})$-module of dimension $k + 1$ and highest weight k, the Lie algebra $\widehat{\mathfrak{g}} = M_k \dot{+} \mathfrak{g}$ is not rigid if and only if $k = 2, 4, 6, 10$. He also shows that in case the half-highest-weight $\frac{k}{2} =: n$ is an odd integer $n > 5$, the Lie algebra cohomology of the Lie algebra $\widehat{\mathfrak{g}}$ is not zero. As a candidate for our Leibniz algebra $\mathfrak{h}$ satisfying the claim of the corollary, we take as before $\mathfrak{h} = I_l \dot{+} \widehat{\mathfrak{g}}$ for some irreducible $\mathfrak{sl}_2(\mathbb{C})$-modules $I_l \neq \mathfrak{sl}_2(\mathbb{C})$ and M_k such that the half-highest-weight $n := \frac{k}{2} > 5$ is an odd integer and $l > 3$ is odd. By Theorem 3.11, $\mathfrak{h}$ is then rigid.

On the other hand, by Proposition 3.14, $H^2(\widehat{\mathfrak{g}}, \widehat{\mathfrak{g}})$ injects into $HL^2(\mathfrak{h}, \mathfrak{h})$, thus we obtain that this Leibniz cohomology space is not zero. □

3.8 LIE-RIGIDITY VERSUS LEIBNIZ-RIGIDITY

We record in this section further results on the question whether $H^2(\mathfrak{g},\mathfrak{g}) = 0$ implies $HL^2(\mathfrak{g},\mathfrak{g}) = 0$. The base field is fixed to be the field $\mathbb{C}$ of complex numbers.

We have seen in Corollary 3.2 that all (non nilpotent) solvable Lie algebras $\mathfrak{g}$ with $H^2(\mathfrak{g},\mathfrak{g}) = 0$ and $\dim Q > 1$ satisfy $HL^2(\mathfrak{g},\mathfrak{g}) = 0$. In the case where the Lie algebra $\mathfrak{g}$ is only (Lie-)rigid, but does not necessarily satisfy $H^2(\mathfrak{g},\mathfrak{g}) = 0$, we cannot conclude that $HL^2(\mathfrak{g},\mathfrak{g}) = 0$. But we will see in Theorems 3.12 and 3.13 below that for Lie algebras of a special form, one can still conclude in this situation that $\mathrm{Cent}(\mathfrak{g}) = 0$.

Recall that by Carles [51], any rigid Lie algebra $\mathfrak{g}$ is algebraic, i.e., it is isomorphic to the Lie algebra of an algebraic group. As the algebraicity implies the decomposability of the algebra [51], it follows that for a rigid Lie algebra $\mathfrak{g}$, we have a decomposition $\mathfrak{g} = \mathcal{S}\dot{+}\mathcal{R}$, where $\mathcal{S}$ is a Levi subalgebra, $\mathcal{R} = \mathcal{N}\dot{+}Q$ is the solvable radical of $\mathfrak{g}$, $\mathcal{N}$ is the nilradical and Q is an exterior torus of derivations in the sense of Malcev; that is, Q is an abelian subalgebra of $\mathfrak{g}$ such that $\mathrm{ad}(x)$ is semisimple for all $x \in Q$. Note that over the complex numbers, the semisimplicity of $\mathrm{ad}(x)$ means that there exists a basis of $\mathcal{N}$ such that for any $x \in Q$, the operator $\mathrm{ad}(x)_{|\mathcal{N}}$ has diagonal form and $\mathrm{ad}(x)_{|Q} = 0$.

Remark. Since in the decomposition $\mathfrak{g} = \mathcal{S}\dot{+}\mathcal{R}$ we have $[\mathcal{S},\mathcal{R}] \subseteq \mathcal{R}$, we can view $\mathcal{R}$ as an $\mathcal{S}$-module. The fact that $\mathcal{S}$ is semisimple implies that $\mathcal{R}$ can be decomposed into a direct sum of irreducible submodules. Let $\mathfrak{r}_0$ be the sum of all one-dimensional submodules (so these are trivial submodules) and $\mathcal{R}_1$ be the sum of the non-trivial irreducible submodules. Then

$$[\mathcal{S},\mathcal{R}_0] = 0, \quad [\mathcal{S},\mathcal{R}_1] = \mathcal{R}_1, \quad [\mathcal{R}_0,\mathcal{R}_0] \subseteq \mathcal{R}_0, \quad \mathcal{R}_1 \subseteq \mathcal{N}, \quad Q \subseteq \mathcal{R}_0.$$

Remark. Above, we have already considered the direct sum Lie algebra $\mathfrak{g} = \mathfrak{sl}_2(\mathbb{C}) \oplus \mathbb{C}$. A short computation with the Hochschild-Serre spectral sequence shows that $H^2(\mathfrak{g},\mathfrak{g}) = 0$, but (cf. proof of Corollary 3 in [71]) $\dim HL^2(\mathfrak{g},\mathfrak{g}) = 1$. Therefore, the Lie algebra $\mathfrak{g}$ is an example of a complex rigid Lie algebra of the form $\mathfrak{g} = \mathcal{S}\dot{+}\mathcal{N}$ where $\dim(\mathfrak{g}/[\mathfrak{g},\mathfrak{g}]) = 1$ and where $\mathrm{Cent}(\mathfrak{g})$ is non-trivial. According to Theorem 3.8, the center is always non-trivial in the case where $H^2(\mathfrak{g},\mathfrak{g})$ and $HL^2(\mathfrak{g},\mathfrak{g})$ differ.

Proposition 3.15. *Let $\mathcal{R} = \mathcal{R}_1 \oplus \mathbb{C}^k$ be a split solvable Lie algebra. Then $\mathcal{R}$ is not rigid.*

Proof. Since $\mathcal{R}$ is solvable, $\mathcal{R}_1$ is also solvable and for this algebra we have $\mathcal{R}_1 = \mathcal{N}_1 \oplus Q_1$, where $\mathcal{N}_1$ is the nilradical of $\mathcal{R}_1$ and Q_1 is a supplementary subspace. Let us fix elements $x \in Q_1$ and $c = (c_1, c_2, \dots, c_k) \in \mathbb{C}^k$ and set $\varphi(x, c_i) = -\varphi(c_i, x) = c_i, 1 \le i \le k$. Clearly, $\varphi \in Z^2(\mathcal{R}, \mathcal{R})$. Consider the infinitesimal deformation $\mathcal{R}_t = \mathcal{R} + t\varphi$ of the algebra $\mathcal{R}$ and let us show that $\varphi \notin B^2(\mathcal{R}, \mathcal{R})$.

Consider

$$(df)(x, c_i) = [f(x), c_i] + [x, f(c_i)] - f([x, c_i]) = [x, f(c_i)], \quad 1 \le i \le k.$$

This equation implies that $\forall f \in C^1(\mathcal{R}, \mathcal{R})$, we have $(df)(x, c_i) \neq c_i$, because $[\mathcal{R}, \mathcal{R}] \subset \mathcal{N}_1$. Therefore, $\varphi \notin B^2(\mathcal{R}, \mathcal{R})$ and $\mathcal{R}_t = \mathcal{R} + t\varphi$ is a non-trivial deformation of the Lie algebra $\mathcal{R}$. Therefore the Lie algebra $\mathcal{R}$ is not rigid. □

Corollary 3.4. *Let $\mathfrak{g} = \mathcal{S} \dot{+} (\mathcal{N} \oplus Q) \oplus \mathbb{C}^k$ be a Lie algebra. Then $\mathfrak{g}$ is not rigid.*

Proof. Due to Remark 3.8, we have that $Q \cap [\mathfrak{g}, \mathfrak{g}] = 0$. So, taking $\mathfrak{g}_t = \mathfrak{g} + t\varphi$ with $\varphi(x, c_i) = -\varphi(c_i, x) = c_i, 1 \le i \le k$ similarly as in Proposition 3.15, we obtain a non-trivial deformation of $\mathfrak{g}$. Therefore, $\mathfrak{g}$ is not rigid. □

Let $\mathcal{R} = \mathcal{N} \oplus Q$ be a rigid solvable Lie algebra with basis $\{x_1, x_2, \dots, x_k\}$ of Q and basis $N := \{e_1, e_2, \dots, e_n\}$ of $\mathcal{N}$. Due to the arguments above, we can assume that its the table of multiplication has the following form:

$$\begin{cases} [e_i, e_j] = \sum\limits_{t=1}^{n} \gamma_{i,j}^t e_t, & 1 \le i, j \le n, \\ [e_i, x_j] = \alpha_{i,j} e_i, & 1 \le i \le n, \ 1 \le j \le k, \ 1 \le k. \end{cases}$$

Note that for any $j \in \{1, \dots, k\}$, there exists i such that $\alpha_{i,j} \neq 0$.

In the first part, we consider the center of rigid solvable Lie algebras of the form $\mathcal{R} = \mathcal{N} \oplus Q$. This study will cumulate in Theorem 3.12 below.

Proposition 3.16. *Let $\mathcal{R} = \mathcal{N} \oplus Q$ be a rigid solvable Lie algebra such that* $\dim Q > 1$. *Then* $\mathrm{Cent}(\mathcal{R}) = \{0\}$.

Proof. First of all, we divide the basis N of $\mathcal{N}$ into two subsets. Namely,

$$N_1 = \{e_i \in N \mid \exists j \in \{1, \dots, k\} \text{ such that } \alpha_{i,j} \neq 0\},$$

$$N_2 = \{e_i \in N \mid \alpha_{i,j} = 0 \ \ \forall j \in \{1, \dots, k\}\}.$$

For convenience, we shall denote the bases N_1 and N_2 as follows:

$$N_1 = \{e_1, e_2, \dots, e_{n_1}\}, \quad N_2 = \{f_1, f_2, \dots, f_{n_2}\}.$$

Evidently, $n = n_1 + n_2$.

Now let us suppose that $\mathrm{Cent}(\mathcal{R}) \neq \{0\}$. Then there exists an element $0 \neq c \in \mathrm{Cent}(\mathcal{R})$. We put

$$c = \sum_{t=1}^{n_1} \beta_t e_t + \sum_{t=1}^{n_2} c_t f_t + \sum_{t=1}^{k} b_t x_t.$$

Then $ad_c \equiv 0$ and $x_1, \dots, x_k$ act diagonally on $\mathcal{N}$, we derive $b_t = 0, \ 1 \leq t \leq k$.

Case 1. Let $N_2 = \emptyset$. Then $n = n_1$.

Consider

$$0 = [c, x_i] = \left[\sum_{t=1}^{n_1} \beta_t e_t, x_i\right] = \sum_{t=1}^{n} \beta_t \alpha_{t,i} e_t.$$

From this we deduce that $\beta_t = 0$ for any $1 \leq t \leq n_1$, as all $\alpha_{t,i}$ are non-zero. Therefore, $c = 0$. So, we have a contradiction with $\mathrm{Cent}(\mathcal{R}) \neq \{0\}$.

Case 2. Let $N_2 \neq \emptyset$. We set $\varphi(x_1, x_2) = c$, which is possible as by our hypothesis $\dim Q > 1$. It is easy to see that φ is a 2-cocycle.

Now we consider the associated infinitesimal deformation of the solvable Lie algebra $\mathcal{R}_t = \mathcal{R} + t\varphi$. Let us check that this deformation $\mathcal{R}_t$ is not equivalent to $\mathcal{R}$. In fact, it is enough to check that φ is not 2-coboundary of $\mathcal{R}$.

Consider

$$\begin{aligned}(df)(x_1, x_2) &= [f(x_1), x_2] + [x_1, f(x_2)] - f([x_1, x_2]) \\ &= [f(x_1), x_2] + [x_1, f(x_2)].\end{aligned}$$

Since $\mathrm{ad}(x_1)$, $\mathrm{ad}(x_2)$ are diagonal and $c \in \mathrm{Cent}(\mathcal{R})$, we conclude that $(df)(x_1, x_2) \neq c$ for any $f \in C^1(\mathcal{R}, \mathcal{R})$. This means, $\varphi \notin B^2(\mathcal{R}, \mathcal{R})$.

Thus, the deformation $\mathcal{R}_t = \mathcal{R} + \varphi$ is a non-trivial deformation of the rigid algebra $\mathcal{R}$. This is a contradiction to the assumption that $\mathrm{Cent}(\mathcal{R}) \neq \{0\}$. □

Remark. Note that the rigidity of the solvable Lie algebra $\mathfrak{g} = \mathcal{N} \oplus Q$ does not always imply that $\dim Q > 1$. Indeed, there are examples of rigid solvable Lie algebras with $\dim Q = 1$ (see [94]).

Recall the commutation relations and notations which we introduced earlier: $[e_i, x_j] = \alpha_{i,j} e_i$ for all $1 \leq i \leq n$, all $1 \leq j \leq k$, and all $1 \leq k$, and $\{e_1, \ldots, e_s\}$ are chosen Lie algebra generators of $\mathcal{N}$. In the following proposition, we neglect the index j in $\alpha_{i,j}$ as Q is one dimensional and we have thus always $j = 1$.

Proposition 3.17. *Let $\mathcal{R} = \mathcal{N} \oplus Q$ be a rigid solvable Lie algebra such that* $\dim Q = 1$. *Then $\alpha_i \neq 0$ for any $s + 1 \leq i \leq n$.*

Proof. Let $\langle x \rangle = Q$. Without loss of generality, one can assume that the basis elements $\{e_1, e_2, \ldots, e_n\}$ of $\mathcal{N}$ are homogeneous products of the generator basis elements, that is, for any $e_j \in [\mathcal{N}, \mathcal{N}]$, we have

$$e_j = [\ldots[e_{j_1}, e_{j_2}], \ldots e_{j_{t_j}}] \quad s + 1 \leq j \leq n, \quad 1 \leq j_i \leq s.$$

Then because of

$$[e_j, x] = (\alpha_{j_1} + \alpha_{j_2} + \cdots + \alpha_{j_{t_j}}) e_j,$$

we obtain

$$\alpha_j = \alpha_{j_1} + \alpha_{j_2} + \cdots + \alpha_{j_{t_j}} \quad s + 1 \leq j \leq n, \quad 1 \leq j_i \leq s.$$

To the algebra $\mathcal{R}$ as above and its regular element x, we associate a linear system $S(x)$ of equations of $n-1$ variables $z_1, \ldots, z_{n-1}$ consisting of equalities $z_i + z_j = z_k$ if and only if the vector $[e_i, e_j]$ contains e_k with a non-zero coefficient. It is clear that $\{\alpha_1, \alpha_2, \ldots, \alpha_n\}$ is one of the solutions of this system.

If the system $S(x)$ has a unique fundamental solution, then we can obtain that $z_p = k_p z_{i_0}$ with $k_p > 0$. Therefore, in this case all elements of the solution are strictly positive. Since $\{\alpha_1, \alpha_2, \ldots, \alpha_n\}$ is a solution, $\alpha_i \neq 0$ for any $s + 1 \leq i \leq n$.

So it remains to consider the case of a system of equations which has a space of fundamental solutions which is of dimension at least two.

Without loss of generality, we can assume $\alpha_1 \neq 0$ and let us assume that there exists some $p \geq s + 1$ such that $\alpha_p = 0$. It follows then that

$$\alpha_p = \alpha_{p_1} + \alpha_{p_2} + \cdots + \alpha_{p_{t_p}} = 0.$$

Let $\{z_1, z_2, \ldots, z_h\}$ be another solution of $S(x)$, linearly independent from $\{\alpha_1, \alpha_2, \ldots, \alpha_n\}$. As $\alpha_1 \neq 0$, we can assume $z_1 = 0$.

Consider the following solution $\{0, 1, 1, \ldots, 1, z_{h+1}, \ldots, z_s, z_{s+1}, z_n\}$, that is, here we get $z_1 = 0, z_2 = 1, \ldots, z_h = 1$.

Then we consider the following cochain:

$$\varphi(e_2, x) = e_2, \quad \varphi(e_3, x) = e_3, \quad \ldots, \quad \varphi(e_h, x) = e_h,$$

$$\varphi(e_{h+1}, x) = z_{h+1}e_{h+1}, \quad \varphi(e_{h+2}, x) = z_{h+2}e_{h+2}, \quad \ldots, \quad \varphi(e_s, x) = z_s e_s,$$

$$\varphi(e_{s+1}, x) = z_{s+1}e_{s+1}, \quad \varphi(e_{s+2}, x) = z_{s+2}e_{s+2}, \quad \ldots, \quad \varphi(e_n, x) = z_n e_n.$$

It is easy to check that $\varphi \in Z^2(\mathcal{R}, \mathcal{R})$.

However, $\varphi \notin B^2(\mathcal{R}, \mathcal{R})$. Indeed, if we had $\varphi \in B^2(\mathcal{R}, \mathcal{R})$, then $\varphi = df$ for some $f \in C^1(\mathcal{R}, \mathcal{R})$.

Consider

$$e_2 = \varphi(e_2, x) = f([e_2, x]) - [f(e_2), x] - [e_2, f(x)] =$$

$$f(\alpha_2 e_2) - [c_2 x + \sum_{i=1}^{n} a_{2,i}e_i, x] - [e_2, d_0 x + \sum_{i=1}^{n} d_i e_i] =$$

$$= \alpha_2(c_2 x + \sum_{i=1}^{n} a_{2,i}e_i) - \sum_{i=1}^{n} \alpha_i a_{2,i} e_i + d_0\alpha_2 e_2 + (*) = d_0\alpha_2 e_2 + (**).$$

Hence, we obtain $d_0\alpha_2 = 1$

On the other hand, we have

$$0 = \varphi(e_1, x) = f([e_1, x]) - [f(e_1), x] - [e_1, f(x)] =$$

$$f(\alpha_1 e_1) - [c_1 x + \sum_{i=1}^{n} a_{1,i}e_i, x] - [e_1, d_0 x + \sum_{i=1}^{n} d_i e_i] =$$

$$= \alpha_1(c_1 x + \sum_{i=1}^{n} a_{1,i}e_i) - \sum_{i=1}^{n} \alpha_i a_{1,i} e_i + d_0\alpha_1 e_1 + (*) = d_0\alpha_1 e_1 + (**).$$

Therefore, $d_0\alpha_1 = 0$. Since $\alpha_1 \neq 0$, we have $d_0 = 0$. This is a contradiction to the assumption that $\varphi \in B^2(\mathcal{R}, \mathcal{R})$. Finally, the deformation $\mathcal{R}_t = \mathcal{R} + t\varphi$ is a non-trivial deformation of $\mathcal{R}$, which contradicts the rigidity of $\mathcal{R}$. Therefore, we obtain that $\alpha_i \neq 0$ for any $1 \leq i \leq n$. □

As a synthesis of Propositions 3.16 and 3.17, we have the following theorem.

Theorem 3.12. *Let $\mathcal{R}$ be a rigid solvable Lie algebra of the form $\mathcal{R} = \mathcal{N} \oplus Q$. Then* $\mathrm{Cent}(\mathcal{R}) = 0$.

Proof. The assertion of the theorem for the case when $\dim Q > 1$ follows from Proposition 3.16. Consider therefore the case $\dim Q = 1$. Let us assume that $\mathrm{Cent}(\mathcal{R}) \neq 0$. Then there exists $0 \neq c \in \mathrm{Cent}(\mathcal{R})$. If $c \in \mathcal{N} \setminus [\mathcal{N}, \mathcal{N}]$, then we obtain that $\mathcal{R} = \mathcal{R}_1 \oplus \mathbb{C}$ and due to Proposition 3.15, we conclude that $\mathcal{R}$ is not rigid.

Let now $c \in [\mathcal{N}, \mathcal{N}]$. We set

$$c = \sum_{i=s+1}^{n} \beta_i e_i.$$

Taking into account Proposition 3.17 and the following equality

$$0 = [c, x] = \sum_{i=s+1}^{n} \beta_i [e_i, x] = \sum_{i=s+1}^{n} \beta_i \alpha_i e_i,$$

we conclude that $c = 0$. The proof is complete. □

Conclusion 1. Let $\mathcal{R}$ be a solvable Lie algebra of the form $\mathcal{R} = \mathcal{N} \oplus Q$ such that $H^2(\mathcal{R}, \mathcal{R}) = 0$, then $\mathcal{R}$ is rigid. By Theorem 3.12, we have $\mathrm{Cent}(\mathcal{R}) = \{0\}$. Now applying Theorem 3.8, we conclude that $HL^2(\mathcal{R}, \mathcal{R}) = 0$, which implies by [29] that $\mathcal{R}$ is rigid as a Leibniz algebra.

In a second part, we will now study the center of general rigid Lie algebras of the form $\mathfrak{g} = \mathcal{S} \dot{+} (\mathcal{N} \oplus Q)$ using the same methods as before.

Proposition 3.18. *Let* $\mathfrak{g} = \mathcal{S} \dot{+} (\mathcal{N} \oplus Q)$ *be a rigid Lie algebra such that* $\dim Q > 1$. *Then* $\mathrm{Cent}(\mathfrak{g}) = \{0\}$.

Proof. Let us suppose that $\mathrm{Cent}(\mathfrak{g}) \neq 0$ and $0 \neq c \in \mathrm{Cent}(\mathfrak{g})$. Clearly, $c \in \mathcal{N}$. From Remark 3.8 we conclude that $[\mathfrak{g}, \mathfrak{g}] \cap Q = 0$. Then similarly as in the Proposition 3.16 we conclude that $\mathfrak{g}_t = \mathfrak{g} + t\varphi$ with $\varphi(x_1, x_2) = c$ is a non-trivial deformation of $\mathfrak{g}$. This implies that $\mathrm{Cent}(\mathfrak{g}) = \{0\}$. □

As recalled in the beginning, if $\mathfrak{g} = \mathcal{S} \dot{+} (\mathcal{N} \oplus Q)$ is rigid, then $\mathcal{S}$ is a Levi subalgebra of $\mathfrak{g}$, $\mathcal{N}$ is the nilradical and Q is an abelian subalgebra with diagonal operators $\mathrm{ad}(x)_{|\mathcal{N}}$ for any $x \in Q$. Moreover,

$[\mathfrak{h}, \mathfrak{h}] = [Q, Q] = 0$ for a Cartan subalgebra $\mathfrak{h}$ of $\mathcal{S}$. Now due to Remark 3.8 (indeed, since $Q \subseteq \mathcal{R}_0$, we obtain $[\mathcal{S}, Q] \subseteq [\mathcal{S}, \mathcal{R}_0] = 0$), we conclude that $\mathfrak{h} \oplus Q$ is toroidal subalgebra in $\mathfrak{g}$ which acts on

$$\mathcal{N} \oplus (\oplus_i \mathfrak{l}_{\beta_i}) \oplus (\oplus_i \mathfrak{l}_{-\beta_i}),$$

where

$$\mathcal{S} = \mathfrak{h} \oplus (\oplus_i \mathfrak{l}_{\beta_i}) \oplus (\oplus_i \mathfrak{l}_{-\beta_i}).$$

Therefore, we have the following toroidal decomposition:

$$\mathfrak{g} = (\mathfrak{h} \oplus Q) \oplus (\mathfrak{g}_{\gamma_1} \oplus \mathfrak{g}_{\gamma_2} \oplus \cdots \oplus \mathfrak{g}_{\gamma_t})$$

such that $[g, h + x] = \gamma(h + x)g$ for any $h \in \mathfrak{h}, x \in Q$ and $g \in \mathfrak{g}_\gamma$.

Let $\dim Q = 1$ with $Q = \langle x \rangle$, $\dim \mathfrak{h} = q$ and $\mathfrak{l}_{\beta_i} = \langle s_{\beta_i} \rangle$ and $\mathfrak{l}_{-\beta_i} = \langle s_{-\beta_i} \rangle$. Then we have the following products in $\mathfrak{g}$:

$$\begin{cases} [e_i, e_j] = \sum\limits_{t=1}^{n} \gamma_{i,j}^t e_t, & 1 \leq i, j \leq n, \\ [e_i, x] = \alpha_i e_i, & 1 \leq i \leq n, \\ [e_i, h_p] = \theta_{i,p} e_i, & 1 \leq i \leq n, \quad 1 \leq p \leq q, \\ [x, \mathcal{S}] = 0. \end{cases}$$

Proposition 3.19. *Let* $\mathfrak{g} = \mathcal{S} \dot{+} (\mathcal{N} \oplus Q)$ *be a rigid Lie algebra such that* $\dim Q = 1$. *Then* $\mathrm{Cent}(\mathfrak{g}) = \{0\}$.

Proof. Similarly as in the proof of Proposition 3.17 we consider the system of linear equations with respect to $\alpha_1, \alpha_2, \dots, \alpha_s$:

$$\alpha_j = \alpha_{j_1} + \alpha_{j_2} + \cdots + \alpha_{j_{t_j}} \quad s + 1 \leq j \leq n, \quad 1 \leq j_i \leq s.$$

Let us assume that there exists some $p \geq s + 1$ such that $\alpha_p = 0$. It follows that

$$\alpha_p = \alpha_{p_1} + \alpha_{p_2} + \cdots + \alpha_{p_{t_p}} = 0.$$

To the above relations, we associate the system $S(z)$ of linear equations with respect to $z_1, z_2, \dots, z_s$. That is, the system contains the relations

$$z_j = z_{j_1} + z_{j_2} + \cdots + z_{j_{t_j}} \quad s + 1 \leq j \leq n, \quad 1 \leq j_i \leq s.$$

Note that the fundamental solution of the system $S(z)$ does not consist of a unique solution. Indeed, if it had a unique fundamental solution, then we would obtain that $0 = z_p = k_p z_1$ with $k_p \neq 0$, which implies $z_1 = 0$ and hence, $\alpha_1 = \alpha_2 = \cdots = \alpha_n = 0$ which is impossible because in this case $\mathrm{ad}(x)$ acts trivially to $\mathcal{N}$.

So, we conclude that the system $S(z)$ of equations has a fundamental solution which is a vector space of dimension at least two. Let us suppose that $\{z_1, z_2, \ldots, z_h\}$ is another solution of $S(z)$, linearly independent of $\{\alpha_1, \ldots, \alpha_h\}$. Moreover, we may assume that α_1 appears in the set $\{\alpha_{p_1}, \alpha_{p_2}, \ldots, \alpha_{p_{t_p}}\}$, i.e., e_1 appears in the following product

$$e_p = [\ldots[e_{p_1}, e_{p_2}], \ldots e_{p_{t_p}}]$$

Consider following solution $\{0, 1, 1, \ldots, 1, z_{h+1}, \ldots, z_s, z_{s+1}, \ldots, z_n\}$, that is, here we get $z_1 = 0, z_2 = 1, \ldots, z_h = 1$.

We have

$$[[e_i, s_\beta], x] = [[e_i, x], s_\beta] + [e_i, [s_\beta, x]] = [[e_i, x], s_\beta] = \alpha_i[e_i, s_\beta].$$

Then we consider the following cochain:

$$\varphi(e_2, x) = e_2, \quad \varphi(e_3, x) = e_3, \quad \ldots, \quad \varphi(e_h, x) = e_h,$$

$$\varphi(e_{h+1}, x) = z_{h+1}e_{h+1}, \quad \varphi(e_{h+2}, x) = z_{h+2}e_{h+2}, \quad \ldots, \quad \varphi(e_s, x) = z_s e_s,$$

$$\varphi(e_{s+1}, x) = z_{s+1}e_{s+1}, \quad \varphi(e_{s+2}, x) = z_{s+2}e_{s+2}, \quad \ldots, \quad \varphi(e_n, x) = z_n e_n.$$

$$\varphi(e_i, h_p) = \theta_{i,p}e_i, \quad \varphi(h_p, x) = 0, \quad 1 \le i \le n, \ 1 \le p \le q,$$

$$\varphi(e_i, s_\beta) = \varphi(x, s_\beta) = \varphi(e_i, e_j) = 0, \quad \varphi([e_i, s_\beta], x) = z_i[e_i, s_\beta],$$

$$1 \le i, j \le n, \ 1 \le p \le q.$$

By straightforward computations we check that φ is a 2-cocycle of $\mathfrak{g}$. However, $\varphi \notin B^2(\mathfrak{g}, \mathfrak{g})$. Indeed, if $\varphi \in B^2(\mathfrak{g}, \mathfrak{g})$, then $\varphi = df$ for some $f = f_{\mathcal{R}} + f_{\mathcal{S}} \in C^1(\mathfrak{g}, \mathfrak{g})$ with $f_{\mathcal{R}} : \mathfrak{g} \to \mathcal{R}$ and $f_{\mathcal{S}} : \mathfrak{g} \to \mathcal{S}$.

Consider

$$0 = \varphi(\mathcal{S}, x) = f([\mathcal{S}, x]) - [f(\mathcal{S}), x] - [\mathcal{S}, f(x)]$$

$$= -[f(\mathcal{S}), x] - [\mathcal{S}, f_{\mathcal{R}}(x)] - [\mathcal{S}, f_{\mathcal{S}}(x)].$$

This implies $[f(\mathcal{S}), x] + [\mathcal{S}, f_{\mathcal{R}}(x)] = 0$ and $[\mathcal{S}, f_{\mathcal{S}}(x)] = 0$.

Taking into account that for any $s \in \mathcal{S}$, there exists $s' \in \mathcal{S}$ such that $[s, s'] \neq 0$, we conclude that $f_{\mathcal{S}}(x) = 0$.

Consider

$$e_2 = \varphi(e_2, x) = f([e_2, x]) - [f(e_2), x] - [e_2, f(x)] =$$

$$f(\alpha_2 e_2) - [c_2 x + \sum_{i=1}^{n} a_{2,i} e_i + f_S(e_2), x] - [e_2, d_0 x + \sum_{i=1}^{n} d_i e_i] =$$

$$= \alpha_2(c_2 x + \sum_{i=1}^{n} a_{2,i} e_i + f_S(e_2)) - \sum_{i=1}^{n} \alpha_i a_{2,i} e_i + d_0 \alpha_2 e_2 + (*) = d_0 \alpha_2 e_2 + (**).$$

Hence, we obtain $d_0 \alpha_2 = 1$

On the other hand, we have

$$0 = \varphi(e_1, x) = f([e_1, x]) - [f(e_1), x] - [e_1, f(x)] =$$

$$f(\alpha_1 e_1) - [c_1 x + \sum_{i=1}^{n} a_{1,i} e_i + f_S(e_1), x] - [e_1, d_0 x + \sum_{i=1}^{n} d_i e_i] =$$

$$= \alpha_1(c_1 x + \sum_{i=1}^{n} a_{1,i} e_i + f_S(e_1)) - \sum_{i=1}^{n} \alpha_i a_{1,i} e_i + d_0 \alpha_1 e_1 + (*) = d_0 \alpha_1 e_1 + (**).$$

Therefore, $d_0 \alpha_1 = 0$. Since $\alpha_1 \neq 0$, we have $d_0 = 0$. This is a contradiction to assumption that $\varphi \in B^2(\mathfrak{g}, \mathfrak{g})$.

Finally, the deformation $\mathfrak{g}_t = \mathfrak{g} + t\varphi$ is a non-trivial deformation of $\mathfrak{g}$, which contradicts the rigidity of $\mathfrak{g}$.

Thus, we have proved that if $[e_p, x] = 0$ for some $e_p \in \mathcal{N}$, then $\mathfrak{g}$ is non-rigid. Similar as in the proof of Theorem 3.12, we obtain that $\mathrm{Cent}(\mathfrak{g}) = \{0\}$. □

Again, we perform a synthesis of the preceding two propositions in the following theorem.

Theorem 3.13. *Let* $\mathfrak{g} = \mathcal{S} \dotplus (\mathcal{N} \oplus \mathcal{Q})$ *be a rigid Lie algebra such that* $\mathcal{Q} \neq 0$. *Then* $\mathrm{Cent}(\mathfrak{g}) = 0$.

Proof. The assertion of the theorem for the case when $\dim \mathcal{Q} > 1$ follows from Proposition 3.18. In case $\dim \mathcal{Q} = 1$, we have that $\mathrm{Cent}(\mathfrak{g}) = 0$ due to Proposition 3.19. □

Conclusion 2. Let $\mathfrak{g} = \mathcal{S} \dotplus (\mathcal{N} \oplus \mathcal{Q})$ be a Lie algebra such that $H^2(\mathfrak{g}, \mathfrak{g}) = 0$, then $\mathfrak{g}$ is rigid. From Theorem 3.13 we obtain that $\mathrm{Cent}(\mathfrak{g}) = \{0\}$. Now applying Theorem 3.8, we conclude that $HL^2(\mathfrak{g}, \mathfrak{g}) = 0$, which implies by [29] that $\mathfrak{g}$ is rigid as a Leibniz algebra.

3.8.1 Necessary criteria for rigidity of Leibniz algebras

An optimal way of exhaustive study of degenerations in a set of algebras includes intensive usage of necessary criteria based on degeneration invariants. The invariants are preserved under degenerations. In the next section for the further references we collect some degeneration invariants. The proofs can be found more or less in the literature, see [166].

For a given Leibniz algebra L we define:

1. $\mathrm{SA}(L)$– the maximal abelian subalgebra of L;
2. $\mathrm{Com}(L)$– the maximal commutative subalgebra of L;
3. $\mathrm{SLie}(L)$– the maximal Lie subalgebra of L;
4. $HL^i(L, L)$– the i-th Leibniz cohomology group with coefficients itself.

3.8.1.1 Invariance Argument 1

Theorem 3.14. *For any $m, r \in \mathbb{N}$ the following subsets of LB_n are closed with respect to the Zariski topology:*

1. $\{L \in LB_n \mid \dim L^m \leq r\}$;
2. $\{L \in LB_n \mid \dim \mathrm{Ann}_r(L) \geq m\}$;
3. $\{L \in LB_n \mid \dim \mathrm{Ann}_l(L) \geq m\}$;
4. $\{L \in LB_n \mid \dim \mathrm{Cent}(L) \geq m\}$;
5. $\{L \in LB_n \mid \dim \mathrm{Aut}(L) > m\}$;
6. $\{L \in LB_n \mid \dim \mathrm{SA}(L) \geq m\}$;
7. $\{L \in LB_n \mid \dim \mathrm{Com}(L) \geq m\}$;
8. $\{L \in LB_n \mid \dim \mathrm{SLie}(L) \geq m\}$;
9. $\{L \in LB_n \mid \dim HL^i(L, L) \geq m\}$.

The proof is an easy consequence of the following fact from algebraic group theory. *Let G be a complex reductive algebraic group acting rationally on an algebraic set X and let B be a Borel subgroup of G. Then $\overline{G} = G * \overline{B}$* (see [13], also [99]).

Corollary 3.5. *An algebra L does not degenerate to an algebra L′ if one of the following conditions is valid:*

1. $\dim L^m < \dim L'^m$ *for some m,*
2. $\dim \mathrm{Ann}_r(L) > \dim \mathrm{Ann}_r(L')$,
3. $\dim \mathrm{Ann}_l(L) > \dim \mathrm{Ann}_l(L')$,
4. $\dim \mathrm{Cent}(L) > \dim \mathrm{Cent}(L')$,
5. $\dim \mathrm{Aut}(L) \geq \dim \mathrm{Aut}(L')$,
6. $\dim \mathrm{SA}(L) > \dim \mathrm{SA}(L')$,
7. $\dim \mathrm{Com}(L) > \dim \mathrm{Com}(L')$,
8. $\dim \mathrm{SLie}(L) > \dim \mathrm{SLie}(L')$.
9. $\dim HL^i(L, L) > \dim HL^i(L', L')$.

The invariance arguments below are stated in a general setting and the Leibniz algebras case is deduced from these as a special case.

3.8.1.2 Invariance Argument 2

Let A be an n-dimensional algebra over a field $\mathbb{F}$ and $\{e_1, e_2, \ldots, e_n\}$ be a basis on it. Then the element $x = x_1 \otimes e_1 + x_2 \otimes e_2 + \cdots + x_n \otimes e_n \in \mathbb{F}[x_1, x_2, \ldots, x_n] \otimes_{\mathbb{F}} A$, where $x_1, x_2, \ldots, x_n$ are independent variables, is called *the generic element* of A. Denote by $f_A(R_x)$ a Cayley-Hamilton polynomial of the right multiplication operator to the generic element x in the algebra $\widehat{A} = \mathbb{F}[x_1, x_2, \ldots, x_n] \otimes_{\mathbb{F}} A$. It is known that $f_A(R_x)$ does not depend on choosing of a basis in A.

Proposition 3.20. *If an algebra A degenerates to algebra B, then* $f_A(R_x) = 0$ *in* $\widehat{B}$.

3.8.1.3 Invariance Argument 3

Let $\{e_1, e_2, \ldots, e_n\}$ *be a basis of A and* $\mathrm{tr}(R_{e_i}) = 0$ *for all* $1 \leq i \leq n$. *If there exists a basis* $\{f_1, f_2, \ldots, f_n\}$ *of B such that* $\mathrm{tr}(R_{f_i}) \neq 0$ *for some* $i \in \{1, \ldots, n\}$, *then A does not degenerate to B.*

3.8.1.4 Invariance Argument 4

Let an algebra A be given by the structure constants $\gamma_1, \gamma_2, \ldots, \gamma_r$ and for a pair (i, j) of positive integers define

$$c_{i,j} = \frac{\mathrm{tr}(R_x)^i \mathrm{tr}(R_y)^j}{\mathrm{tr}((R_x)^i \circ (R_y)^j)}.$$

Then $c_{i,j}$ is a polynomial of $\gamma_1, \gamma_2, \ldots, \gamma_r$ and it does not depend on the elements x, y of A.

If neither of these polynomials is zero, we call $c_{i,j}$ *as* (i, j)*-invariant of* A*. Suppose that* A *has an* (i, j)*-invariant* $c_{i,j}$*. Then all* $B \in \overline{\mathrm{Orb}(A)}$ *have the same* (i, j)*-invariant* $c_{i,j}$*.*

3.8.1.5 Invariance Argument 5

Assume that in the previous invariance argument either $\mathrm{tr}(R_x)^i \mathrm{tr}(R_y)^j = 0$ *or* $\mathrm{tr}((R_x)^i \circ (R_y)^j) = 0$ *for all* $x, y \in A$ *and some pair* (i, j)*. Then these equations hold for all* $B \in \overline{\mathrm{Orb}(A)}$.

The proofs, except for the argument on dimension of the group of cohomology, can be found in the literature, see for example [39], [40], [41], [99]. The main argument relies on the following more general fact from the theory of algebraic groups. Let G be a complex reductive algebraic group acting rationally on an algebraic set X. Let B be a Borel subgroup of G. Then $\overline{G \cdot x} = G \cdot \overline{B \cdot x}$ for all $x \in X$.

We shall use it in the following context. Let B be a Borel subgroup of GL_n and L and L' be two elements of a class of algebras $\mathfrak{A}$. If $L \to L'$ and L lies in a B-stable closed subset $\mathfrak{R}$ of $\mathfrak{A}$, then L' must also be represented by a structure in $\mathfrak{R}$.

Let us now consider the argument on the group of cohomologies. It is clear that $\dim ZL^i(L, L) \leq \dim ZL^i(L', L')$ for a natural i.

One has

$$\dim HL^i(L, L) = \dim ZL^i(L, L) - \dim CL^{i+1}(L, L) + \dim ZL^{i+1}(L, L).$$

This implies that $\dim HL^i(L, L) \leq \dim HL^i(L', L')$, since $\dim CL^{i+1}(L, L) = \dim CL^{i+1}(L', L')$.

3.8.2 Applications of invariance arguments to varieties of low-dimensional Leibniz algebras

3.8.2.1 *Two-dimensional Leibniz algebras*

In Table 3.2 we give the orbit closures of two-dimensional Leibniz algebras under the action of GL_2 as above.

Notation	Leibniz brackets	Orbit closure
L_1	$[e_1, e_2] = e_1, \ \ [e_2, e_2] = e_1$	L_1, L_2, L_4
L_2	$[e_2, e_2] = e_1$	L_2, L_4
L_3	$[e_1, e_2] = -[e_2, e_1] = e_2$	L_3, L_4
L_4	-	L_4

Table 3.2 Orbit closures of two-dimensional Leibniz algebras

It is easy to see here that the algebra L_1 is rigid. Non degeneracy $L_1 \not\to L_3$ and $L_2 \not\to L_3$ occur due to Corollary 3.5 item 2). The matrix $\begin{pmatrix} t^2 & 0 \\ 0 & t \end{pmatrix}$ is the base change giving the degeneration $L_1 \to L_2$ and the algebra L_3 being a Lie algebra does not degenerate to the Leibniz algebra L_1. The Lie algebra L_3 itself also is Leibniz rigid, the fact which follows from a cohomological version of the result given in [146]. Therefore, the variety of two-dimensional Leibniz algebras has two irreducible components, they are generated by the algebras L_1 and L_3.

3.8.3 Variety of three-dimensional nilpotent Leibniz algebras

Let LN_n be the variety of n-dimensional nilpotent Leibniz algebras. In this section we describe LN_n for $n \leq 3$. The varieties of one and two-dimensional Leibniz algebras easily can be described as follows. The variety LN_1 consists of one point (it is the trivial Leibniz algebra). The variety LN_2 consists of two points: one of them is the trivial algebra and another one is the Leibniz algebra L given by the table of multiplication $[e_1, e_1] = e_2$, on a basis $\{e_1, e_2\}$. The algebra L degenerates to abelian. Hence, LN_2 is irreducible and its irreducible component is generated by the rigid algebra L, i.e., $LN_2 = \overline{\mathrm{Orb}(L)}$ and $\dim LN_2 = 2$.

In the dimension three there are five isolated and one parametric family of pairwise non isomorphic algebras. The values of invariants of these algebras are given in Table 3.1.

By using degeneration arguments stated in the table we write down all possible degenerations for nilpotent three-dimensional Leibniz algebras:

$$\begin{aligned} \lambda_4\ (\alpha = 0) &\to \lambda_1, \lambda_2, \\ \lambda_4\ (\alpha \neq 0) &\to \lambda_1, \lambda_2, \lambda_3, \\ \lambda_5\ &\to \lambda_1, \lambda_2, \\ \lambda_6\ &\to \lambda_1, \lambda_2, \lambda_4(\alpha = 0). \end{aligned}$$

Later we shall prove (see Proposition 3.21) that all non Lie Leibniz algebras degenerate to λ_2. Moreover, the algebra λ_6 degenerates to the algebra λ_4 $(\alpha = 0)$ via the following family of matrices: $g_t(e_1) = e_2 - e_3$, $g_t(e_2) = t^{-1}e_1$, $g_t(e_3) = t^{-1}e_3$.

Algebra	Multiplication table of λ	Orbit closure
λ_1	abelian	λ_1
λ_2	$[e_1, e_1] = e_2$	λ_1, λ_2
λ_3	$[e_2, e_3] = e_1$ $[e_3, e_2] = -e_1$ $[e_2, e_2] = e_1$	λ_1, λ_3
$\lambda_4(\alpha \neq 0)$	$[e_3, e_3] = \alpha e_1$ $[e_2, e_3] = e_1$ $[e_2, e_2] = e_1$	$\lambda_1, \lambda_2, \lambda_3, \lambda_4(\alpha \neq 0)$
$\lambda_4(\alpha = 0)$	$[e_2, e_3] = e_1$ $[e_2, e_2] = e_1$	$\lambda_1, \lambda_2, \lambda_4(\alpha = 0)$
λ_5	$[e_3, e_2] = e_1$ $[e_2, e_3] = e_1$	$\lambda_1, \lambda_2, \lambda_5$
λ_6	$[e_3, e_3] = e_1$ $[e_1, e_3] = e_2$	$\lambda_1, \lambda_2, \lambda_4(\alpha = 0), \lambda_6$

Table 3.3 Geometric descriptions of the algebras

$$LN_3(\mathbb{C}): \quad \begin{array}{ccccccc} & \lambda_4(\alpha \neq 0) & & & \lambda_5 & & \lambda_6 \\ \swarrow & & \searrow & \swarrow & & \swarrow & \\ \lambda_3 & & & \lambda_2 & \longleftarrow & \lambda_4(\alpha = 0) & \\ & \searrow & \swarrow & & & & \\ & & \lambda_1 & & & & \end{array}$$

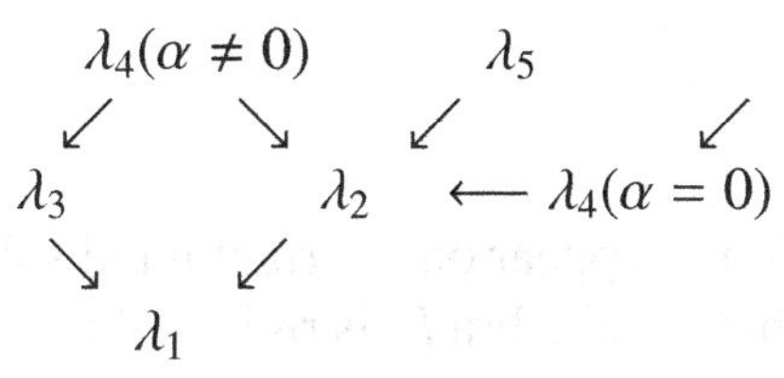

The following Hasse diagram gives us the complete description of the set of the algebras given in Table 3.3.

Summing up all the above, we conclude that the variety $LN_3(\mathbb{C})$ is the union of three irreducible components $\mathfrak{C}_i$, $i = 1, 2.3$ as follows

$$\mathfrak{C}_1 = \overline{\bigcup_{\alpha\neq 0} \mathrm{Orb}(\lambda_4)}, \quad \mathfrak{C}_2 = \overline{\mathrm{Orb}(\lambda_5)}, \quad \mathfrak{C}_3 = \overline{\mathrm{Orb}(\lambda_6)}$$

and $\dim LN_3(\mathbb{C}) = 6$.

3.8.4 Three-dimensional rigid Leibniz algebras

This section deals with Leibniz algebras in dimension three. We explore rigidity problems here. One needs the algebraic classification of all three-dimensional Leibniz algebras from Table 3.1. By using the invariance arguments, we find all possible degenerations of three-dimensional complex Leibniz algebras that may occur:

$L_1 \to L_2,\ L_5,\ L_6, L_7, L_8, L_{12}(\alpha = 0), L_{14}, L_{15}, L_{18};$

$L_2 \to L_7, L_8, L_{12}(\alpha = 0), L_{14}, L_{15}, L_{18};$

$L_3 \to L_{10}, L_{17}, L_{18};$

$L_4(\alpha = 0) \to L_7, L_8, L_{10}, L_{12}(\alpha = 0), L_{13}, L_{14}, L_{15}, L_{17}, L_{18};$

$L_4(\alpha \neq 0) \to L_4(\alpha = 0),\ L_5,\ L_6,\ L_7,\ L_8,\ L_{10},\ L_{12}(\alpha = 0),\ L_{13},\ L_{14},\ L_{15},\ L_{17},\ L_{18};$

$L_5 \to L_7, L_8, L_{10}, L_{12}(\alpha = 0), L_{15}, L_{18};$

$L_6 \to L_7, L_8, L_{12}(\alpha = 0), L_{14}, L_{15}, L_{18}$

$L_7 \to L_8, L_{12}(\alpha = 0), L_{15}, L_{18};$

$L_8 \to L_8, L_{12}(\alpha = 0), L_{15}, L_{18};$

$L_9 \to L_{10}, L_{11}, L_{16}, L_{17}, L_{18};$

$L_{10} \to L_{17}, L_{18};$

$L_{11} \to L_{16}, L_{17}, L_{18};$

$L_{12}(\alpha = 0) \to L_{15}, L_{18}$

$L_{12}(\alpha \neq 0) \to L_{12}(\alpha = 0), L_{15}, L_{18};$

$L_{13} \to L_{15}, L_{17}, L_{18};$

$L_{14} \to L_{17}, L_{18};$

$L_{15} \to L_{18}$

$L_{16} \to L_{17}, L_{18}$

$L_{17} \to L_{18}$

$L_{18} \to L_{18}$

Some algebras do not appear on the right hand side of this list after arrows, this means that the algebra L_3 is rigid and the parametric family of algebras $L_1(\alpha), L_4(\alpha), L_9(\alpha), L_{12}(\alpha)$ are rigid families, i.e., they are not degeneration of other Leibniz algebras in dimension three. The final result can be written as follows:

Theorem 3.15. *The algebra L_3 and the continuous parametric families of algebras $L_1(\alpha), (\alpha \neq 0), L_4(\alpha), L_9(\alpha), (|\alpha| < 1, \alpha \neq 0), L_{12}(\alpha), (\alpha \neq 0)$ generate rigid irreducible components of $LB_3(\mathbb{C})$ with the dimensions:*

$$\mathfrak{C}_1 = \overline{\mathrm{Orb}(L_3)}, \quad \dim \mathfrak{C}_1 = 6,$$

$$\mathfrak{C}_2 = \overline{\cup_\alpha \mathrm{Orb}(L_1(\alpha))}, \quad \dim \mathfrak{C}_2 = 7,$$

$$\mathfrak{C}_3 = \overline{\cup_{|\alpha|<1, \alpha\neq 0} \mathrm{Orb}(L_9(\alpha))}, \quad \dim \mathfrak{C}_3 = 7,$$

$$\mathfrak{C}_4 = \overline{\cup_\alpha \mathrm{Orb}(L_4(\alpha))}, \quad \dim \mathfrak{C}_4 = 6,$$

$$\mathfrak{C}_5 = \overline{\cup_{\alpha\neq 0} \mathrm{Orb}(L_{12}(\alpha))}, \quad \dim \mathfrak{C}_5 = 5.$$

$$dim LB_3(\mathbb{C}) = max\{\dim \mathfrak{C}_1, \dim \mathfrak{C}_2, \dim \mathfrak{C}_3, \dim \mathfrak{C}_4, \dim \mathfrak{C}_5\} = 7.$$

3.8.5 Rigid nilpotent Leibniz algebras in dimension four

First we prove a few facts on degenerations of nilpotent Leibniz algebras.

Proposition 3.21. *Let λ be a non Lie algebra in $LN_n(\mathbb{C})$. Then $\lambda \to \mu \oplus \mathbb{C}^{n-2}$, where μ is two-dimensional non abelian nilpotent Leibniz algebra.*

Proof. Since λ is non Lie Leibniz algebra, there exists x such that $[x, x] = y$, where $y \neq 0$. These two elements are linearly independent. Thus, x and y can be included to the basis $\{e_1 = x, e_2 = y, e_3, \dots, e_n\}$ of λ. Then taking the following family g_t $(t \in \mathbb{C})$ in $GL_n(\mathbb{C}(t))$:

$$g_t(e_1) = e_1, \; g_t(e_2) = e_2, \; g_t(e_i) = te_i, \quad \text{where } 3 \leq i \leq n$$

we obtain that $\lambda \to \mu \oplus \mathbb{C}^{n-2}$, with μ defined by the following table of multiplications: $[e_1, e_1] = e_2$. □

Note that μ is an example of null-filiform Leibniz algebra in dimension two. Moreover, it can be easily proved that the null-filiform Leibniz algebra is rigid in $LN_n(\mathbb{C})$.

Proposition 3.22. *The n-dimensional null-filiform Leibniz algebra NF_n, non Lie filiform Leibniz algebra from classes* FLb_n *and* SLb_n *degenerates to the algebra* $v \oplus \mathbb{C}^{n-3}$, *where v is a three-dimensional non abelian nilpotent Leibniz algebra with the following table of multiplications:*

$$[e_2, e_2] = e_1, \quad [e_2, e_3] = e_1.$$

Proof. First we consider the null-filiform algebras case. According to Theorem 3.2 a complex n-dimensional null-filiform Leibniz algebra can be represented in a basis $\{e_1, e_2, \ldots, e_n\}$ by the following table of multiplications:

$$[e_i, e_1] = e_{i+1}, \quad 1 \leq i \leq n-1.$$

It is easy to check that the following family g_t $(t \in \mathbb{C})$ in $GL_n(\mathbb{C}(t))$:

$$g_t(e_1) = t^{-2}e_2, \ g_t(e_2) = t^{-4}e_1, \ g_t(e_3) = t^{-3}e_2 - t^{-3}e_3,$$

$$g_t(e_4) = t^{-5}e_1 + t^{-4}e_4, \ g_t(e_i) = t^{-1}e_i \ (5 \leq i \leq n)$$

gives the corresponding degeneration.

Now, let us consider filiform Leibniz algebras of the cases FLb_n, SLb_n. Verifying the following family g_t $(t \in \mathbb{C})$ of transformations from $GL_n(\mathbb{C}(t))$:

$g_t(e_0) = t^{-1}e_2,$
$g_t(e_1) = t^{-1}e_2 - t^{-1}e_3,$
$g_t(e_2) = t^{-2}e_1,$
$g_t(e_3) = t^{-1}e_0,$
$g_t(e_i) = t^{-1}e_i, \ 4 \leq i \leq n-1,$

it is easy to conclude that the first class degenerates to the $v \oplus \mathbb{C}^{n-3}$.

The fact that the second class degenerates to the $v \oplus \mathbb{C}^{n-3}$ is obtained by applying consecutively the families of transformations g_t and f_t in $GL_n(\mathbb{C}(t))$ $(t \in \mathbb{C})$

$g_t(e_0) = t^{-3}e_1,$
$g_t(e_1) = t^{-2}e_2,$
$g_t(e_2) = t^{-3}e_3,$
$g_t(e_3) = t^{-1}e_0,$
$g_t(e_i) = te_i, \ \ 4 \leq i \leq n,$

and

$$f_t(e_0) = t^{-2}e_0 + t^{-1}e_1,$$
$$f_t(e_1) = e_0 + t^{-1}e_2,$$
$$f_t(e_0) = -e_0,$$
$$f_t(e_i) = e_i, \ \ 4 \leq i \leq n.$$

□

This section concerns the rigid four-dimensional nilpotent Leibniz algebras. For this purpose we make use the list of isomorphism classes of nilpotent Leibniz algebras in dimension four from Theorem 3.3.

Non rigidities of some algebras from the list are obtained from the degenerations given below:

$\mathcal{R}_4(\alpha = 0) \to \mathcal{R}_6$, $g_t(e_1) = t^{-2}e_1, g_t(e_2) = t^{-3}e_2$, $g_t(e_3) = t^{-4}e_3$, $g_t(e_4) = t^{-6}e_4$,

$\mathcal{R}_9 \to \mathcal{R}_7$, $g_t(e_1) = -t^{-1}e_2, g_t(e_2) = t^{-1}e_1 + t^{-3}e_2$, $g_t(e_3) = t^{-2}e_3$, $g_t(e_4) = t^{-2}e_4$,

$\mathcal{R}_{10} \to \mathcal{R}_8$, $g_t(e_1) = -\frac{1}{2}ie_1 + 2t^{-1}e_2, g_t(e_2) = 2it^{-1}e_2$, $g_t(e_3) = t^{-1}(e_3 - \frac{1}{2}e_4)$, $g_t(e_4) = -\frac{1}{2}it^{-1}e_4$,

$\mathcal{R}_{10} \to \mathcal{R}_9$, $g_t(e_1) = te_1, g_t(e_2) = t^2e_2$, $g_t(e_3) = t^3e_3$, $g_t(e_4) = t^4e_4$,

$\mathcal{R}_{13}(\alpha = 1) \to \mathcal{R}_{19}$, $g_t(e_1) = t^{-2}e_1, g_t(e_2) = t^{-1}e_2$, $g_t(e_3) = -t^{-3}e_4$, $g_t(e_4) = -t^{-2}e_3$,

$\mathcal{R}_{26} \to \mathcal{R}_{22}$, $g_t(e_1) = e_1, g_t(e_2) = t^{-1}e_3$, $g_t(e_3) = e_2$, $g_t(e_4) = e_4$,

$\mathcal{R}_{14}(\alpha \neq 0) \to \mathcal{R}_{25}$, $g_t(e_1) = e_1, g_t(e_2) = -\alpha e_1 + 2\alpha e_2 - ut^{-1}e_4$, $g_t(e_3) = t^{-1}e_4$, $g_t(e_4) = e_3$, where $u^2 = 4\alpha^2\beta - \alpha^2 - 1$,

$\mathcal{R}_{16} \to \mathcal{R}_{26}$, $g_t(e_1) = e_1, g_t(e_2) = t^{-1}e_4$, $g_t(e_3) = ie_1 - ie_2$, $g_t(e_4) = e_3$,

$\mathcal{R}_7 \to \mathcal{R}_{27}$, $g_t(e_1) = -t^1e_1, g_t(e_2) = t^2e_2$, $g_t(e_3) = t^3e_3$, $g_t(e_4) = t^4e_4$,

$\mathcal{R}_3 \to \mathcal{R}_{28}$, $g_t(e_1) = e_1, g_t(e_2) = t^{-1}e_4$, $g_t(e_3) = e_2$, $g_t(e_4) = e_3$,

$\mathcal{R}_{10} \to \mathcal{R}_{16}$, $g_t(e_1) = t^{-1}e_1, g_t(e_2) = t^{-1}e_3$, $g_t(e_3) = -t^{-1}e_2$, $g_t(e_4) = t^{-2}e_4$,

$\mathcal{R}_4(\alpha = 0) \to \mathcal{R}_4(\alpha = 1)$, $g_t(e_1) = te_1 - te_2, g_t(e_2) = t^2e_2$, $g_t(e_3) = t^3e_3 - t^3e_4$, $g_t(e_4) = t^4e_4$,

$\mathcal{R}_{14}\left(\frac{1}{t}\right) \to \mathcal{R}_{15}$, $g_t(e_1) = t^{-2}e_1 + e_3, g_t(e_2) = t^{-1}e_2 - it^{-2}e_3$, $g_t(e_3) = ie_3$, $g_t(e_4) = t^{-2}e_4$,

$\mathcal{R}_6 \to \mathcal{R}_{14}(i)$, $g_t(e_1) = \frac{1}{2}t^{-1}(e_1 - ie_2), g_t(e_2) = t^{-1}e_3$, $g_t(e_3) = \frac{1}{2}t^{-1}(e_1 + ie_2)$, $g_t(e_4) = t^{-2}e_4$,

$\mathcal{R}_5 \to \mathcal{R}_{20}(\alpha)$, $g_t(e_1) = t^{-1}e_2$, $g_t(e_2) = \frac{1-\alpha}{1+\alpha}(t^{-1}e_1 - t^{-2}e_3)$, $g_t(e_3) = t^{-2}e_3$, $g_t(e_4) = t^{-2}e_4$,

$\mathcal{R}_6 \to \mathcal{R}_{23},\ g_t(e_1) = t^{-1}e_1, g_t(e_2) = e_3,\ g_t(e_3) = t^{-2}e_2,\ g_t(e_4) = e_4,$
$\mathcal{R}_{12} \to \mathcal{R}_{24},\ g_t(e_1) = e_3, g_t(e_2) = t^{-1}e_2,\ g_t(e_3) = -t^{-1}e_1,$
$g_t(e_4) = t^{-1}e_1 + e_4,$
$\mathcal{R}_{15} \to \mathcal{R}_{27},\ g_t(e_1) = e_2 - t^{-1}e_4,\ g_t(e_2) = t^{-1}e_4,\ g_t(e_3) = e_3,$
$g_t(e_4) = e_1,$
$\mathcal{R}_4(\alpha = 1) \to \mathcal{R}_{29},\ g_t(e_1) = e_3,\ g_t(e_2) = t^{-1}e_4,\ g_t(e_3) = e_1,$
$g_t(e_4) = e_2,$
$\mathcal{R}_4(\alpha = 0) \to \mathcal{R}_2,\ g_t(e_1) = t^{-1}e_2,\ g_t(e_2) = e_1+(t^{-1}-1)e_2+(t^{-1}+1)e_3,$
$g_t(e_3) = -t^{-1}e_3,\ g_t(e_4) = t^{-1}e_4,$
$\mathcal{R}_2 \to \mathcal{R}_3,\ g_t(e_1) = te_1,\ g_t(e_2) = te_2,\ g_t(e_3) = t^2e_3,\ g_t(e_4) = t^3e_4,$
$\mathcal{R}_4(\alpha \neq 1) \to \mathcal{R}_4(\alpha = 1),\ g_t(e_1) = te_1 + (t^2 - t)e_2 + t^3e_3 + (t^4 - t^3)e_4,$
$g_t(e_2) = t^2e_2,\ g_t(e_3) = t^3e_3 + (t^4 - t^3)e_4,\ g_t(e_4) = t^4e_4,$
$\mathcal{R}_2 \to \mathcal{R}_5,\ g_t(e_1) = t^{-1}e_1,\ g_t(e_2) = t^{-2}e_2,\ g_t(e_3) = t^{-2}e_3,$
$g_t(e_4) = t^{-3}e_4,$
$\mathcal{R}_4(\alpha = 1) \to \mathcal{R}_{11},\ g_t(e_1) = \sqrt{2}t^{-1}e_1,\ g_t(e_2) = \frac{1}{\sqrt{2}}t^{-1}e_1 - \frac{1}{\sqrt{2}}t^{-1}e_2,$
$g_t(e_3) = 2t^{-2}e_4,\ g_t(e_4) = -t^{-2}e_3 + t^{-2}e_4,$
$\mathcal{R}_2 \to \mathcal{R}_{12},\ g_t(e_1) = it^{-1}e_2,\ g_t(e_2) = -it^{-1}e_1,\ g_t(e_3) = t^{-2}e_3,$
$g_t(e_4) = t^{-2}e_4,$
$\mathcal{R}_4(\alpha) \to \mathcal{R}_{13}(\alpha),\ g_t(e_1) = t^{-1}e_2,\ g_t(e_2) = -t^{-1}e_1,\ g_t(e_3) = -t^{-2}e_4,$
$g_t(e_4) = t^{-2}e_3,$
$\mathcal{R}_2 \to \mathcal{R}_{17},\ g_t(e_1) = t^{-3}e_1,\ g_t(e_2) = t^{-2}e_2,\ g_t(e_3) = t^{-5}e_4,$
$g_t(e_4) = t^{-5}e_3,$
$\mathcal{R}_4(\alpha = 1) \to \mathcal{R}_{19},\ g_t(e_1) = t^{-1}e_1,\ g_t(e_2) = e_2,\ g_t(e_3) = t^{-1}e_4,$
$g_t(e_4) = e_3,$
$\mathcal{R}_{29} \to \mathcal{R}_{20}(\alpha = -1),\ g_t(e_1) = e_1 + t^{-2}e_3,\ g_t(e_2) = e_3 + t^{-1}e_4,$
$g_t(e_3) = t^{-1}e_2,\ g_t(e_4) = e_4,$
$\mathcal{R}_{16} \to \mathcal{R}_{21},\ g_t(e_1) = t^{-1}e_1,\ g_t(e_2) = e_2,\ g_t(e_3) = t^{-\frac{1}{2}}e_3,$
$g_t(e_4) = t^{-1}e_4,$
$\mathcal{R}_{11} \to \mathcal{R}_{18},\ g_t(e_1) = t^{-1}e_1,\ g_t(e_2) = e_2,\ g_t(e_3) = t^{-1}e_3,$
$g_t(e_4) = 2t^{-1}e_3 + e_4.$

Applying the same invariance degeneration arguments as for dimensions three and their values given in [14] and [109] we obtain the rigidity of the algebras $\mathcal{R}_1, \mathcal{R}_{10}, \mathcal{R}_4(0)$ and the rigidity of the family of algebras $\mathcal{R}_{14}(\alpha)$ with $(\alpha \neq i)$. Therefore, the final result of this section is given as follows.

Theorem 3.16. $\mathcal{R}_1, \mathcal{R}_{10}, \mathcal{R}_4(0)$ *are rigid and the family of algebras* $\mathcal{R}_{14}(\alpha)$ *is a rigid family in* $NL_4(\mathbb{C})$ *and they generate irreducible components of* $LN_4(\mathbb{C})$.

$$\begin{aligned}
\dim \mathfrak{C}_1 &= \dim \overline{\mathrm{Orb}(\mathcal{R}_1)} = 12,\\
\dim \mathfrak{C}_2 &= \dim \overline{\mathrm{Orb}(\mathcal{R}_{10})} = 13,\\
\dim \mathfrak{C}_3 &= \dim \overline{\mathrm{Orb}(\mathcal{R}_4(0))} = 13,\\
\dim \mathfrak{C}_4 &= \dim \overline{\cup_{\alpha\in\mathbb{C}\setminus\{i\}} \mathrm{Orb}(\mathcal{R}_{14}(\alpha)} = 11.
\end{aligned}$$

Therefore,

$$\dim LN_4(\mathbb{C}) = \max\{\dim \mathfrak{C}_1, \dim \mathfrak{C}_2, \dim \mathfrak{C}_3, \dim \mathfrak{C}_4, \} = 13.$$

CHAPTER 4

ON SOME CLASSES OF LEIBNIZ ALGEBRAS

In the chapter we study complex non Lie filiform Leibniz algebras. For Lie algebras, the notion of p-filiformity ($p \in \mathbb{N} \cup \{0\}$) makes sense for $p \geq 1$ and loses sense for $p = 0$, since a Lie algebra has at least two generators. In the case of Leibniz algebras, this notion is meaningful for $p = 0$; so the introduction of null-filiform algebra is quite justified. We give some equivalent conditions for a Leibniz algebra to be filiform and describe naturally graded complex filiform Leibniz algebras.

4.1 IRREDUCIBLE COMPONENTS OF A SUBCLASS OF NILPOTENT LEIBNIZ ALGEBRAS

This section is devoted to description of the irreducible components of the class of nilpotent Leibniz algebras containing an algebra of maximal nilindex.

Let x be an element from the set $L \setminus L^2$ such that R_x is a nilpotent operator. Denote by $C(x) = (n_1, n_2, \ldots, n_k)$ the decreasing sequence which consists of the dimensions of the Jordan blocks of the R_x. On the set of such sequences we consider the lexicographic order, i.e. $C(x) = (n_1, n_2, \ldots, n_k) \leq C(y) = (m_1, m_2, \ldots, m_s)$ means that there exists $i \in \mathbb{N}$ such that $n_j = m_j$ for any $j < i$ and $n_i < m_i$.

Definition 4.1. The sequence $C(L) = \max_{x \in L \setminus L^2} C(x)$ is called the *characteristic sequence* of the Leibniz algebra L.

As in the Lie algebras case it can be shown that the characteristic sequence is an invariant under isomorphisms.

Example 4.1. *Let L be an algebra with $C(L) = (1, 1, ..., 1)$. Then L is abelian.*

Example 4.2. *Let L be an n-dimensional Leibniz algebra. According to Theorem 3.2, L is a null-filiform algebra if and only if $C(L) = (n)$.*

Since the set $\{L \in LB_n(K) \mid \dim \mathrm{Ann}_r(L) \geq n - 1\}$ is closed with respect to the Zariski topology, we have

$$\overline{\mathrm{Orb}_n(NF_n)} \subseteq LN_n \cap \{L \in LB_n(K) : \dim \mathrm{Ann}_r(L) \geq n - 1\}.$$

For a convenience, we introduce the notation

$$N_nZ = LN_n \cap \{L \in LB_n(K) \mid \dim \mathrm{Ann}_r(L) = n - 1\}.$$

The case $\dim \mathrm{Ann}_r(L) = n$ is not interesting, since in this case L is abelian.

Lemma 4.1. *Let L be an algebra in N_nZ with the characteristic sequence $C(L) = (m, n - m)$. Then for $m = \frac{n}{2}$ the algebra L is isomorphic to the algebra:*

$$[e_i, e_1] = e_{i+1}, \quad [e_{\frac{n}{2}+i}, e_1] = e_{\frac{n}{2}+i+1}, \; 1 \leq i \leq \frac{n}{2} - 1$$

and for $m > n/2$ it is isomorphic to one of the two non-isomorphic algebras:

$$[e_i, e_1] = e_{i+1}, \; 1 \leq i \leq m - 1, \quad [e_{m+i}, e_1] = e_{m+i+1}, \; 1 \leq i \leq n - m - 1,$$

$$[e_i, e_1] = e_{i+1}, \; 1 \leq i \leq n-m-1, \quad [e_{n-m+i}, e_1] = e_{n-m+i+1}, \; 1 \leq i \leq m-1.$$

Proof. Let $L \in N_nZ$ and $C(L) = (m, n - m)$. Then there is a basis $\{e_1, \ldots, e_n\}$ of L such that for $x \in L \setminus L^2$ the matrix form of the operator R_x can be represented as follows:

$$R_x :\simeq \begin{pmatrix} J_m & 0 \\ 0 & J_{n-m} \end{pmatrix}, \quad R_x :\simeq \begin{pmatrix} J_{n-m} & 0 \\ 0 & J_m \end{pmatrix},$$

i.e., in the first case we have products:

$$[e_i, x] = e_{i+1},\ 1 \le i \le m-1,\ [e_{m+i}, x] = e_{m+i+1},\ 1 \le i \le n-m-1,$$
$$[e_m, x] = [e_n, x] = 0,$$

whereas in the second case we have

$$[e_i, x] = e_{i+1},\ 1 \le i \le n-m-1,\ [e_{n-m+i}, x] = e_{n-m+i+1},\ 1 \le i \le m-1,$$
$$[e_{n-m}, x] = [e_n, x] = 0.$$

Consider the first case.

Since $L^2 \subseteq \text{Ann}_r(L)$, the basis vectors $e_2, \dots, e_m, e_{m+2}, \dots, e_n$ are in $\text{Ann}_r(L)$. Consequently, the element x is not in the linear span of the basis vectors $\{e_2, \dots, e_m, e_{m+2}, \dots, e_n\}$. Therefore, instead of x we can chose either e_1 or e_{m+1} (due to $\dim \text{Ann}_r(L) = n-1$).

If $x = e_1$, we get the algebra L_1 :

$$[e_i, e_1] = e_{i+1},\ 1 \le i \le m-1,\ [e_{m+i}, e_1] = e_{m+i+1},\ 1 \le i \le n-m-1.$$

If $x = e_{m+1}$, we obtain L_2 :

$$[e_i, e_{m+1}] = e_{i+1},\ 1 \le i \le m-1,\ [e_{m+i}, e_{m+1}] = e_{m+i+1},\ 1 \le i \le n-m-1.$$

Consider now the second case:

$$[e_i, x] = e_{i+1},\ 1 \le i \le n-m-1,\ [e_{n-m+i}, x] = e_{n-m+i+1},\ 1 \le i \le m-1,$$
$$[e_{n-m}, x] = [e_n, x] = 0.$$

Applying the similar arguments as in the first case we can suppose that either $x = e_1$ or $x = e_{n-m+1}$. So, we derive the following two algebras L_1' and L_2' :

$$L_1' : \begin{array}{l} [e_i, e_1] = e_{i+1},\ 1 \le i \le n-m-1, \\ [e_{n-m+i}, e_1] = e_{n-m+i+1},\ 1 \le i \le m-1, \end{array}$$

$$L_2' : \begin{array}{l} [e_i, e_{n-m+1}] = e_{i+1},\ 1 \le i \le n-m-1, \\ [e_{n-m+i}, e_{n-m+1}] = e_{n-m+i+1},\ 1 \le i \le m-1. \end{array}$$

It is not difficult to check that the algebras L_1 and L_1' are isomorphic to L_2' and L_2, respectively. It is easy to see that in the case $m \neq \frac{n}{2}$ the algebras L_1 and L_1' are non isomorphic, otherwise they coincide. □

For a convenience, in the case of $\dim \text{Ann}_r(L) = n-1$ we henceforth express the table of multiplication of L via the right multiplication operator of an basis element $e_1 \in L \setminus \text{Ann}_r(L)$.

Lemma 4.2. *Assume that $L \in N_nZ$ and $C(L) = (n_1, \dots, n_s)$. Then L is isomorphic to one of the algebras*

$$R_{e_1,\sigma} := \begin{pmatrix} J_{n_{\sigma(1)}} & 0 & \dots & 0 & 0 \\ 0 & J_{n_{\sigma(2)}} & \dots & 0 & 0 \\ \vdots & \vdots & \ddots & \vdots & \vdots \\ 0 & 0 & \dots & 0 & J_{n_{\sigma(s)}} \end{pmatrix},$$

where σ is an element of the symmetric group S_s and $J_{n_1}, \dots, J_{n_s}$ are Jordan blocks with dimensions $n_1, \dots, n_s$, respectively. In particular, $R_{e_1,\sigma_1} \cong R_{e_2,\sigma_2}$ if and only if $n_{\sigma_1} = n_{\sigma_2}$.

Proof. Suppose that L satisfies the conditions of the lemma. By permutation of corresponding basis elements it is easy to see that the interchange of the second and the third Jordan blocks do not change the multiplication of the algebra L. Continuing this procedure, we conclude that permutations of Jordan blocks $J_{n_{\sigma(2)}}, \dots, J_{n_{\sigma(s)}}$ do not change the multiplication of the algebra R_{e_1,σ_1}. The arguments similar to those in Lemma 4.1 complete the proof of the corollary. □

Under the assumptions of Lemma 4.2, we also have

Corollary 4.1. *The number of non isomorphic algebras in N_nZ with characteristic sequence $(n_1, \dots, n_s)$ equals the cardinality of different numbers in the set $\{n_1, \dots, n_s\}$.*

Lemma 4.3. *The set $\overline{\mathrm{Orb}_n(NF_n)}$ consists of the algebras of the form:*

$$R_{e_1} := \begin{pmatrix} J_{n_1} & 0 & \dots & 0 & 0 \\ 0 & J_{n_2} & \dots & 0 & 0 \\ \vdots & \vdots & \ddots & \vdots & \vdots \\ 0 & 0 & \dots & 0 & J_{n_s} \end{pmatrix},$$

where $n_1 + n_2 + \dots + n_s = n$.

Proof. Since $NF_n \in LN_n \cap \{L \in LB_n(K) : \dim \mathrm{Ann}_r(L) \geq n-1\} = N_nZ \cup \mathbb{C}^n$ and $\overline{N_nZ \cup \mathbb{C}^n}$ is a closed set in Zariski topology we have $\overline{\mathrm{Orb}_n(NF_n)} \subseteq N_nZ \cup \mathbb{C}^n$. Therefore, $\overline{\mathrm{Orb}_n(NF_n)}$ is in the set of the algebras of Lemma 4.2. We show that the set $\overline{\mathrm{Orb}_n(NF_n)}$ coincides with $N_nZ \cup \mathbb{C}^n$.

Let L be an element of $N_nZ \cup \mathbb{C}^n$ and

$$R_{e_1} := \begin{pmatrix} J_{n_1} & 0 & \dots & 0 & 0 \\ 0 & J_{n_2} & \dots & 0 & 0 \\ \vdots & \vdots & \ddots & \vdots & \vdots \\ 0 & 0 & \dots & 0 & J_{n_s} \end{pmatrix}.$$

Consider the family of the matrices $(g_t)_{t\in\mathbb{R}\setminus\{0\}}$ defined as follows:

$$\begin{array}{ll} g_t(e_i) & = t^{-i}e_i \qquad \text{for } 1 \leq i \leq n_1, \\ g_t(e_i) & = t^{-(i-1)}e_i \quad \text{for } n_1+1 \leq i \leq n. \end{array}$$

Passing to the limit of this family as $t \to 0$, i.e., $\lim\limits_{t\to 0} g_t^{-1}[g_t(e_i), g_t(e_j)]$, we obtain

$$g_t * NF_n \underset{t\to 0}{\longrightarrow} R_{e_1} := \begin{pmatrix} J_{n_1} & 0 \\ 0 & J_{n-n_1} \end{pmatrix}.$$

Now, take the family of matrices $(f_t)_{t\in\mathbb{R}\setminus\{0\}}$ defined by

$$\begin{array}{ll} f_t(e_i) & = t^{-i}e_i, \qquad \text{for } 1 \leq i \leq n_1+n_2, \\ f_t(e_i) & = t^{-(i-1)}e_i, \quad \text{for } n_1+n_2+1 \leq i \leq n. \end{array}$$

Taking the limit of this family as $t \to 0$, i.e., $\lim\limits_{t\to 0} f_t^{-1}[f_t(e_i), f_t(e_j)]$, we obtain

$$f_t * NF_n \underset{t\to 0}{\longrightarrow} R_{e_1} := \begin{pmatrix} J_{n_1} & 0 & 0 \\ 0 & J_{n_2} & 0 \\ 0 & 0 & J_{n-n_1-n_2} \end{pmatrix}.$$

Repeating this procedure s times, we conclude that the algebra defined by the operator

$$R_{e_n} := \begin{pmatrix} J_{n_1} & 0 & \dots & 0 & 0 \\ 0 & J_{n_2} & \dots & 0 & 0 \\ \vdots & \vdots & \ddots & \vdots & \vdots \\ 0 & 0 & \dots & 0 & J_{n_s} \end{pmatrix}.$$

is in $\overline{\mathrm{Orb}_n(NF_n)}$. □

Since the orbit of a null-filiform algebra is an open set in LN_n, we conclude that its closure is an irreducible component of LN_n (see [170]) and the following theorem holds.

Theorem 4.1. *An irreducible component of the variety LN_n, containing the null-filiform algebra, consists of the following algebras:*

$$R_{e_1} :\simeq \begin{pmatrix} J_{n_1} & 0 & \dots & 0 & 0 \\ 0 & J_{n_2} & \dots & 0 & 0 \\ \vdots & \vdots & \ddots & \vdots & \vdots \\ 0 & 0 & \dots & 0 & J_{n_s} \end{pmatrix}$$

where $n_1 + \cdots + n_s = n$.

Moreover, two algebras

$$R_{e_1} :\simeq \begin{pmatrix} J_{n_1} & 0 & \dots & 0 & 0 \\ 0 & J_{n_2} & \dots & 0 & 0 \\ \vdots & \vdots & \ddots & \vdots & \vdots \\ 0 & 0 & \dots & 0 & J_{n_s} \end{pmatrix}, \quad R'_{e_1} :\simeq \begin{pmatrix} J_{m_1} & 0 & \dots & 0 & 0 \\ 0 & J_{m_2} & \dots & 0 & 0 \\ \vdots & \vdots & \ddots & \vdots & \vdots \\ 0 & 0 & \dots & 0 & J_{m_s} \end{pmatrix}$$

are isomorphic if and only if $n_1 = m_1$ and the set $\{m_2, m_3, \dots, m_s\}$ is obtained by permutation of $\{n_2, n_3, \dots, n_s\}$.

Proof. The proof follows from Lemmas 4.3 and 4.2. □

Remark 1. Theorem 4.1 implies that the number of non isomorphic algebras in the irreducible component of LN_n containing the algebra NF_n equals $p(n)$, where $p(n)$ is the number of integer solutions of the equation $x_1 + x_2 + \cdots + x_n = n$, with $x_1 \geq x_2 \geq \cdots \geq x_n \geq 0$. The asymptotic value of $p(n)$, given in [102] by the expression $p(n) \approx \frac{1}{4n\sqrt{3}} e^{A\sqrt{n}}$ with $A = \pi\sqrt{\frac{2}{3}}$, (where $a(n) \approx b(n)$ means that $\lim\limits_{n\to\infty} \frac{a(n)}{b(n)} = 1$) shows how small is the set of non isomorphic Leibniz algebras in the irreducible component of LN_n containing the algebra NF_n, i.e., the number of orbits in this component is finite for every value of n.

4.2 CLASSIFICATION OF NATURALLY GRADED COMPLEX FILIFORM LEIBNIZ ALGEBRAS

Let L be a nilpotent Leibniz algebra with nilindex s.

Consider $L_i = L^i / L^{i+1}$, $1 \leq i \leq s-1$, and $grL = L_1 \oplus L_2 \oplus \cdots \oplus L_{s-1}$. Then $[L_i, L_j] \subseteq L_{i+j}$ and we obtain the graded algebra grL.

Definition 4.2. *If a Leibniz algebra L' is isomorphic to a naturally graded filiform algebra grL, then L' is said to be naturally graded filiform Leibniz algebra.*

Lemma 4.4. *Let L be an n-dimensional Leibniz algebra. Then the following statements are equivalent:*

(a) $C(L) = (n-1, 1)$;

(b) *L is a filiform Leibniz algebra;*

(c) $L^{n-1} \neq 0$ *and* $L^n = 0$.

Proof. The implications (a) $\Rightarrow$ (b) $\Rightarrow$ (c) are obvious.

(b) $\Rightarrow$ (a): Let $\{e_1, \dots, e_n\}$ be a basis for a filiform algebra L such that

$$\{e_3, \dots, e_n\} \subseteq L^2, \ \{e_4, \dots, e_n\} \subseteq L^3, \ \dots \ , \{e_n\} \subseteq L^{n-1}.$$

Consider the products

$[x, e_1 + \alpha e_2] = \gamma_1 e_3 + \alpha\beta_1 e_3, \ [e_3, e_1 + \alpha e_2] = \gamma_2 e_4 + \alpha\beta_2 e_4,$
$[e_4, e_1 + \alpha e_2] = \gamma_3 e_5 + \alpha\beta_3 e_5, \ \dots \ , [e_n, e_1 + \alpha e_2] = 0,$

where x is an arbitrary element of L and $(\gamma_i, \beta_i) \neq (0, 0)$ for any i. Choose α such that $\gamma_i + \alpha\beta_i \neq 0$ for any i. Then $z = e_1 + \alpha e_2 \in L \setminus [L, L]$ and $C(z) = (n-1, 1)$.

(c) $\Rightarrow$ (b): Assume that $L^n = 0$. Then we obtain a decreasing chain of subalgebras $L \supset L^2 \supset L^3 \supset \cdots \supset L^{n-1} \supset L^n = 0$ of the length n. Obviously, $\dim L^2 = n-1$ or $\dim L^2 = n-2$ (otherwise $L^{n-1} = 0$). The case $\dim L^2 = n-1$ is not possible because of Proposition 3.3. Therefore, $\dim L^2 = n-2$. Obviously, for $i = 2, 3, \dots, n$ we have $\dim L^i = n - i$, i.e., L is a filiform Leibniz algebra. □

Let L be an $(n+1)$-dimensional complex filiform Leibniz algebra. By the arguments similar to those are in [178], over an infinite field we can find a basis $e_0, e_1 \in L_1, \ e_i \in L_i \ (i \geq 2)$ for L such that $[e_i, e_0] = e_{i+1}$ and $[e_n, e_0] = 0, \ 1 \leq i \leq n$.

Case 1. Assume that $[e_0, e_0] = \alpha e_2, \ \alpha \neq 0$. Then $e_2 \in \mathrm{Ann}_r(L)$. Hence, $\{e_3, \dots, e_n\} \subseteq \mathrm{Ann}_r(L)$. Changing the basis as below

$$\overline{e}_1 = \alpha e_1, \ \overline{e}_2 = \alpha e_2, \ \overline{e}_3 = \alpha e_3, \ \dots, \ \overline{e}_n = \alpha e_n,$$

we may assume that α is equal to 1. Thus, $[e_0, e_0] = e_2, [e_i, e_0] = e_{i+1}$, and $[e_n, e_0] = 0$. Suppose that $[e_0, e_1] = \beta e_2$ and $[e_1, e_1] = \gamma e_2$. Then

$$[e_0, [e_1, e_0]] = [[e_0, e_1], e_0] - [[e_0, e_0], e_1] \Rightarrow \beta e_3 = [e_2, e_1]$$

and

$$[e_1,[e_0,e_1]] = [[e_1,e_0],e_1] - [[e_1,e_1],e_0] \Rightarrow \gamma e_3 = [e_2,e_1].$$

It follows that $\beta = \gamma$. Applying the induction with respect to the dimension and using the equality

$$[e_i,[e_0,e_1]] = [[e_i,e_0],e_1] - [[e_i,e_1],e_0],$$

one can easily prove that $[e_i,e_1] = \beta e_{i+1}$, i.e., in Case 1 we obtain the algebra

$$[e_0,e_0] = e_2, [e_i,e_0] = e_{i+1}, [e_1,e_1] = \beta e_2, [e_i,e_1] = \beta e_{i+1},$$

$$[e_0,e_1] = \beta e_2.$$

Case 2. Let $[e_0,e_0] = 0$ and $[e_1,e_1] = \alpha e_2$, $\alpha \neq 0$. In this case $e_2 \in \mathrm{Ann}_r(L)$. Hence, $\{e_3,\dots,e_n\} \subseteq \mathrm{Ann}_r(L)$. Putting

$$\overline{e}_0 = \alpha e_0,\ \ \overline{e}_2 = \alpha e_2,\ \ \overline{e}_3 = \alpha^2 e_3,\dots,\overline{e}_n = \alpha^{n-1} e_n,$$

we may assume that $\alpha = 1$, i.e., $[e_1,e_1] = e_2$, $[e_i,e_0] = e_{i+1}$. Put $[e_0,e_1] = \beta e_2$. Then

$$[e_0,[e_1,e_0]] = [[e_0,e_1],e_0] - [[e_0,e_0],e_1] \Rightarrow [[e_0,e_1],e_0] = 0,$$

i.e., $\beta[e_2,e_0] = \beta e_3 = 0 \Rightarrow \beta = 0$. Again applying the induction with respect to the dimension and using the equality $[e_i,[e_0,e_1]] = [[e_{i+1},e_0],e_1] - [[e_i,e_1],e_0]$, we can easily show that $[e_i,e_1] = e_{i+1}$, i.e., in Case 2 we obtain the algebra

$$[e_i,e_0] = e_{i+1}, [e_i,e_1] = e_{i+1}(i > 1).$$

Changing the basis as $\overline{e}_0 := e_0 - e_1$, $\overline{e}_1 := e_1$ and others are left unchanged, we get the algebra $[\overline{e}_i,\overline{e}_1] = \overline{e_{i+1}}$. It is easy to see that this algebra is isomorphic to the algebra of Case 1 for $\beta = 1$ $(e_0' := e_0 - e_1, e_1' := e_1)$.

Case 3. Let $[e_0,e_0] = 0$ and $[e_1,e_1] = 0$. We set $[e_0,e_1] = \alpha e_2$.

Subcase 1. Assume that $[e_0,e_1] = \alpha e_2$ with $\alpha \neq 1$. Then $e_2 \in \mathrm{Ann}_r(L)$. Again $\{e_3,\ \dots\ ,e_n\} \subseteq \mathrm{Ann}_r(L)$. Since $\alpha \neq 1$, on putting $\overline{e}_1 = e_1 + e_0$ we obtain $\overline{e}_1^2 = (\alpha + 1)e_2$ and $[\overline{e}_1,e_0] = e_2$, i.e., we come to Case 2.

Subcase 2. $[e_0, e_1] = -e_2$.

Before treating this subcase, we prove the following

Lemma 4.5. *Let L be an* $(n+1)$*-dimensional naturally graded filiform Leibniz algebra with a basis* $\{e_0, e_1, \ldots, e_n\}$ *satisfying the conditions:*

$$[e_1, e_1] = [e_0, e_0] = 0, [e_0, e_1] = -e_2, \ [e_i, e_0] = e_{i+1}.$$

Then L is a Lie algebra.

Proof. Again applying the induction with respect to the dimension and using the relation $[e_0, [e_i, e_0]] = [[e_0, e_i], e_0] - [[e_0, e_0], e_i]$, one can show that $[e_0, e_i] = -[e_i, e_0]$ for $1 \leq i \leq n$.

The equality $[e_1, [e_1, e_0]] = [[e_1, e_1], e_0] - [[e_1, e_0], e_1]$ implies $[e_1, e_2] = -[e_2, e_1]$.

Applying the chain of the equalities

$$\begin{aligned}
[e_1, e_{i+1}] &= [e_1, [e_i, e_0]] \\
&= [[e_1, e_i], e_0] - [[e_1, e_0], e_i] \\
&= -[[e_i, e_1], e_0] - [e_2, e_i] \\
&= [e_0, [e_i, e_1]] - [e_2, e_i] \\
&= [[e_0, e_i], e_1] - [[e_0, e_1], e_i] - [e_2, e_i] \\
&= [[e_0, e_i], e_1] + [e_2, e_i] - [e_2, e_i] \\
&= -[[e_i, e_0], e_1] = -[e_{i+1}, e_1]
\end{aligned}$$

and the induction hypothesis we obtain $[e_1, e_i] = -[e_i, e_1]$ for $1 \leq i \leq n$. Thus,

$$[e_1, e_i] = -[e_i, e_1] \text{ and } [e_0, e_i] = -[e_i, e_0] \text{ for } 0 \leq i \leq n.$$

Let us prove the equality $[e_i, e_j] = -[e_j, e_i]$ for all i, j. We proceed by the induction on i for a fixed j. Observe that j may be assumed to be greater than i. Using the chain of the equalities

$$\begin{aligned}
[e_{i+1}, e_j] &= [[e_i, e_0], [e_{j-1}, e_0]] \\
&= [[[e_i, e_0], e_{j-1}], e_0] - [[[e_i, e_0], e_0], e_{j-1}] \\
&= -[e_0, [[e_i, e_0], e_{j-1}]] + [[e_0, [e_i, e_0]], e_{j-1}] \\
&= [e_0, [[e_0, e_i], e_{j-1}] - [[e_0, [e_0, e_i]], e_{j-1}] \\
&= [[e_0, [e_0, e_i]], e_{j-1}] - [[e_0, e_{j-1}, [e_0, e_i]] \\
&\quad -[[[e_0, e_0], e_i], e_{j-1}] + [[e_0, e_i], e_0], e_{j-1}] \\
&= [[[e_0, e_0], e_i], e_{j-1}] \\
&\quad -[[[e_0, e_i], e_0], e_{j-1}] - [[[e_0, e_0], e_i], e_{j-1}] \\
&\quad -[[e_{j-1}, e_0], [e_i, e_0]] + [[[e_0, e_i], e_0], e_{j-1}] \\
&= -[e_j, e_{i+1}],
\end{aligned}$$

we obtain anti-commutativity of the basis elements of the algebra L. □

Thus, the naturally graded filiform Leibniz algebras which are not Lie algebras are given as follows:

$$[e_0, e_0] = e_2, [e_i, e_0] = e_{i+1}, [e_i, e_1] = \beta e_{i+1}, [e_0, e_1] = \beta e_2.$$

Let us suppose that $\beta \neq 1$. Changing the basis as follows

$$\overline{e}_0 = (1-\beta)e_0,\ \overline{e}_1 = -\beta e_0 + e_1,\ \overline{e}_2 = (1-\beta)^2 e_2,\ \dots, \overline{e}_n = (1-\beta)^n e_n,$$

we may assume that $\beta = 0$.

Now, consider the case $\beta = 1$, i.e.,

$$[e_0, e_0] = e_2, [e_i, e_1] = e_{i+1}, [e_0, e_1] = e_2 \ \ (1 \leq i \leq n).$$

Making the basis change $\overline{e}_1 = e_1 - e_0$, we obtain

$$[e_0, e_0] = e_2, [e_i, e_0] = e_{i+1} \ (1 \leq i \leq n).$$

We show that the obtained algebras

$$L_1 := \quad [e_0, e_0] = e_2, [e_i, e_0] = e_{i+1} \ (1 \leq i \leq n-1),$$

and

$$L_2 := \quad [e_0, e_0] = e_2, [e_i, e_0] = e_{i+1} \ (2 \leq i \leq n-1)$$

are not isomorphic.

Assume the contrary and let φ be such an isomorphism, i.e., $\varphi : L_1 \to L_2$ and

$$\varphi(e_i) = \sum_{j=0}^{n} \alpha_{ij} e_j.$$

We have

$$\left[\varphi(e_0), \varphi(e_0)\right] = \left[\sum_{j=0}^{n} \alpha_{0j} e_j, \alpha_{00} e_0\right] = \alpha_{00}\Big(\alpha_{00} e_2 + \alpha_{02} e_3 + \cdots + \alpha_{0,n-1} e_n\Big).$$

On the other hand,

$$\varphi([e_0, e_0]) = \varphi(e_2) = \sum_{j=0}^{n} \alpha_{2j} e_j.$$

Comparing the two equalities above, we conclude that

$$\alpha_{20} = \alpha_{21} = 0, \alpha_{22} = \alpha_{00}^2, \alpha_{2,k} = \alpha_{00}\alpha_{0,k-1} \tag{4.1}$$

for $3 \leq k \leq n$.

Consider the product

$$[\varphi(e_i), \varphi(e_0)] = \left[\sum_{j=0}^{n} \alpha_{ij}e_j, \alpha_{00}e_0\right] = \alpha_{00}\sum_{j=0}^{n}\alpha_{ij}[e_j, e_0]$$
$$= \alpha_{00}(\alpha_{i,0}e_2 + \alpha_{i,2}e_3 + \cdots + \alpha_{i,n-1}e_n)$$

and the element

$$\varphi([e_i, e_0]) = \varphi(e_{i+1}) = \sum_{j=0}^{n}\alpha_{i+1,j}x_j \text{ for } 1 \le i \le n-1.$$

Comparing the two equalities, we deduce that

$$\alpha_{i+1,0} = \alpha_{i+1,1} = 0, \alpha_{i+1,2} = \alpha_{00}\alpha_{i,0}, \alpha_{i+1,k} = \alpha_{00}\alpha_{i,k-1} \qquad (4.2)$$

for $3 \le k \le n, 1 \le i \le n-1$.

It follows from (4.2) that $\alpha_{22} = \alpha_{00}\alpha_{10}$. Since $\alpha_{00} \neq 0$ (otherwise φ is degenerate) the relation (4.1) implies that $\alpha_{00} = \alpha_{10}$.

We have $\varphi([e_0, e_1]) = \varphi(0) = 0$. On the other hand,

$$[\varphi(e_0), \varphi(e_1)] = \left[\sum_{j=0}^{n}\alpha_{0j}e_j, \alpha_{10}e_0\right] = \alpha_{10}\sum_{j=0}^{n}\alpha_{0j}[e_j, e_0]$$
$$= \alpha_{10}(\alpha_{00}e_0 + \alpha_{02}e_3 + \cdots + \alpha_{0,n-1}e_n) = 0.$$

Hence, $\alpha_{10}\alpha_{00} = 0$ and so $\alpha_{10} = 0$, i.e., the first column of the matrix $[\varphi]$ of φ is zero. Therefore, φ is degenerate. Thus we have proved the following

Proposition 4.1. *There are exactly two non isomorphic naturally graded complex non-Lie filiform Leibniz algebras NGF_1 and NGF_2 of dimension $n+1$, where*

$$NGF_1 := \quad [e_0, e_0] = e_2, [e_i, e_0] = e_{i+1} \text{ for } 1 \le i \le n-1,$$

$$NGF_2 := \quad [e_0, e_0] = e_2, [e_i, e_0] = e_{i+1} \text{ for } 2 \le i \le n-1,$$

the other products vanish.

Remark. The naturally graded complex filiform Lie algebras have been described by M. Vergne in [178].

The following theorem summarizes the results of Proposition 4.1 and Vergne [178].

Theorem 4.2. *Any complex $(n+1)$-dimensional naturally graded filiform Leibniz algebra is isomorphic to one of the following pairwise non isomorphic algebras:*

$$NGF_1 := \begin{cases} [e_0, e_0] = e_2, \\ [e_i, e_0] = e_{i+1}, \quad 1 \le i \le n-1 \end{cases}$$

$$NGF_2 := \begin{cases} [e_0, e_0] = e_2, \\ [e_i, e_0] = e_{i+1}, \quad 2 \le i \le n-1 \end{cases}$$

$$NGF_3 := \begin{cases} [e_i, e_0] = -[e_0, e_i] = e_{i+1}, & 1 \le i \le n-1 \\ [e_i, e_{n-i}] = -[e_{n-i}, e_i] = \alpha(-1)^{i+1} e_n, & 1 \le i \le n-1, \\ \alpha \in \{0, 1\} \textit{ for odd } n \textit{ and } \alpha = 0 \textit{ for even } n. \end{cases}$$

It is clear that NGF_3 is a Lie algebra, but NGF_1 and NGF_2 are non Lie algebras.

Theorem 4.3. *An arbitrary $(n+1)$–dimensional complex filiform Leibniz algebra admits a basis $\{e_0, e_1, \ldots, e_n\}$ called adapted, such that the table of multiplication of the algebra has one of the following forms:*

$$\mathrm{FLb}_{n+1} := \begin{cases} [e_0, e_0] = e_2, \\ [e_i, e_0] = e_{i+1}, \quad 1 \le i \le n-1, \\ [e_0, e_1] = \alpha_3 e_3 + \alpha_4 e_4 + \cdots + \alpha_{n-1} e_{n-1} + \theta e_n, \\ [e_j, e_1] = \alpha_3 e_{j+2} + \alpha_4 e_{j+3} + \cdots + \alpha_{n+1-j} e_n, \\ 1 \le j \le n-2, \text{ and } \alpha_3, \alpha_4, \ldots, \alpha_n, \theta \in \mathbb{C}. \end{cases}$$

$$\mathrm{SLb}_{n+1} := \begin{cases} [e_0, e_0] = e_2, \\ [e_i, e_0] = e_{i+1}, \quad 2 \le i \le n-1, \\ [e_0, e_1] = \beta_3 e_3 + \beta_4 e_4 + \cdots + \beta_n e_n, \\ [e_1, e_1] = \gamma e_n, \\ [e_j, e_1] = \beta_3 e_{j+2} + \beta_4 e_{j+3} + \cdots + \beta_{n+1-j} e_n, \\ 2 \le j \le n-2 \text{ and } \beta_3, \beta_4, \ldots, \beta_n, \gamma \in \mathbb{C}. \end{cases}$$

$$\mathrm{TLb}_{n+1} := \begin{cases} [e_0, e_0] = \gamma e_n, \\ [e_1, e_1] = \alpha e_n, \\ [e_i, e_0] = e_{i+1}, \quad 1 \le i \le n-1 \\ [e_0, e_1] = -e_2 + \beta e_n, \\ [e_0, e_i] = -e_{i+1}, \quad 2 \le i \le n-1 \\ [e_i, e_j] = -[e_j, e_i] \in \mathrm{Span}\{e_{i+j+1}, e_{i+j+2}, \ldots, e_n\}, \\ 1 \le i \le n-3, \text{ and } 2 \le j \le n-1-i \\ [e_{n-i}, e_i] = -[e_i, e_{n-i}] = (-1)^i \delta e_n, \quad 1 \le i \le n-1 \end{cases}$$

where $\delta \in \{0, 1\}$ for odd n, $\delta = 0$ for even n and the bracket of TLb_{n+1} must satisfy the Leibniz identity.

Proof. From Theorem 4.2 we have that the natural gradation of a filiform Leibniz algebra may be an algebra from one of NGF_i for $i = 1, 2, 3$.

By straightforward verification we can make sure that all the above algebras are Leibniz algebras. By Proposition 4.1, every $(n + 1)$-dimensional complex non-Lie filiform Leibniz algebra (L, μ) is isomorphic to the algebra $\mu_0^n + \beta$, where $NGF_1 = (V, \mu_0^n)$ and

$$\begin{array}{ll} \beta(e_0, e_0) & = 0, \\ \beta(e_i, e_0) & = 0 \text{ for } 1 \le i \le n-1, \\ \beta(e_i, e_j) & \in \text{Span}\{e_{i+j+1}, \ldots, e_n\} \text{ for } i \neq 0, \\ \beta(e_0, e_j) & \in \text{Span}\{e_{j+2}, \ldots, e_n\} \text{ for } 1 \le j \le n-2, \end{array}$$

or to the algebra $\mu_1^n + \beta$, where $NGF_2 = (V, \mu_1^n)$ and

$$\begin{array}{ll} \beta(e_0, e_0) & = 0, \\ \beta(e_i, e_0) & = 0, \text{ for } 2 \le i \le n-1, \\ \beta(e_i, e_j) & \in \text{Span}\{e_{i+j+1}, \ldots, e_n\} \text{ for } i, j \neq 0, \\ \beta(e_0, e_j) & \in \text{Span}\{e_{j+2}, \ldots, e_n\} \text{ for } 1 \le j \le n-2. \end{array}$$

Case 1. Assume that $\mu :\cong \mu_0^n + \beta$. Then $[e_0, e_0] = e_2$ and $[e_i, e_0] = e_{i+1}$ for $1 \le i \le n-1$; hence $\text{Span}\{e_2, e_3, \ldots, e_n\} = \text{Ann}_r(L)$, so that $[e_i, e_j] = 0$ for $2 \le j \le n, 0 \le i \le n$.

Put $[e_1, e_1] = \alpha_3 e_3 + \alpha_4 e_4 + \cdots + \alpha_n e_n$. Consider

$$[e_i, [e_0, e_1]] = [[e_i, e_0], e_1] - [[e_i, e_1], e_0].$$

Since $[e_0, e_1] \in \text{Ann}_r(L)$, we have $[e_i, [e_0, e_1]] = 0$ and so $[[e_i, e_0], e_1] = [[e_i, e_1], e_0$ for all $i \ge 1$. Thus, $[e_i, e_1] = \alpha_3 e_{i+2} + \alpha_4 e_{i+3} + \cdots + \alpha_{n+1-i} e_n$ for $1 \le i \le n$.

Let $[e_0, e_1] = \theta_3 e_3 + \theta_4 e_4 + \cdots + \theta_n e_n$. Consider

$$[e_0, [e_1, e_0]] = [[e_0, e_1], e_0] - [[e_0, e_0], e_1].$$

We have

$$[[e_0, e_1], e_0] = [[e_0, e_0], e_1].$$

However, $[e_0, e_0] = e_2$ and $[e_i, e_0] = e_{i+1}$. Therefore,

$$\theta_3 e_4 + \theta_4 e_5 + \cdots + \theta_{n-1} e_n = \alpha_3 e_4 + \alpha_4 e_5 + \cdots + \alpha_{n-1} e_n;$$

whence

$$[e_0, e_1] = \alpha_3 e_3 + \alpha_4 e_4 + \cdots + \alpha_{n-1} e_{n-1} + \alpha_n e_n.$$

Thus, in Case 1 we obtain the following family:

$$\begin{aligned}
&[e_0, e_0] = e_2,\\
&[e_i, e_0] = e_{i+1},\\
&[e_0, e_1] = \alpha_3 e_3 + \alpha_4 e_4 + \cdots + \alpha_{n-1} e_{n-1} + \theta_n e_n,\\
&[e_i, e_1] = \alpha_3 e_{i+2} + \alpha_4 e_{i+3} + \cdots + \alpha_{n+1-i} e_n
\end{aligned}$$
for $1 \leq i \leq n$.

Case 2. $\mu :\cong \mu_1^n + \beta$. In this case $[e_0, e_0] = e_2$ and $[e_i, e_0] = e_{i+1}$ for $2 \leq i \leq n-1$, whence $\{e_2, e_3, \ldots, e_n\} \subseteq \mathrm{Ann}_r(L)$ and so $[e_i, e_j] = 0$ for $2 \leq j \leq n$, $0 \leq i \leq n$.

Let $\beta(e_1, e_0) = \alpha_3 e_3 + \alpha_4 e_4 + \cdots + \alpha_n e_n$.

Making the change $\overline{e}_1 := e_1 - \alpha_3 e_2 - \alpha_4 e_3 - \cdots - \alpha_n e_{n-1}$, we may assume that $[e_1, e_0] = 0$.

Let $[e_0, e_1] = \beta_3 e_3 + \beta_4 e_4 + \cdots + \beta_n e_n$. Consider the product

$$[e_0, [e_1, e_0]] = [[e_0, e_1], e_0] - [[e_0, e_0], e_1].$$

Since $[e_1, e_0] \in \mathrm{Ann}_r(L)$, we have $[[e_0, e_1], e_0] = [[e_0, e_0], e_1]$. Therefore, $[[e_0, e_1], e_0] = [e_2, e_1]$, i.e., $[e_2, e_1] = \beta_3 e_4 + \beta_4 e_5 + \cdots + \beta_{n-1} e_n$.

Consider the equality

$$[e_1, [e_0, e_1]] = [[e_1, e_0], e_1] - [[e_1, e_1], e_0].$$

In view of $[e_0, e_1] \in \mathrm{Ann}_r(L)$ and $[e_1, e_0] = 0$, we have $[[e_1, e_1], e_0] = 0$. However, e_0 annihilates from the left only e_n. Therefore, $[e_1, e_1] = e_n$.

Considering the equality

$$[e_i, [e_0, e_1]] = [[e_i, e_0], e_1] - [[e_i, e_1], e_0]$$

for $2 \leq i \leq n-1$ and taking into account $[e_0, e_1] \in \mathrm{Ann}_r(L)$ we derive $[[e_i, e_0], e_1] = [[e_i, e_1], e_0]$. Therefore

$$[e_{i+1}, e_1] = [[e_i, e_1], e_0], \text{ i.e., } [e_i, e_1] = \beta_3 e_{i+2} + \beta_4 e_{i+3} + \cdots + \beta_{n+1-i} e_n$$

for $2 \leq i \leq n-1$. Thus, in Case 2 we obtain the following family

$$[e_0, e_0] = e_2,$$
$$[e_i, e_0] = e_{i+1},$$
$$[e_0, e_1] = \beta_3 e_3 + \beta_4 e_4 + \cdots + \beta_n e_n = \gamma e_n.$$
$$[e_i, e_1] = \beta_3 e_{i+2} + \beta_4 e_{i+3} + \cdots + \beta_{n+1-i} e_n, \quad 2 \leq i \leq n.$$

Case 3. Let $L \in \mathrm{TLb}_{n+1}$ and let $\{e_0, e_1, \ldots, e_n\}$ be a basis of L. Then due to Theorem 4.3

$$[e_i, e_j] \in \mathrm{Span}\{e_{i+j+1}, \ldots, e_n\} \text{ for any } i, j \neq 0.$$

Then

$$[e_i, e_0] = e_{i+1} + (*)e_{i+2} + \cdots + (*)e_n, \ 1 \leq i \leq n-1.$$

Putting $e'_1 = e_1,\ e'_0 = e_0,\ e'_{i+1} := [e'_i, e'_0]$ we may assume that $[e_i, e_0] = e_{i+1},\ 1 \leq i \leq n-1$.

Now consider

$$[e_0, e_i] = -e_{i+1} + \alpha_{0,i}^{i+2} e_{i+2} + \alpha_{0,i}^{i+3} e_{i+3} + \cdots + \alpha_{0,i}^{n} e_n, \ 1 \leq i \leq n-1.$$

Then we get

$$[e_i, e_0] + [e_0, e_i] = \alpha_{0,i}^{i+2} e_{i+2} + \alpha_{0,i}^{i+3} e_{i+3} + \cdots + \alpha_{0,i}^{n} e_n, \ 1 \leq i \leq n-1. \tag{4.3}$$

From Leibniz identity we have $[x, y]+[y, x] \in \mathrm{Ann}_r(L)$, for any $x, y \in L$. Therefore, if we multiply by e_0 the both sides of (4.3) from the right-hand side $(n-i-2)$ times, we obtain $\alpha_{0,i}^{i+2} = 0$. Substituting it in (4.3) and repeating this action sufficient times, we get

$$\alpha_{0,i}^{i+k} = 0, \ 2 \leq k \leq n-1-i.$$

Applying the above to $[e_i, e_i]$ for $i:\ 0 \leq i \leq \left[\frac{n}{2}\right]$ we conclude that $[e_i, e_i] = \alpha_{i,i}^{n} e_n$.

The following chain of equalities

$$[e_0, e_i] = [e_0, [e_{i-1}, e_0]] = [[e_0, e_{i-1}], e_0] - [[e_0, e_0], e_{i-1}] =$$
$$= [-e_i + \alpha_{0,i-1}^{n} e_n, e_0] = -[e_i, e_0] = -e_{i+1}$$

leads to $[e_i, e_0] = -[e_0, e_i] = e_{i+1}$ for $2 \leq i \leq n-1$, i.e., $[e_0, x] = -[x, e_0]$ for any $x \in L^2$.

We claim that

$$[e_i, e_j] = -[e_j, e_i], \quad 1 \leq i < j \leq n-1. \tag{4.4}$$

Indeed, the induction by i for any j and the following chain of equalities:

$$\begin{aligned} [e_i, e_{j+1}] &= [e_i, [e_j, e_0]] \\ &= [[e_i, e_j], e_0] - [[e_i, e_0], e_j] \end{aligned}$$

(since $[e_i, e_j] \in L^2$)

$$\begin{aligned} &= -[e_0, [e_i, e_j]] + [[e_0, e_i] - \alpha^n_{0,i} e_n, e_j] \\ &= -[e_0, [e_i, e_j]] + [[e_0, e_i], e_j] \\ &= -[[e_0, e_i], e_j] + [[e_0, e_j], e_i] + [[e_0, e_i], e_j] \\ &= -[e_{j+1}, e_i], \quad 1 \leq j \leq n-1 \end{aligned}$$

gives (4.4).

The above observations lead to the required table of multiplication of $L \in \mathrm{TLb}_{n+1}$.

□

Remark. It should be noted that a Leibniz algebra from the family TLb_{n+1} is non Lie algebra if and only if $(\alpha, \beta, \gamma) \neq (0, 0, 0)$. In fact, this inequality can be replaced by the condition $\gamma = 1$.

For $L \in \mathrm{TLb}_{n+1}$ the subspace $\mathrm{Span}\{e_n\}$ spanned by $\{e_n\}$ is an ideal of L and the quotient algebra $L/\mathrm{Span}\{e_n\}$ is the n-dimensional filiform Lie algebra with the composition law

$$\begin{aligned} [e_i, e_0] &= e_{i+1}, \quad i = 1, 2, \ldots, n-1, \\ [e_i, e_j] &= a^1_{i,j} e_{i+j+1} + \cdots + a^{n-(i+j+1)}_{i,j} e_{n-1}, \quad 1 \leq i < j \leq n-1. \end{aligned}$$

Lemma 4.6. *Let $L \in \mathrm{TLb}_{n+1}$. Then*

$$\sum_{s=1}^{n-(i+j+k+1)} a^s_{j,k} b_{i,j+k+s} = \sum_{s=1}^{n-(i+j+k+1)} (a^s_{i,j} b_{i+j+s,k} - a^s_{i,k} b_{i+k+s,j}). \tag{4.5}$$

Proof. The Leibniz identity for e_i, e_j and e_k gives the required relations between the structure constants. Indeed,

$$[e_i,[e_j,e_k]] = \left[e_i, \sum_{s=1}^{n-(j+k+1)} a_{j,k}^s e_{j+k+s} + b_{j,k}e_n\right]$$
$$= \sum_{s=1}^{n-(i+j+k+1)} a_{j,k}^s \left(\sum_{t=1}^{n-(i+j+k+s+1)} a_{i,j+k+s}^t e_{i+j+k+s+t} + b_{i,j+k+s}e_n\right),$$
$$[[e_i,e_j],e_k] = \left[\sum_{s=1}^{n-(i+j+1)} a_{i,j}^s e_{i+j+s} + b_{i,j}e_n, e_k\right]$$
$$= \sum_{s=1}^{n-(i+j+k+1)} a_{i,j}^s \left(\sum_{t=1}^{n-(i+j+k+s+1)} a_{i+j+s,k}^t e_{i+j+k+s+t} + b_{i+j+s,k}e_n\right),$$
$$[[e_i,e_k],e_j] = \left[\sum_{s=1}^{n-(i+k+1)} a_{i,k}^s e_{i+k+s} + b_{i,k}e_n, e_j\right]$$
$$= \sum_{s=1}^{n-(i+j+k+1)} a_{i,k}^s \left(\sum_{t=1}^{n-(i+j+k+s+1)} a_{i+k+s,j}^t e_{i+j+k+s+t} + b_{i+k+s,j}e_n\right),$$

and this implies that

$$\sum_{s=1}^{n-(i+j+k+1)} a_{j,k}^s b_{i,j+k+s} = \sum_{s=1}^{n-(i+j+k+1)} (a_{i,j}^s b_{i+j+s,k} - a_{i,k}^s b_{i+k+s,j}).$$

□

Here are several useful remarks regarding (4.5) which can significantly simplify the multiplication table of TLb_{n+1} :

1. It is symmetric with respect to i, j, k (since $a_{s,t}^k = -a_{t,s}^k$ and $b_{s,t} = -b_{t,s}$ for any s and t, except for $(s,t) = (0,0),(1,1),(0,1),(1,0)$).

2. In the case when $(i,j,k) = (0,j,k)$ we get

$$\sum_{s=1}^{n-(j+k+1)} a_{j,k}^s b_{0,j+k+s} = \sum_{s=1}^{n-(j+k+1)} (a_{0,j}^s b_{j+s,k} - a_{0,k}^s b_{k+s,j}),$$

where $j \neq 0, k \neq 0$.

3. Since $a_{0,t}^s = 0$ as $s \neq 1$ and $a_{0,t}^1 = -1$, we get

$$a_{j,k}^1 b_{0,j+k+1} + a_{j,k}^2 b_{0,j+k+2} + \cdots + a_{j,k}^{n-(j+k+1)} b_{0,n-1} = -b_{j+1,k} + b_{k+1,j}.$$

4. Since $b_{0,t} = 0$ as $t = 2,\ldots,n-2$ and $b_{0,n-1} = -1$, one has that

$$a_{j,k}^{n-(j+k+1)} = b_{j+1,k} - b_{k+1,j},$$

for $k = j+1, j+2, \ldots, n-j-2$ and $j = 1,2,\ldots,\left[\frac{n-3}{2}\right]$.

Lemma 4.7. *Let $L \in \mathrm{TLb}_{n+1}$. Then*

$$[e_i, e_{j+k}] = \sum_{s=0}^{k} (-1)^{k-s} \binom{k}{s} [e_{i+k-s}, e_j] R_{e_0}^s, \tag{4.6}$$

where $1 \leq i,\ j,\ k \leq n$.

Proof. We proceed by the induction on k. Let $k = 1$. Then $[e_i, e_{j+1}] = [e_i, [e_j, e_0]] = -[e_{i+1}, e_j] + [[e_i, e_j], e_0]$, i.e., the equality (4.6) holds at $k = 1$. This is the base of the induction. Then the following chain of equalities lead to the claim:

$$
\begin{aligned}
&[e_i, e_{j+k+1}] \\
&\quad = [e_i, [e_{j+k}, e_0]] \\
&\quad = [[e_i, e_{j+k}], e_0] - [[e_i, e_0], e_{j+k}] \\
&\quad = \sum_{s=0}^{k} (-1)^{k-s} \binom{k}{s} [e_{i+k-s}, e_j] R_{e_0}^{s+1} \\
&\quad\quad - \sum_{s=0}^{k} (-1)^{k-s} \binom{k}{s} [e_{i+k+1-s}, e_j] R_{e_0}^{s} \\
&\quad = - \sum_{s=1}^{k+1} (-1)^{k-s} \binom{k}{s-1} [e_{i+k+1-s}, e_j] R_{e_0}^{s} \\
&\quad\quad - \sum_{s=0}^{k} (-1)^{k-s} \binom{k}{s} [e_{i+k+1-s}, e_j] R_{e_0}^{s} \\
&\quad = \sum_{s=1}^{k} (-1)^{k+1-s} \left(\binom{k}{s-1} + \binom{k}{s} \right) [e_{i+k+1-s}, e_j] R_{e_0}^{s} \\
&\quad\quad + [e_{i+k+1}, e_j] R_{e_0}^{k+1} - (-1)^k [e_{i+1+k}, e_k] \\
&\quad = \sum_{s=0}^{k+1} (-1)^{k+1-s} \binom{k+1}{s} [e_{i+k+1-s}, e_j] R_{e_0}^{s}.
\end{aligned}
$$

□

4.3 CLASSIFICATION OF SOME SOLVABLE LEIBNIZ ALGEBRAS

In this section we give classification of solvable Leibniz algebras with given nilradicals.

Let us first consider a solvable Leibniz algebra whose nilradical is the null-filiform Leibniz algebra NF_n.

Due to Theorem 3.4 we need to describe the derivations of NF_n. Their description is given below.

Proposition 4.2. *Any derivation of the algebra NF_n has the following matrix form:*

$$
\begin{pmatrix}
a_1 & a_2 & a_3 & \dots & a_n \\
0 & 2a_1 & a_2 & \dots & a_{n-1} \\
0 & 0 & 3a_1 & \dots & a_{n-2} \\
\vdots & \vdots & \vdots & \vdots & \vdots \\
0 & 0 & 0 & \dots & na_1
\end{pmatrix}.
$$

Proof. The proof is straightforward verification using the table of multiplication of NF_n. □

Corollary 4.2. *The maximal number of nil-independent derivations of the n-dimensional null-filiform Leibniz algebra* NF_n *is 1.*

Proof. Let

$$D_i := \begin{pmatrix} a_1^i & a_2^i & a_3^i & \dots & a_n^i \\ 0 & 2a_1^i & a_2^i & \dots & a_{n-1}^i \\ 0 & 0 & 3a_1^i & \dots & a_{n-2}^i \\ \vdots & \vdots & \vdots & \vdots & \vdots \\ 0 & 0 & 0 & \dots & na_1^i \end{pmatrix}, \quad i = 1, 2, \dots, p,$$

be derivations of NF_n. If $p > 1$, then $\left(D_i - \frac{a_1^i}{a_1^1} D_1\right)^n = 0$ with non-trivial scalars. Hence the system $\{D_1, D_2, \dots, D_p\}$ is not nil-independent. □

Corollary 4.3. *The dimension of a solvable Leibniz algebra with nil-radical* NF_n *is equal to* $n + 1$.

Proof. Let us assume that the solvable Leibniz algebra is decomposed as $R = NF_n \oplus Q$. Then by Corollary 4.2 and Theorem 3.4 we have $1 \leq \dim Q \leq 1$. Hence, $dim_K Q = 1$. □

Theorem 4.4. *Let R be a solvable Leibniz algebra whose nilradical is* NF_n. *Then there exists a basis* $\{e_1, e_2, \dots, e_n, x\}$ *of the algebra R such that the multiplication table of R with respect to this basis has the following form:*

$$\begin{cases} [e_i, e_1] = e_{i+1}, & 1 \leq i \leq n-1, \\ [x, e_1] = e_1, \\ [e_i, x] = -ie_i, & 1 \leq i \leq n. \end{cases}$$

Proof. According to Theorem 3.2 and Corollary 4.3 there exists a basis $\{e_1, e_2, \dots, e_n, x\}$ such that all products of elements of the basis are known, except of the products $[e_i, x]$, $[x, e_1]$ and $[x, x]$. The products $[e_i, x]$ can be derived from the equalities

$$[e_{i+1}, x] = [[e_i, e_1], x] = [e_i, [e_1, x]] + [[e_i, x], e_1], \quad 1 \leq i \leq n-1$$

and $[e_1, x]$.

Let us introduce the notations

$$[x, e_1] = \sum_{i=1}^{n} \alpha_i e_i, \quad [e_1, x] = \sum_{i=1}^{n} \beta_i e_i, \quad [x, x] = \sum_{i=1}^{n} \gamma_i e_i,$$

where $\{e_1, e_2, \dots, e_n\}$ is a basis of NF_n and $\{x\}$ is a basis of Q.

Now we consider the following two possible cases.

Case 1. Let $\alpha_1 \neq 0$. Then taking the change of basis:

$$e_i' = \frac{1}{\alpha_1}\sum_{j=i}^{n}\alpha_{j-i+1}e_j, \quad 1 \leq i \leq n, \qquad x' = \frac{1}{\alpha_1}x,$$

we can assume that $[x, e_1] = e_1$ and other products by redesignation of parameters can be assumed not changed.

From the products

$$0 = [x,[x,x]] = \left[x, \sum_{i=1}^{n}\gamma_i e_i\right] = \sum_{i=1}^{n}\gamma_i[x, e_i] = \gamma_1 e_1,$$

we can deduce that $\gamma_1 = 0$.

On the other hand, from the Leibniz identity

$$[x,[e_1,x]] = [[x,e_1],x] - [[x,x],e_1]$$

we get

$$\beta_1[x,e_1] = [e_1,x] - \sum_{i=3}^{n}\gamma_{i-1}e_i, \text{ i.e.,}$$

$$\beta_1 e_1 = \sum_{i=1}^{n}\beta_i e_i - \sum_{i=3}^{n}\gamma_{i-1}e_i.$$

Comparing the coefficients at the elements of the basis, we obtain $\beta_2 = 0$ and $\gamma_i = \beta_{i+1}$ for $2 \leq i \leq n-1$. From the equality $[e_1,[e_1,x]] = -[e_1,[x,e_1]]$, we derive that $\beta_1 = -1$.

Thus, we have

$$[e_1,x] = -e_1 + \sum_{i=3}^{n}\beta_i e_i, \qquad [x,x] = \sum_{i=2}^{n-1}\beta_{i+1}e_i + \gamma_n e_n.$$

Now we prove the following identity

$$[e_i,x] = -ie_i + \sum_{j=i+2}^{n}\beta_{j-i+1}e_j, \tag{4.7}$$

for $1 \leq i \leq n$. It is obvious that (4.7) is true for $i = 1$. Assume that (4.7) holds for each i, $1 \leq i < k \leq n$. Then

$$\begin{aligned}
[e_k, x] &= [[e_{k-1}, e_1], x] = [e_{k-1}, [e_1, x]] + [[e_{k-1}, x], e_1] \\
&= [e_{k-1}, -e_1] + \left[-(k-1)e_{k-1} + \sum_{j=k+1}^{n} \beta_{j-k+2} e_j, e_1\right] \\
&= -e_k - (k-1)e_k + \sum_{j=k+1}^{n} \beta_{j-k+2}[e_j, e_1] = -ke_k + \sum_{j=k+2}^{n} \beta_{j-k+1} e_j.
\end{aligned}$$

By induction, we see that indeed (4.7) holds for all i, $1 \leq i \leq n$.

Thus, the multiplication table of the algebra R is written as follows

$$\begin{cases}
[e_i, e_1] = e_{i+1}, & 1 \leq i \leq n-1, \\
[x, e_1] = e_1, & \\
[e_i, x] = -ie_i + \sum\limits_{j=i+2}^{n} \beta_{j-i+1} e_j, & 1 \leq i \leq n, \\
[x, x] = \sum\limits_{i=2}^{n-1} \beta_{i+1} e_i + \gamma_n e_n. &
\end{cases} \tag{4.8}$$

Let us make the following change of basis:

$$e_i' = e_i + \sum_{j=i+2}^{n} A_{j-i+1} e_j, \ 1 \leq i \leq n, \qquad x' = \sum_{i=2}^{n-1} A_{i+1} e_i + B_n e_n + x,$$

where A_i, B_n are given as follows

$$\begin{aligned}
A_3 &= \tfrac{1}{2}\beta_3, \\
A_4 &= \tfrac{1}{3}\beta_4 \\
A_i &= \tfrac{1}{i-1}\left(\sum_{j=3}^{i-2} A_{i-j+1}\beta_j + \beta_i\right), \quad 5 \leq i \leq n, \\
B_n &= \tfrac{1}{n}\left(\sum_{j=3}^{n-1} A_{n-j+2}\beta_j + \gamma_n\right).
\end{aligned}$$

Then taking into account the multiplication table (4.8) we compute the products in the new basis

$$[e_i', e_1'] = \left[e_i + \sum_{j=i+2}^{n} A_{j-i+1} e_j, e_1\right] = e_{i+1} + \sum_{j=i+3}^{n} A_{j-i} e_j = e_{i+1}',$$
where $1 \leq i \leq n-1$,

$$
\begin{aligned}
[x', e_1'] &= \left[\sum_{i=2}^{n-1} A_{i+1}e_i + B_n e_n + x, e_1\right] \\
&= \sum_{i=3}^{n} A_i e_i + [x, e_1] = e_1 + \sum_{i=3}^{n} A_i e_i = e_1', \\
[x', x'] &= \left[\sum_{i=2}^{n-1} A_{i+1}e_i + B_n e_n + x, x\right] \\
&= \sum_{i=2}^{n-1} A_{i+1}[e_i, x] + B_n[e_n, x] + [x, x] \\
&= \sum_{i=2}^{n-1} A_{i+1}\left(-ie_i + \sum_{j=i+2}^{n} \beta_{j-i+1}e_j\right) - nB_n e_n \\
&\qquad + \sum_{i=2}^{n-1} \beta_{i+1}e_i + \gamma_n e_n \\
&= -\sum_{i=2}^{n-1} iA_{i+1}e_i + \sum_{i=2}^{n-3} A_{i+1} \sum_{j=i+2}^{n-1} \beta_{j-i+1}e_j + \sum_{i=2}^{n-1} \beta_{i+1}e_i \\
&\qquad + \sum_{i=2}^{n-2} A_{i+1}\beta_{n-i+1}e_n - B_n e_n + \gamma_n e_n \\
&= \sum_{i=2}^{n-1}(-iA_{i+1} + \beta_{i+1})e_i + \sum_{i=4}^{n-1}\sum_{j=3}^{i-1} A_{i-j+2}\beta_j e_i \\
&\qquad + \left(-nB_n + \gamma_n + \sum_{i=2}^{n-1} A_{i+1}\beta_{n-i+1}\right)e_n \\
&= (-2A_3 + \beta_3)e_2 + (-3A_4 + \beta_4)e_3 \\
&\qquad + \sum_{i=4}^{n-1}\sum_{j=3}^{i-1}(-iA_{i+1} + \beta_{i+1} + A_{i-j+2}\beta_j)e_i = 0,
\end{aligned}
$$

$$
\begin{aligned}
&[e_1', x'] = \left[e_1 + \sum_{i=3}^{n} A_i e_i, x\right] = [e_1, x] + \sum_{i=3}^{n} A_i[e_i, x] \\
&= -e_1 + \sum_{i=3}^{n} \beta_i e_i + \sum_{i=3}^{n} A_i\left(-ie_i + \sum_{j=i+2}^{n} \beta_{j-i+1}e_j\right) \\
&= -e_1 + \sum_{i=3}^{n} \beta_i e_i - \sum_{i=3}^{n} iA_i e_i + \sum_{i=3}^{n} A_i \sum_{j=i+2}^{n} \beta_{j-i+1}e_j \\
&= -e_1 - \sum_{i=3}^{n} A_i e_i - \sum_{i=3}^{n}(i-1)A_i e_i + \sum_{i=3}^{n} \beta_i e_i + \sum_{i=3}^{n}\left(\sum_{j=3}^{i-2} A_{i-j+1}b_j\right)e_i \\
&= -e_1 - \sum_{i=3}^{n} A_i e_i + \sum_{i=3}^{n}\left(-(i-1)A_i + \beta_i\right)e_i + \sum_{i=5}^{n}\sum_{j=3}^{i-2} A_{i-j+1}\beta_j e_i \\
&= -e_1 - \sum_{i=3}^{n} A_i e_i + (-2A_3 + \beta_3)e_3 + (-3A_4 + \beta_4)e_4 \\
&+ \sum_{i=5}^{n}\sum_{j=3}^{i-2}\left(-(i-1)A_i + \beta_i + A_{i-j+1}\beta_j\right)e_i = -e_1 - \sum_{i=3}^{n} A_i e_i = -e_1'.
\end{aligned}
$$

Similarly to that in (4.7) we obtain $[e'_i, x'] = -ie'_i$, $1 \leq i \leq n$. Thus, we get the required multiplication table for R.

Case 2. Let $\alpha_1 = 0$. Then from the equalities $[e_1, [e_1, x]] = -[e_1, [x, e_1]]$ and $0 = [x, [x, x]]$ we get $\beta_1 = 0$ and $\gamma_1 = 0$, respectively.

Thus, we have the following products:

$$\begin{cases} [e_i, e_1] = e_{i+1}, & 1 \leq i \leq n-1, \\ [x, e_1] = \sum\limits_{i=2}^{n} \alpha_i e_i, \\ [e_1, x] = \sum\limits_{i=2}^{n} \beta_i e_i, \\ [x, x] = \sum\limits_{i=2}^{n} \gamma_i e_i. \end{cases}$$

Similarly to that in (4.7) we can prove the equality: $[e_i, x] = \sum\limits_{j=i+1}^{n} \beta_{j-i+1} e_j$. Consequently, we have $[e_i, x] \in \operatorname{Span}\{e_{i+1}, e_{i+2}, \ldots, e_n\}$, i.e., $R^i \subseteq \operatorname{Span}\{e_i, e_{i+1}, \ldots, e_n\}$. Thus, $R^{n+1} = 0$ which contradicts the assumption of non nilpotency of the algebra R. This implies that, in the case of $\alpha_1 = 0$, there is no non nilpotent solvable Leibniz algebra with nilradical NF_n. □

Now we clarify the situation when the nilradical is represented as a direct sum of its two null-filiform ideals and the complementary space to the nilradical is one-dimensional.

Theorem 4.5. *Let R be a solvable Leibniz algebra such that $R = NF_k \oplus NF_s + Q$, where $NF_k \oplus NF_s$ is the nilradical of R, NF_k and NF_s are ideals of the nilradical and* $\dim Q = 1$. *Then NF_k and NF_s are also ideals of the algebra R.*

Proof. Let $\{e_1, e_2, \ldots, e_k\}$ be a basis of NF_k and $\{f_1, f_2, \ldots, f_s\}$ a basis of NF_s and let $\{x\}$ be a basis of Q. We can assume, without loss of generality, that $k \geq s$.

Theorem 3.2 implies the inclusion

$$\{e_2, e_3, \ldots, e_k, f_2, f_3, \ldots, f_s\} \subseteq \operatorname{Ann}_r(R)$$

and the following equalities:

$$[e_i, e_1] = e_{i+1}, \ \ 1 \le i \le k-1, \qquad [f_i, f_1] = f_{i+1}, \ \ 1 \le i \le s-1.$$

Let us introduce the notations:

$$\begin{cases} [x, e_1] = \sum_{i=1}^{k} \alpha_i e_i + \sum_{i=1}^{s} \beta_i f_i, & [x, f_1] = \sum_{i=1}^{k} \delta_i e_i + \sum_{i=1}^{s} \gamma_i f_i, \\ [e_1, x] = \sum_{i=1}^{k} \lambda_i e_i + \sum_{i=1}^{s} \sigma_i f_i, & [f_1, x] = \sum_{i=1}^{k} \tau_i e_i + \sum_{i=1}^{s} \mu_i f_i, \\ [x, x] = \sum_{i=1}^{k} \rho_i e_i + \sum_{i=1}^{s} \xi_i f_i. \end{cases}$$

From the products

$$0 = [x, [e_1, f_1]] = [[x, e_1], f_1] - [[x, f_1], e_1] = \sum_{i=2}^{s} \beta_{i-1} f_i - \sum_{i=2}^{k} \delta_{i-1} e_i,$$

we obtain $\beta_i = 0, \ 1 \le i \le s-1$ and $\delta_i = 0, \ 1 \le i \le k-1$.

The equalities $[e_1, [e_1, x]] = -[e_1, [x, e_1]]$ and $[f_1, [f_1, x]] = -[f_1, [x, f_1]]$ imply that $\lambda_1 = -\alpha_1, \mu_1 = -\gamma_1$.

From the equalities $0 = [e_1, [x, x]] = \rho_1 e_2$ and $0 = [f_1, [x, x]] = \xi_1 f_2$, we get $\rho_1 = \xi_1 = 0$.

In a similar way as in the proof of Theorem 4.4, the following equalities can be proved:

$$[e_i, x] = -i\alpha_1 e_i + \sum_{j=i+1}^{k} \lambda_{j-i+1} e_j, \qquad 2 \le i \le k,$$

$$[f_i, x] = -i\gamma_1 f_i + \sum_{j=i+1}^{s} \mu_{j-i+1} f_j, \qquad 2 \le i \le s.$$

Summarizing, we obtain the following multiplication table for the algebra R:

$$\begin{cases} [e_i, e_1] = e_{i+1}, \quad 1 \leq i \leq k-1, \\ [f_i, f_1] = f_{i+1}, \quad 1 \leq i \leq s-1, \\ [x, e_1] = \sum_{i=1}^{k} \alpha_i e_i + \beta_s f_s, \\ [x, f_1] = \delta_k e_k + \sum_{i=1}^{s} \gamma_i f_i, \\ [e_1, x] = -\alpha_1 e_1 + \sum_{i=2}^{k} \lambda_i e_i + \sum_{i=1}^{s} \sigma_i f_i, \\ [f_1, x] = \sum_{i=1}^{k} \tau_i e_i - \gamma_1 f_1 + \sum_{i=2}^{s} \mu_i f_i, \\ [e_i, x] = -i\alpha_1 e_i + \sum_{j=i+1}^{k} \lambda_{j-i+1} e_j, \ 2 \leq i \leq k, \\ [f_i, x] = -i\gamma_1 e_i + \sum_{j=i+1}^{s} \mu_{j-i+1} f_j, \ 2 \leq i \leq s, \\ [x, x] = \sum_{i=2}^{k} \rho_i e_i + \sum_{i=2}^{s} \xi_i f_i. \end{cases} \tag{4.9}$$

Below, we analyze the different cases that can appear in terms of the possible values of α_1 and γ_1.

Case 1. Let $\alpha_1 = \gamma_1 = 0$. Then the multiplication table (4.9) implies

$$\begin{array}{ll} [e_i, x] & \in \mathrm{Span}\{e_{i+1}, e_{i+2}, \ldots, e_k\}, \\ [f_i, x] & \in \mathrm{Span}\{f_{i+1}, f_{i+2}, \ldots, f_s\}, \\ [e_1, x] & \in \mathrm{Span}\{e_2, e_3, \ldots, e_k, f_1, f_2, \ldots, f_s\}, \\ [f_1, x] & \in \mathrm{Span}\{e_1, e_2, \ldots, e_k, f_2, f_3, \ldots, f_s\}. \end{array}$$

The above relations mean that the algebra R is nilpotent, so we get a contradiction with the assumption of non nilpotency of R. Therefore, this case is impossible.

Case 2. Let $\alpha_1 \neq 0$ and $\gamma_1 = 0$. Using the following change of basis:

$$
\begin{aligned}
e_1' &= \frac{1}{\alpha_1}\left(\sum_{i=1}^{k} \alpha_i e_i + \beta_s f_s\right),\\
e_i' &= \frac{1}{\alpha_1}\sum_{j=i}^{k} \alpha_{j-i+1} e_j,\ 2 \le i \le k,\\
x' &= \frac{1}{\alpha_1} x,
\end{aligned}
$$

we may assume that

$$[x, e_1] = e_1.$$

From the identity

$$[x, [x, e_1]] = [[x, x], e_1] - [[x, e_1], x]$$

we have that

$$e_1 = \sum_{i=2}^{k} \rho_i [e_i, e_1] - [e_1, x] = \sum_{i=3}^{k} \rho_{i-1} e_i + e_1 - \sum_{i=2}^{k} \lambda_i e_i - \sum_{i=1}^{s} \sigma_i f_i.$$

Consequently, $\lambda_2 = \sigma_i = 0$ for $1 \le i \le s$ and $\rho_i = \lambda_{i+1}$ for $2 \le i \le k-1$.

From the identity

$$[f_1, [x, e_1]] = [[f_1, x], e_1] - [[f_1, e_1], x]$$

we conclude that $0 = [[f_1, x], e_1] = \sum\limits_{i=2}^{k} \tau_{i-1} e_i \Rightarrow \tau_i = 0,\ 1 \le i \le k-1.$

From the identity

$$[x, [x, f_1]] = [[x, x], f_1] - [[x, f_1], x],$$

we obtain

$$
\begin{aligned}
0 &= \sum_{i=3}^{s} \xi_{i-1} f_i - \sum_{i=2}^{s} \gamma_i [f_i, x] + \delta_k [e_k, x]\\
&= \sum_{i=3}^{s} \xi_{i-1} f_i - \sum_{i=2}^{s} \gamma_i \left(\sum_{j=i+1}^{s} \mu_{j-i+1} f_j\right) - k\delta_k e_k\\
&= \sum_{i=3}^{s} \xi_{i-1} f_i - \sum_{i=3}^{s} \left(\sum_{j=3}^{i} \gamma_{j-1} \mu_{i-j+2}\right) f_i - k\delta_k e_k\\
&= \sum_{i=3}^{s} \left(\xi_{i-1} - \sum_{j=3}^{i} \gamma_{j-1} \mu_{i-j+2}\right) f_i - k\delta_k e_k.
\end{aligned}
$$

By comparison of coefficients at the elements of the basis we deduce that:

$$\xi_i = \sum_{j=3}^{i+1} \gamma_{j-1}\mu_{i-j+3}, \ 2 \le i \le s-1 \text{ and } \delta_k = 0.$$

Now we consider the following change of basis:

$$\begin{array}{ll} f'_1 & = f_1 + \frac{\tau_k}{k} e_k, \\ f'_i & = f_i, \ 2 \le i \le s. \end{array}$$

Then we obtain

$$[f'_1, x] = [f_1 + \frac{\tau_k}{k} e_k, x] = \sum_{i=2}^{s} \mu_i f_i + \tau_k e_k - \tau_k e_k = \sum_{i=2}^{s} \mu_i f_i = \sum_{i=2}^{s} \mu_i f'_i$$

and

$$[x, f'_1] = [x, f_1 + \frac{\tau_k}{k} e_k] = [x, f_1] = \sum_{i=2}^{s} \gamma_i f_i = \sum_{i=2}^{s} \gamma_i f'_i.$$

Thus, we have the following multiplication table of the algebra R:

$$\begin{cases} [e_i, e_1] = e_{i+1}, & 1 \le i \le k-1, \\ [f_i, f_1] = f_{i+1}, & 1 \le i \le s-1, \\ [x, e_1] = e_1, & \\ [x, f_1] = \sum\limits_{i=2}^{s} \gamma_i f_i, & \\ [e_1, x] = -e_1 + \sum\limits_{i=2}^{k} \lambda_i e_i, & \\ [f_1, x] = \sum\limits_{i=2}^{s} \mu_i f_i, & \\ [e_i, x] = -ie_i + \sum\limits_{j=i+2}^{k} \lambda_{j-i+1} e_j, \ 2 \le i \le k, & \\ [f_i, x] = \sum\limits_{j=i+1}^{s} \mu_{j-i+1} f_j, \ 2 \le i \le s, & \\ [x, x] = \sum\limits_{i=2}^{k} \rho_i e_i + \sum\limits_{i=2}^{s} \xi_i f_i. & \end{cases}$$

From the above multiplication table the following inclusions can be immediately derived:

$$[x, NF_k] \subseteq NF_k, \quad [NF_k, x] \subseteq NF_k,$$

$$[x, NF_s] \subseteq NF_s, \quad [NF_s, x] \subseteq NF_s.$$

This completes the proof of the assertion of the theorem for this case.

Case 3. Let $\alpha_1 = 0$ and $\gamma_1 \neq 0$. Due to symmetry of Cases 2 and 3, the proof of the theorem follows similar arguments as in Case 2.

Case 4. Let $\alpha_1 \neq 0$ and $\gamma_1 \neq 0$. Consider the following change of basis:

$$e_1' = \frac{1}{\alpha_1}\left(\sum_{i=1}^{k} \alpha_i e_i + \beta_s f_s\right), \quad e_i' = \frac{1}{\alpha_1}\sum_{j=i}^{k} \alpha_{j-i+1} e_j, \quad 2 \leq i \leq k,$$

$$f_1' = \frac{1}{\gamma_1}\left(\sum_{i=1}^{s} \gamma_i f_i + \delta_k e_k\right), \quad f_i' = \frac{1}{\gamma_1}\sum_{j=i}^{s} \gamma_{j-i+1} f_j, \quad 2 \leq i \leq s, \; x' = \frac{1}{\alpha_1}x.$$

Then we derive

$$[x', e_1'] = \left[\frac{1}{\alpha_1}x, \frac{1}{\alpha_1}\left(\sum_{i=1}^{k} \alpha_i e_i + \beta_s f_s\right)\right] = \frac{1}{\alpha_1^2}\alpha_1[x, e_1] = \frac{1}{\alpha_1}[x, e_1] = e_1',$$

$$[x', f_1'] = \left[\frac{1}{\alpha_1}x, \frac{1}{\gamma_1}\left(\sum_{i=1}^{s} \gamma_i f_i + \delta_k e_k\right)\right] = \frac{1}{\alpha_1\gamma_1}\gamma_1[x, f_1] = \frac{\gamma_1}{\alpha_1}f_1'.$$

From the identity $[x, [x, e_1]] = [[x, x], e_1] - [[x, e_1], x]$ we deduce:

$$e_1 = \sum_{i=2}^{k} \rho_i[e_i, e_1] - [e_1, x] = \sum_{i=3}^{k} \rho_{i-1} e_i + \alpha_1 e_1 - \sum_{i=2}^{k} \lambda_i e_i - \sum_{i=1}^{s} \sigma_i f_i.$$

Therefore, $\alpha_1 = 1, \lambda_1 = -1, \; \lambda_2 = \sigma_i = 0, \; 1 \leq i \leq s$ and $\rho_i = \lambda_{i+1}, \; 2 \leq i \leq k-1$.

Expanding the identity $[x, [x, f_1]] = [[x, x], f_1] - [[x, f_1], x]$, we derive the equalities:

$$\left(\frac{\gamma_1}{\alpha_1}\right)^2 f_1 = \sum_{i=2}^{s} \xi_i[f_i, f_1] - \frac{\gamma_1}{\alpha_1}[f_1, x] = \sum_{i=3}^{s} \xi_{i-1} f_i - \frac{\gamma_1}{\alpha_1}\sum_{i=1}^{s} \mu_i f_i - \frac{\gamma_1}{\alpha_1}\sum_{i=1}^{k} \tau_i e_i$$

from which we have $\mu_1 = -\frac{\gamma_1}{\alpha_1}, \; \mu_2 = \tau_i = 0, \; 1 \leq i \leq k$ and $\xi_i = \frac{\gamma_1}{\alpha_1}\mu_{i+1}, \; 2 \leq i \leq s-1$.

Finally, we obtain the following products of basis elements in the algebra R:

$$\begin{cases} [e_i, e_1] = e_{i+1}, & 1 \le i \le k-1, \\ [f_i, f_1] = f_{i+1}, & 1 \le i \le s-1, \\ [x, e_1] = e_1, & \\ [x, f_1] = \frac{\gamma_1}{\alpha_1} f_1, & \\ [e_1, x] = -e_1 + \sum_{i=3}^{k} \lambda_i e_i, & \\ [f_1, x] = -\frac{\gamma_1}{\alpha_1} f_1 + \sum_{i=3}^{s} \mu_i f_i, & \\ [x, x] = \sum_{i=2}^{k} \rho_i e_i + \sum_{i=2}^{s} \xi_i f_i. & \end{cases}$$

These products are sufficient in order to check the inclusions:

$$[x, NF_k] \subseteq NF_k, \quad [NF_k, x] \subseteq NF_k,$$

$$[x, NF_s] \subseteq NF_s, \quad [NF_s, x] \subseteq NF_s.$$

Thus, the ideals NF_k and NF_s of the nilradical are also ideals of the algebra. □

Now we study the solvable Leibniz algebras with nilradical $NF_k \oplus NF_s$ and with one-dimensional complementary vector space. Due to Theorem 4.5 we can assume that NF_k and NF_s are ideals of the algebra.

Theorem 4.6. *Let R be a solvable Leibniz algebra such that $R = NF_k \oplus NF_s + Q$, where $NF_k \oplus NF_s$ is the nilradical of R and* $\dim Q = 1$. *Let us assume that $\{e_1, e_2, \dots, e_k\}$ is a basis of NF_k, $\{f_1, f_2, \dots, f_s\}$ is a basis of NF_s and $\{x\}$ is a basis of Q. Then the algebra R is isomorphic to one* *of the following pairwise non-isomorphic algebras:*

$$R(\alpha) := \begin{cases} [e_i, e_1] = e_{i+1}, & 1 \leq i \leq k-1, \\ [f_i, f_1] = f_{i+1}, & 1 \leq i \leq s-1, \\ [x, e_1] = e_1, & \\ [x, f_1] = \alpha f_1, & \alpha \neq 0, \\ [e_i, x] = -ie_i, & 1 \leq i \leq k, \\ [f_i, x] = -i\alpha f_i, & 1 \leq i \leq s, \end{cases}$$

$$R(\beta_2, \beta_3, \ldots, \beta_s, \gamma) := \begin{cases} [e_i, e_1] = e_{i+1}, & 1 \leq i \leq k-1, \\ [f_i, f_1] = f_{i+1}, & 1 \leq i \leq s-1, \\ [x, e_1] = e_1, & \\ [f_i, x] = \sum\limits_{j=i+1}^{s} \beta_{j-i+1} f_j, & 1 \leq i \leq s, \\ [e_i, x] = -ie_i, & 1 \leq i \leq k, \\ [x, x] = \gamma f_s. & \end{cases}$$

In the second family of algebras the first non-zero element of the vector $(\beta_2, \beta_3, \ldots, \beta_s, \gamma)$ can be assumed to be equal to 1.

Proof. Firstly, we note that the algebras $NF_k + Q$ and $NF_s + Q$ are not simultaneously nilpotent. Indeed, if they are both nilpotent, then we have:

$$\begin{array}{ll} [e_i, e_1] \in \text{Span}\{e_{i+1}, \ldots, e_k\}, & 1 \leq i \leq k-1, \\ [f_i, f_1] \in \text{Span}\{f_{i+1}, \ldots, f_s\}, & 1 \leq i \leq s-1, \\ [x, e_1] \in \text{Span}\{e_2, e_3, \ldots, e_k\}, & [x, f_1] \in \text{Span}\{f_2, f_3, \ldots, f_s\}, \\ [e_i, x] \in \text{Span}\{e_{i+1}, \ldots, e_k\}, & 1 \leq i \leq k-1, \\ [f_j, x] \in \text{Span}\{f_{j+1}, \ldots, f_s\}, & 2 \leq i \leq s-1. \end{array}$$

From the equalities $0 = [e_1, [x, x]]$, $0 = [f_1, [x, x]]$ we conclude that:

$$[x, x] \in \text{Span}\{e_2, e_3, \ldots, e_k, f_2, f_3, \ldots, f_s\}.$$

Therefore, $R^2 \subseteq \text{Span}\{e_2, e_3, \ldots, e_k, f_2, f_3, \ldots, f_s\}$. Moreover, we have

$$R^i \subseteq \text{Span}\{e_i, e_{i+1}, \ldots, e_k, f_i, f_{i+1}, \ldots, f_s\},$$

which implies that $R^{\max\{k,s\}+1} = \{0\}$. Thus, we have a contradiction to the assumption that R is not nilpotent. Hence, the algebras $NF_k + Q$ and $NF_s + Q$ cannot be nilpotent at the same time.

Without loss of generality, we can assume that the algebra $NF_k + Q$ is non-nilpotent.

We take the quotient algebra by the ideal NF_s, then $R/NF_s \cong \overline{NF}_k + \overline{Q}$. By Theorem 4.4 the structure of the algebra $\overline{NF}_k + \overline{Q}$ is known. Namely,

$$\begin{cases} [\overline{e}_i, \overline{e}_1] = \overline{e}_{i+1}, & 1 \leq i \leq k-1, \\ [\overline{x}, \overline{e}_1] = \overline{e}_1, & \\ [\overline{e}_i, \overline{x}] = -i\,\overline{e}_i, & 1 \leq i \leq k. \end{cases} \tag{4.10}$$

Using the fact that NF_k and NF_s are ideals of R and considering the multiplication table (4.10), we have that:

$$\begin{cases} [e_i, e_1] = e_{i+1}, & 1 \leq i \leq k-1, \\ [f_i, f_1] = f_{i+1}, & 1 \leq i \leq s-1, \\ [x, e_1] = e_1, & \\ [x, f_1] = \sum\limits_{i=1}^{s} \alpha_i f_i, & \\ [e_i, x] = -ie_i, & 1 \leq i \leq k, \\ [f_1, x] = \sum\limits_{i=1}^{s} \beta_i f_i, & \\ [x, x] = \sum\limits_{i=1}^{s} \gamma_i f_i. & \end{cases} \tag{4.11}$$

If $\alpha_1 \neq 0$, then in a similar way as in the Case 1 of Theorem 4.4 we obtain the family of algebras $R(\alpha)$, where $\alpha \neq 0$.

The fact that two algebras in the family $R(\alpha)$ with different values of parameter α are not isomorphic can be easily established by a general change of basis and considering the expansion of the product $[x', f_1']$ in both bases.

Now consider $\alpha_1 = 0$. Then by the change of basis

$$x' = x - (\alpha_2 f_1 + \alpha_3 f_2 + \cdots + \alpha_s f_{s-1}),$$

we can suppose that $[x, f_1] = 0$.

From the identity $[f_1,[f_1,x]] = [[f_1,f_1],x] - [[f_1,x],f_1]$ we get $\beta_1 = 0$.

Similarly to the proof of (4.7), we can show that

$$[f_i, x] = \sum_{m=i+1}^{s} \beta_{m-i+1} f_j, \ 1 \le i \le s.$$

The identity $[x,[f_1,x]] = [[x,f_1],x]-[[x,x],f_1]$ implies the following equalities:

$$0 = -[[x,x],f_1] = -\sum_{m=3}^{s} \gamma_{m-1} f_m.$$

Consequently, $\gamma_i = 0, \ 2 \le i \le s-1$.

Thus, we obtain the products of the family $R(\beta_2,\beta_3,\dots,\beta_s,\gamma)$

$$\begin{cases} [f_i, f_1] = f_{i+1}, & 1 \le i \le s-1, \\ [f_i, x] = \sum_{m=i+1}^{s} \beta_{m-i+1} f_m, & 1 \le i \le s, \\ [x, x] = \gamma_s f_s. \end{cases}$$

Now we are going to study the isomorphism inside the family $R(\beta_2,\beta_3,\dots,\beta_s,\gamma)$.

Taking into account that, under general basis transformation, the products (4.11) should be the same, we conclude that it is sufficient to take the following change of basis:

$$f_i' = A_1^{i-1} \sum_{j=i}^{s} A_{j-i+1} f_j, \ (A_1 \ne 0), \qquad 1 \le i \le s, \qquad x' = x.$$

Then we have

$$[f_1', x'] = \sum_{i=1}^{s} A_i [f_i, x] = \sum_{i=1}^{s-1} A_i \left(\sum_{j=i+1}^{s} \beta_{j-i+1} f_j \right) = \sum_{i=2}^{s} \left(\sum_{j=1}^{i-1} A_j B_{i-j+1} \right) f_i.$$

On the other hand

$$[f_1', x'] = \sum_{i=2}^{s} \beta_i' f_i' = \sum_{i=1}^{s-1} A_1^i \beta_{i+1}' \left(\sum_{j=1}^{s-i} A_j f_{i+j} \right) = \sum_{i=2}^{s} \left(\sum_{j=1}^{i-1} A_1^j A_{i-j} \beta_{j+1}' \right) f_i.$$

Comparing coefficients at the elements of the basis we deduce that:

$$\sum_{i=1}^{k-1} A_i\beta_{k-i+1} = \sum_{i=1}^{k-1} A_1^i A_{k-i}\beta'_{i+1}, \quad k = 2, 3, \ldots, s.$$

From these systems of equations it follows that

$$\beta'_i = \frac{\beta_i}{A_1^{i-1}}, A_1^{i-1}, \quad 2 \le i \le s.$$

If we consider

$$\gamma'_s A_1^s f_s = \gamma'_s f'_s = [x', x'] = [x, x] = \gamma_s f_s,$$

then we obtain

$$\gamma'_s = \frac{\gamma_s}{A_1^s}.$$

It is easy to see that by choosing an appropriate value for the parameter A_1, the first non-zero element of the vector $(\beta_2, \beta_3, \ldots, \beta_s, \gamma)$ can be assumed to be equal to 1.

Therefore, two algebras $R(\beta_2, \beta_3, \ldots, \beta_s, \gamma)$ and $R(\beta'_2, \beta'_3, \ldots, \beta'_s, \gamma')$ with different set of parameters are not isomorphic.

For given parameters α and $\beta_2, \beta_3, \ldots, \beta_s, \gamma$, the algebras $R(\alpha)$ and $R(\beta_2, \beta_3, \ldots, \beta_s, \gamma)$ are not isomorphic because

$$k + s = \dim R(\alpha)^2 \neq \dim R(\beta_2, \beta_3, \ldots, \beta_s, \gamma)^2 = k + s - 1.$$

□

Remark. In the case when all coefficients $(\beta_2, \beta_3, \ldots, \beta_s, \gamma)$ are equal to zero we have the split algebra $(NF_k + Q) \oplus NF_s$. Therefore, in the non split case, we can always assume that $(\beta_2, \beta_3, \ldots, \beta_s, \gamma) \neq (0, 0, 0, \ldots, 0)$.

Now, by an induction process, we are going to generalize Theorem 4.6 to the case when the nilradical is a direct sum of several (greater than 2) copies of null-filiform ideals.

Theorem 4.7. *Let R be a solvable Leibniz algebra such that $R = NF_{n_1} \oplus NF_{n_2} \oplus \cdots \oplus NF_{n_s} + Q$, where $NF_{n_1} \oplus NF_{n_2} \oplus \cdots \oplus NF_{n_s}$ is the nilradical of R and $\dim Q = 1$. There exist $p, q \in \mathbb{N}$ with $p \neq 0$ and $p + q = s$, a basis $\{e_1^i, e_2^i, \ldots, e_{n_i}^i\}$ of NF_{n_i}, for $1 \le i \le p$, a basis*

$\{f_1^k, f_2^k, \dots, f_{n_k}^k\}$ of $NF_{n_{p+k}}$, for $1 \leq k \leq q$, and a basis $\{x\}$ of Q such that the multiplication table of the algebra R is given by:

$$R_{p,q} := \begin{cases} [e_i^j, e_1^j] = e_{i+1}^j, & 1 \leq i \leq n_j - 1, \\ [f_i^k, f_1^k] = f_{i+1}^k, & 1 \leq i \leq n_k - 1, \\ [x, e_1^j] = \delta^j e_1^j, & \delta^j \neq 0 \\ [f_i^k, x] = \sum\limits_{m=i+1}^{n_k} \beta_{m-i+1}^k f_m^k, & 1 \leq i \leq n_k, \\ [e_i^j, x] = -i\delta^j e_i^j, & 1 \leq i \leq n_j, \\ [x, x] = \sum\limits_{m=1}^{k} \gamma^m f_{n_m}, & \end{cases} \tag{4.12}$$

where $1 \leq j \leq p, 1 \leq k \leq q$ and $\delta^1 = 1$. Moreover, the first non-zero component of the vectors $(\beta_2^k, \beta_3^k, \dots, \beta_{n_k}^k, \gamma^k)$ can be assumed to be equal to 1. Moreover, the above algebras are pairwise non-isomorphic.

Proof. We shall prove the theorem by the induction on s:

If $s = 1$, then $p = 1$, $q = 0$, so $R_{1,0}$ is the algebra given in Theorem 4.4.

If $s = 2$, then we have two cases: either $p = 2$, $q = 0$ or $p = 1$, $q = 1$, which were considered in Theorem 4.5. Namely, we have two families of algebras: $R(\alpha)$, which corresponds to $R_{2,0}$, and $R(\beta_2, \beta_3, \dots, \beta_s, \gamma)$, which corresponds to $R_{1,1}$.

Let us assume that the theorem is true for s and we shall prove it for $s + 1$.

Let $R = NF_{n_1} \oplus NF_{n_2} \oplus \cdots \oplus NF_{n_s} \oplus NF_{n_{s+1}} + Q$. We consider the quotient algebra by $NF_{n_{s+1}}$, i.e. $R/NF_{n_{s+1}} \cong \overline{NF}_{n_1} \oplus \overline{NF}_{n_2} \oplus \cdots \oplus \overline{NF}_{n_s} + \overline{Q}$. Then we get the multiplication table given in (4.12).

Note that the multiplication table for the algebra R can be obtained from (4.12) by adding the products

$$[e_i^{s+1}, e_1^{s+1}] = e_{i+1}^{s+1}, \qquad 1 \leq i \leq n_{s+1} - 1,$$

$$[x, e_1^{s+1}] = \sum_{m=1}^{n_{s+1}} \alpha_m^{s+1} e_m^{s+1},$$

$$[e_1^{s+1}, x] = \sum_{m=1}^{n_{s+1}} \beta_m^{s+1} e_m^{s+1},$$

$$[x, x] = \sum_{m=1}^{n_{s+1}} \gamma_m^{s+1} e_m^{s+1}.$$

If $\alpha_1^{s+1} \neq 0$, then in a similar way as in proof of Theorem 4.4, we deduce that

$$\begin{aligned} [e_i^{s+1}, e_1^{s+1}] &= e_{i+1}^{s+1}, && 1 \leq i \leq n_{s+1} - 1, \\ [x, e_1^{s+1}] &= \alpha_{s+1}^{s+1} e_1^{s+1}, && \\ [e_i^{s+1}, x] &= -i\alpha^{s+1} e_i^{s+1}, && 1 \leq i \leq n_{s+1}. \end{aligned}$$

Therefore we get the algebra $R_{p+1,q}$.

If $\alpha_1^{s+1} = 0$, then by similar arguments as in Theorem 4.6, we obtain

$$\begin{aligned} [e_i^{s+1}, e_1^{s+1}] &= e_{i+1}^{s+1}, && 1 \leq i \leq n_{s+1} - 1, \\ [e_i^{s+1}, x] &= \sum_{m=i+1}^{n_{s+1}} \beta_{m-i+1}^{s+1} f_m^{s+1}, && 1 \leq i \leq n_{s+1}, \\ [x, x] &= \sum_{m=1}^{k} \gamma^m f_{n_m} + \gamma^{s+1} f_{n_{s+1}}^{s+1}. && \end{aligned}$$

Setting $f_{i-1}^{q+1} = e_{i-1}^{s+1}$ we get the family of algebras $R_{p,q+1}$.

The proof that two algebras of the family $R_{p,q}$ with different values of parameters are not isomorphic is carried on in the same way as in the proof of Theorem 4.6. □

In fact, due to Theorem 3.4, when the nilradical of a solvable Leibniz algebra is a direct sum of s copies of null-filiform ideals, the complementary vector space has dimension not greater than s. By taking the direct sum of ideals $NF_i + Q_i$ and $NF_k \oplus \cdots \oplus NF_s$, where $1 \leq i \leq k - 1$, $k \leq s$, we can construct a solvable Leibniz algebra whose nilradical is $NF_1 \oplus \cdots \oplus NF_s$ and whose complementary vector space is k-dimensional.

CHAPTER 5

ISOMORPHISM CRITERIA FOR FILIFORM LEIBNIZ ALGEBRAS

5.1 ON BASE CHANGES IN COMPLEX FILIFORM LEIBNIZ ALGEBRAS

Since an arbitrary filiform Leibniz algebra, up to isomorphism, belongs to one of the classes from Theorem 4.3, we conclude that in order to investigate the problem of isomorphisms inside the classes, we need to study the behavior of the parameters (structure constants) under base change inside each family.

Let L be an $(n+1)$-dimensional complex filiform Leibniz algebra obtained from a naturally graded filiform non Lie Leibniz algebra and let $\{e_0, e_1, \ldots, e_n\}$ be its basis.

Definition 5.1. *The basis $\{e_0, e_1, \ldots, e_n\}$ of L is said to be adapted if the table of multiplications of the algebra with respect to $\{e_0, e_1, \ldots, e_n\}$ has one of the forms* FLb_{n+1}, SLb_{n+1} *and* TLb_{n+1}.

Let L be a Leibniz algebra defined on a vector space V and let $\{e_0, e_1, \ldots, e_n\}$ be an adapted basis of L.

Definition 5.2. *A basis transformation $f \in GL(V)$ is said to be adapted with respect to the table of multiplications of L if the basis $\{f(e_0), f(e_1), \ldots, f(e_n)\}$ also is adapted.*

Obviously, the composition of the adapted base changes is an adapted transformation. The closed subgroup of $GL(V)$ consisting of the adapted transformations is denoted by $GL_{ad}(V)$.

Further we shall need the following identity, the proof of which is straightforward.

$$\sum_{i=k}^{n} a_i \sum_{j=i+p}^{n} b_{i,j}e_j = \sum_{j=k+p}^{n} \sum_{i=k}^{j-p} a_i b_{i,j} e_j, \qquad 0 \le p \le n-k, \ 3 \le k \le n. \tag{5.1}$$

The proposition below describes the adapted basis transformations in algebras from the classes FLb_{n+1} and SLb_{n+1}.

Proposition 5.1. *Let $f \in GL_{ad}(V)$ and $\{e_0, e_1, \ldots, e_n\}$ be the adapted basis.*

a) *If $L \in \mathrm{FLb}_{n+1}$ then $f : L \longrightarrow L$ has the following form:*

$$\begin{cases} f(e_0) &= \sum\limits_{i=0}^{n} a_i e_i, \\ f(e_1) &= (a_0 + a_1)e_1 + \sum\limits_{i=2}^{n-2} a_i e_i + (a_{n-1} + a_1(\theta - \alpha_n))e_{n-1} + b_n e_n, \\ f(e_{i+1}) &= [f(e_i), f(e_0)], \quad 1 \le i \le n-1 \\ f(e_2) &= [f(e_0), f(e_0)]. \end{cases}$$

b) *If $L \in \mathrm{SLb}_{n+1}$ then $f : L \longrightarrow L$ has the following form:*

$$\begin{cases} f(e_0) &= \sum\limits_{i=0}^{n} a_i e_i, \\ f(e_1) &= b_1 e_1 - \frac{a_1 b_1 \gamma}{a_0} e_{n-1} + b_n e_n \\ f(e_{i+1}) &= [f(e_i), f(e_0)], \quad 2 \le i \le n-1 \\ f(e_2) &= [f(e_0), f(e_0)]. \end{cases}$$

Proof. Note that an adapted basis change for filiform Lie and Leibniz algebras is defined by its action on the basis vectors e_0 and e_1. Therefore we set

$$f(e_0) = \sum_{i=0}^{n} a_i e_i \text{ and } f(e_1) = \sum_{j=0}^{n} b_j e_j.$$

Case a) Consider the product $f(e_2) = [f(e_0), f(e_0)]$. Then by (5.1) we have

$$
\begin{aligned}
[f(e_0),\ f(e_0)] &= a_0(a_0 + a_1)e_2 + a_0 \sum_{i=3}^{n} a_{i-1}e_i \\
&\quad + a_0a_1\left(\sum_{i=3}^{n-1} \alpha_i e_i + \theta e_n\right) + a_1^2 \sum_{i=3}^{n} \alpha_i e_i \\
&\quad + a_1 \sum_{i=2}^{n-2} a_i \sum_{k=i+2}^{n} \alpha_{k+1-i}e_k \\
&= a_0(a_0 + a_1)e_2 + a_0 \sum_{i=3}^{n} a_{i-1}e_i + a_1(a_0 + a_1) \sum_{i=3}^{n-1} \alpha_i e_i \\
&\quad + a_1(a_0\theta + a_1\alpha_n)e_n + a_1 \sum_{i=4}^{n} a_{i-2} \sum_{k=i}^{n} \alpha_{k+3-i}e_k \\
&= a_0(a_0 + a_1)e_2 + a_0 \sum_{i=3}^{n} a_{i-1}e_i \\
&\quad + a_1(a_0 + a_1) \sum_{i=3}^{n-1} \alpha_i e_i + a_1(a_0\theta + a_1\alpha_n)e_n \\
&\quad + a_1 \sum_{i=4}^{n} \sum_{i=4}^{k} (a_{i-2}\alpha_{k+3-i}e_k) \\
&= a_0(a_0 + a_1)e_2 + (a_0a_2 + a_1(a_0 + a_1)\alpha_3)e_3 \\
&\quad + \sum_{t=4}^{n-1} (a_0a_{t-1} + a_1(a_0 + a_1)\alpha_t \\
&\quad + a_1 \sum_{i=4}^{t} a_{i-2}\alpha_{t+3-i})e_t + (a_0a_{n-1} + a_1(a_0\theta + a_1\alpha_n) \\
&\quad + a_1 \sum_{i=4}^{n} a_{i-2}\alpha_{n+3-i})e_n \\
&= f(e_2).
\end{aligned}
$$

Due to $[f(e_0), f(e_0)] \in L^2$, we get $a_0(a_0 + a_1) \neq 0$.

Consider the product

$$[f(e_0), f(e_1)] = b_0(a_0 + a_1)e_2 + \sum_{i=3}^{n} c_i e_i.$$

Since $[f(e_0), f(e_1)] \notin L^2$ and $a_0 + a_1 \neq 0$, this implies that $b_0 = 0$.

The properties of the adapted transformation give $f(e_2) = [f(e_1), f(e_0)]$.

The product $[f(e_1), f(e_0)]$ in the basis $\{e_0, e_1, \ldots, e_n\}$ is expanded as follows

$$[f(e_1), f(e_0)] = a_0b_1e_2 + (a_0b_2 + a_1b_1\alpha_3)e_3$$

$$+\sum_{t=4}^{n-1}\left(a_0b_{t-1}+a_1b_1\alpha_t+a_1\sum_{i=4}^{t}b_{i-2}\alpha_{t+3-i}\right)e_t$$
$$+\left(a_0b_{n-1}+a_1b_1\alpha_n+a_1\sum_{i=4}^{n}b_{i-2}\alpha_{n+3-i}\right)e_n.$$

Comparing the coefficients at the basis vectors we get the conditions to the coefficients of f:

$$\begin{cases} a_0+a_1=b_1, \quad a_2=b_2, \\ a_0a_{t-1}+a_1\sum_{i=4}^{t}a_{i-2}\alpha_{t+3-i}=a_0b_{t-1}+a_1\sum_{i=4}^{t}b_{i-2}\alpha_{t+3-i}, \\ a_0a_{n-1}+a_1(a_0\theta+a_1\alpha_n)+a_1\sum_{i=4}^{n}a_{i-2}\alpha_{n+3-i} \\ \qquad\qquad =a_0b_{n-1}+a_1b_1\alpha_n+a_1\sum_{i=4}^{n}b_{i-2}\alpha_{n+3-i}. \end{cases}$$

Simplifying we obtain

$$\begin{cases} b_1 &= a_0+a_1, \\ b_i &= a_i, \quad 2\le i\le n-2 \\ b_{n-1} &= a_{n-1}+a_1(\theta-\alpha_n). \end{cases}$$

Case b) is proved similarly.

□

We introduce the notion of elementary transformations for algebras from families FLb_{n+1} and SLb_{n+1} as follows (for Lie algebras case see [87]).

Definition 5.3. *The following types of the adapted transformations are said to be elementary:*

$$\textit{first type} - \tau(a,b,k)=\begin{cases} f(e_0) &= e_0+ae_k, \quad 2\le k\le n \\ f(e_1) &= e_1+be_k, \quad 2\le k\le n \\ f(e_{i+1}) &= [f(e_i),f(e_0)], \quad 1\le i\le n-1, \\ f(e_2) &= [f(e_0),f(e_0)] \end{cases}$$

$$\textit{second type} - \vartheta(a,b)=\begin{cases} f(e_0) &= ae_0+be_1 \\ f(e_1) &= (a+b)e_1+b(\theta-\alpha_n)e_{n-1}, \\ & \qquad\qquad a(a+b)\neq 0 \\ f(e_{i+1}) &= [f(e_i),f(e_0)], \quad 1\le i\le n-1, \\ f(e_2) &= [f(e_0),f(e_0)] \end{cases}$$

$$\text{third type} - \sigma(b,n) = \begin{cases} f(e_0) &= e_0 \\ f(e_1) &= e_1 + be_n, \\ f(e_{i+1}) &= [f(e_i), f(e_0)], \quad 2 \le i \le n-1, \\ f(e_2) &= [f(e_0), f(e_0)] \end{cases}$$

$$\text{fourth type} - \eta(a,k) = \begin{cases} f(e_0) &= e_0 + ae_k \\ f(e_1) &= e_1 \\ f(e_{i+1}) &= [f(e_i), f(e_0)], \ 2 \le i \le n-1, \\ f(e_2) &= [f(e_0), f(e_0)] \end{cases}$$
$$2 \le k \le n$$

$$\text{fifth type} - \delta(a,b,d) = \begin{cases} f(e_0) &= ae_0 + be_1 \\ f(e_1) &= de_1 - \frac{bd\gamma}{a} e_{n-1}, \quad ad \neq 0 \\ f(e_{i+1}) &= [f(e_i), f(e_0)], \quad 2 \le i \le n-1, \\ f(e_2) &= [f(e_0), f(e_0)] \end{cases}$$
$$a, b, d \in \mathbb{C}.$$

Let f be an arbitrary element of the group $GL_{ad}(V)$, then f can be expressed as compositions of the elementary transformations as follows.

Proposition 5.2.

i) *Let f be of the form* a) *of Proposition* 5.1. *Then*

$$f = \tau(a_n, b_n, n) \circ \tau(a_{n-1}, a_{n-1}, n-1) \circ \ldots \circ \tau(a_2, a_2, 2) \circ \vartheta(a_0, a_1)$$

ii) *Let f be of the form* b) *of Proposition* 5.1. *Then*

$$f = \sigma(b_n, n) \circ \eta(a_n, n) \circ \eta(a_{n-1}, n-2) \circ \cdots \circ \eta(a_2, 2) \circ \delta(a_0, a_1, b_1)$$

Proof. Straightforward. □

One has the following proposition.

Proposition 5.3.

1) *The basis transformation*

$$\phi = \tau(a_n, b_n, n) \circ \tau(a_{n-1}, a_{n-1}, n-1) \circ \cdots \circ \tau(a_2, a_2, 2)$$

does not change the structure constants of algebras from FLb_{n+1}.

2) *The basis transformation*

$$\varphi = \sigma(b_n, n) \circ \eta(a_n, n) \circ \eta(a_{n-1}, n-2) \circ \cdots \circ \eta(a_2, 2)$$

does not change the structure constants of algebras from SLb_{n+1}.

Proof. Let us prove the first assertion.

Consider the basis transformation $\tau(a,b,k)$:

$$\tau(a,b,k) = \begin{cases} f(e_0) = e_0 + ae_k, \\ f(e_1) = e_1 + be_k, \\ f(e_{i+1}) = [f(e_i), f(e_0)], \ 1 \leq i \leq n-1 \\ f(e_2) = [f(e_0), f(e_0)] \end{cases} \qquad 2 \leq k \leq n$$

For $2 \leq k \leq n-1$ we put $a = b$ and consider the products which involve the parameters:

$$\begin{aligned} [f(e_0), f(e_1)] &= \sum_{i=3}^{n-k+1} \alpha_i(e_i + ae_{k+i-1}) + \sum_{i=n-k+2}^{n-1} \alpha_i e_i + \theta e_n \\ &= \sum_{i=3}^{n-1} \alpha_i f(e_i) + \theta f(e_n), \\ [f(e_1), f(e_1)] &= \sum_{i=3}^{n} \alpha_i e_i + a \sum_{i=3}^{n-k+1} \alpha_i e_{k+i-1} \\ &= \sum_{i=3}^{n-k+1} \alpha_i(e_i + ae_{k+i-1}) + \sum_{i=n-k+2}^{n} \alpha_i e_i = \sum_{i=3}^{n} \alpha_i f(e_i). \end{aligned}$$

Therefore, the basis transformation $\tau(a,a,k)$, $2 \leq k \leq n-1$, for any $a \in \mathbb{C}$ does not change the structure constants α_i, θ.

Similarly, one can check that $\tau(a,b,n) \in GL_{ad}(V)$ also does not change the structure constants α_i, θ for any value of a.

Since a composition of adapted transformations is again an adapted transformation, we conclude that the transformation

$$\phi = \tau(a_n, b_n, n) \circ \tau(a_{n-1}, a_{n-1}, n-1) \circ \cdots \circ \tau(a_2, a_2, 2)$$

also does not change the structure constants of the family FLb_{n+1}.

The proof of the second part is carried out similarly. □

Thus, the study of all basis transformations is reduced to the second and fifth types of elementary transformations for the families FLb_{n+1} and SLb_{n+1}, respectively.

5.2 CRITERIA OF ISOMORPHISMS OF COMPLEX FILIFORM NON-LIE LEIBNIZ ALGEBRAS

We recall that for any element a of a Leibniz algebra L the right multiplication operator is defined as follows $R_a(x) = [x,a]$.

Set

$$R_a^m(x) := [[\dots[x, \underbrace{a], a], \dots, a]}_{m \text{ times}} \text{ and } R_a^0(x) := x.$$

It is easy to check that for an algebra from the first two families of Theorem 4.3 the following equality is valid:

$$[[e_s, e_1], e_0] = [e_{s+1}, e_1], \ 2 \le s \le n. \tag{5.2}$$

Let L belong to the family FLb_{n+1} (respectively, of the family SLb_{n+1}). Then (5.2) implies that for $m \in \mathbb{N}$, $0 \le p \le n$ (respectively, $0 \le p \le n$, $p \ne 1$) the following identity holds true:

$$R^m_{e_1}(e_p) = R^{p-1}_{e_0}(R^m_{e_1}(e_0)). \tag{5.3}$$

To prove the main result (Theorem 5.1) we need the following lemma.

Lemma 5.1. *Let L belong to the family the first two families from Theorem* 4.3. *Then for* $2 \le m \le \frac{n-1}{2}$ *the following identities hold true*

$$R^m_{e_1}(e_0) = \sum_{i_m=2m+1}^{n} \sum_{i_{m-1}=2m+1}^{i_m} \cdots \sum_{i_1=2m+1}^{i_2} \eta_{i_m+3-i_{m-1}} \cdots \eta_{i_2+3-i_1}\eta_{i_1+3-2(m-1)}e_{i_m},$$

$$\textit{where } \eta_i = \begin{cases} \alpha_i, & \textit{if } L \in \mathrm{FLb}_{n+1} \\ \beta_i, & \textit{if } L \in \mathrm{SLb}_{n+1}. \end{cases}$$

Proof. Let us consider the case $\eta_i = \alpha_i$, whereas the case $\eta_i = \beta_i$ is proved similarly.

We use the method of mathematical induction with respect to m. For $m = 2$ we have

$$\begin{aligned} R^2_{e_1}(e_0) &:= \left[\sum_{i=3}^{n-1} \alpha_i e_i + \theta e_n, e_1\right] = \sum_{i=5}^{n} \alpha_{i-2}[e_{i-2}, e_1] \\ &= \sum_{i=5}^{n} \alpha_{i-2} \sum_{j=i}^{n} \alpha_{j+3-i} e_1 = \sum_{j=5}^{n} \sum_{i=5}^{j} \alpha_{i-2}\alpha_{j+3-i} e_1. \end{aligned}$$

Assume that for m the equality of the lemma is true. Then the following equalities complete the proof

$$R^{m+1}_{e_1}(e_0) = [R^m_{e_1}(e_0), e_1]$$

$$= \left[\sum_{i_k=2m+1}^{n} \sum_{i_{m-1}=2m+1}^{i_m} \cdots \sum_{i_1=2m+1}^{i_2} \alpha_{i_m+3-i_{k-1}} \cdots \alpha_{i_2+3-i_1}\alpha_{i_1+3-(2m+1)}e_{i_m}, e_1\right]$$

$$= \left[\sum_{i_m=2m+3}^{n} \sum_{i_{m-1}=2m+3}^{i_m} \cdots \sum_{i_1=2m+3}^{i_2} \alpha_{i_m+3-i_{m-1}} \dots \alpha_{i_2+3-i_1} \alpha_{i_1+3-(2m+3)} e_{i_m-2}, e_1 \right]$$

$$= \sum_{i_m=2m+3}^{n} \sum_{i_{m-1}=2m+3}^{i_m} \cdots \sum_{i_1=2m+3}^{i_2} \alpha_{i_m+3-i_{m-1}} \dots \alpha_{i_2+3-i_1} \alpha_{i_1+3-(2m+3)} \sum_{i_{m+1}=i_m}^{n} \alpha_{i_{m+1}+3-i_m} e_{i_m+1}$$

$$= \sum_{i_m=2m+3}^{n} \sum_{i_{m+1}=i_m}^{n} \sum_{i_{m-1}=2m+3}^{i_m} \cdots \sum_{i_1=2m+3}^{i_2} \alpha_{i_m+3-i_{m-1}} \dots \alpha_{i_2+3-i_1} \alpha_{i_1+3-(2m+3)} e_{i_m+1}$$

$$= \sum_{i_{m+1}=2m+3}^{n} \sum_{i_m=2m+3}^{i_{m+1}} \sum_{i_{m-1}=2m+3}^{i_m} \cdots \sum_{i_1=2m+3}^{i_2} \alpha_{i_m+3-i_{m-1}} \dots \alpha_{i_2+3-i_1} \alpha_{i_1+3-(2m+3)} e_{i_m+1}.$$

□

The study of adapted transformations for the family FLb_{n+1} is reduced to the study of its action on e_0 and e_1 as follows

$$\begin{cases} e_0' = f(e_0) = Ae_0 + Be_1, & A(A+B) \neq 0, \\ e_1' = f(e_2) = (A+B)e_1 + B(\theta - \alpha_n)e_{n-1}. \end{cases} \tag{5.4}$$

Let us find out the action on whole basis. It is given in the corollary below.

Corollary 5.1. *The adapted base change for the class* FLb_{n+1} *is given as follows*

$$\begin{aligned} e_0' &= Ae_0 + Be_1, \quad A(A+B) \neq 0, \\ e_1' &= (A+B)e_1 + B(\theta - \alpha_n)e_{n-1} \\ e_2' &= A(A+B)e_2 + B(A+B)\sum_{i=3}^{n-1} \alpha_i e_i + B(A\theta + B\alpha_n), \\ e_k' &= (A+B)\left(\sum_{i=0}^{k-2} \binom{k-1}{k-1-i} A^{k-1-i} B^i R_{e_1}^i (e_{k-i}) \right. \\ &\qquad \left. + B^{k-1} R_{e_1}^{k-1}(e_0) \right), \end{aligned} \tag{5.5}$$

where $3 \leq k \leq n$ *and* $\binom{s}{t} = \frac{s!}{(s-t)!t!}$.

Proof. We prove the corollary by induction on k. For $k = 2, 3$ we have

$$\begin{aligned} e_2' &= [e_1', e_0'] = A(A+B)e_2 + B(A+B)\sum_{i=3}^{n-1} \alpha_i e_i + B(A\theta + B\alpha_n), \\ e_3' &= [e_2', e_0'] = A^2(A+B)e_3 + 2AB(A+B)[e_2, e_1] \\ &\qquad + B^2(A+B)[[e_0, e_1], e_1] \\ &= (A+B)(A^2 e_3 + 2ABR_{e_1}(e_2) + B^2 R_{e_1}^2(e_0)). \end{aligned}$$

Suppose that equality (5.5) is true for k. Taking into account the equality (5.2) and the following equalities

$$\begin{aligned}
e'_{k+1} &= [e'_k, e_0] \\
&= \left[(A+B)\left(\sum_{i=0}^{k-2}\binom{k-1}{k-1-i}A^{k-1-i}B^iR^i_{e_1}(e_{k-i}) + B^{k-1}R^{k-1}_{e_1}(e_0)\right), Ae_0 + Be_1\right] \\
&= (A+B)\left(\sum_{i=0}^{k-2}\binom{k-1}{k-1-i}A^{k-i}B^iR^i_{e_1}(e_{k+1-i}) + AB^{k-1}R^{k-1}_{e_1}(e_2)\right. \\
&\quad \left. + \sum_{i=0}^{k-2}\binom{k-1}{k-1-i}A^{k-1-i}B^{i+1}R^{i+1}_{e_1}(e_{k-i}) + B^kR^k_{e_1}(e_0)\right) \\
&= (A+B)\left(\sum_{i=0}^{k-2}\binom{k-1}{k-1-i}A^{k-i}B^iR^i_{e_1}(e_{k+1-i}) + AB^{k-1}R^{k-1}_{e_1}(e_2)\right. \\
&\quad \left. + \sum_{i=1}^{k-1}\binom{k-1}{k-i}A^{k-i}B^iR^i_{e_1}(e_{k+1-i}) + B^kR^k_{e_1}(e_0)\right) \\
&= (A+B)\left(\sum_{i=1}^{k-2}\left(\binom{k-1}{k-1-i}+\binom{k-1}{k-i}\right)A^{k-i}B^iR^i_{e_1}(e_{k+1-i})\right. \\
&\quad \left. + \binom{k-1}{k-1}A^ke_{k+1} + \binom{k-1}{1}AB^{k-1}R^{k-1}_{e_1}(e_2) + AB^{k-1}R^{k-1}_{e_1}(e_2) + B^kR^k_{e_1}(e_0)\right) \\
&= (A+B)\left(\sum_{i=1}^{k-2}\binom{k}{k-i}A^{k-i}B^iR^i_{e_1}(e_{k+1-i}) + \binom{k}{k}A^ke_{k+1}\right. \\
&\quad \left. + \binom{k-1}{1}AB^{k-1}R^{k-1}_{e_1}(e_2) + B^kR^k_{e_1}(e_0)\right) \\
&= (A+B)\left(\sum_{i=0}^{k-1}\binom{k}{k-i}A^{k-i}B^iR^i_{e_1}(e_{k+1-i}) + B^kR^k_{e_1}(e_0)\right)
\end{aligned}$$

we complete the proof of the equality (5.5) for $k+1$. □

Similarly, for the family SLb_{n+1} the result is given by the following corollary.

Corollary 5.2. *The adapted base change for the class* SLb_{n+1} *is given as follows*

$$\begin{cases}
e'_0 &= Ae_0 + Be_1, \quad AD \neq 0, \\
e'_1 &= De_1 - \frac{BD\gamma}{A}e_{n-1}, \\
e'_k &= A\left(\sum_{i=0}^{k-2}\binom{k-1}{k-1-i}A^{k-1-i}B^iR^i_{e_1}(e_{k-i}) + B^{k-1}R^{k-1}_{e_1}(e_0)\right), \\
& \qquad\qquad\qquad\qquad 3 \leq k \leq n.
\end{cases}$$

Proof. The proof is the similar to that of Corollary 5.1. □

We shall denote an algebra from family FLb_{n+1} (respectively, SLb_{n+1}) by $L(\alpha_3, \alpha_4, \ldots, \alpha_n, \theta)$ (respectively, $L(\beta_3, \beta_4, \ldots, \beta_n, \gamma)$).

Theorem 5.1. a) *Two algebras* $L(\alpha_3, \alpha_4, \ldots, \alpha_n, \theta)$ *and* $L'(\alpha'_3, \alpha'_4, \ldots, \alpha'_n, \theta')$ *are isomorphic if and only if there exist* $A, B \in \mathbb{C}$ *such that* $A(A+B) \neq 0$ *and the following conditions hold:*

$$\alpha'_3 = \frac{(A+B)}{A^2}\alpha_3$$

$$\alpha'_t = \frac{1}{A^{t-1}}\Bigg((A+B)\alpha_t - \sum_{k=3}^{t-1}\Bigg(\binom{k-1}{k-2}A^{k-2}B\alpha_{t+2-k}$$
$$+\binom{k-1}{k-3}A^{k-3}B^2\sum_{i_1=k+2}^{t}\alpha_{t+3-i_1}\alpha_{i_1+1-k}$$
$$+\binom{k-1}{k-4}A^{k-4}B^3\sum_{i_2=k+3}^{t}\sum_{i_1=k+3}^{i_2}\alpha_{t+3-i_2}\,\alpha_{i_2+3-i_1}\cdot\alpha_{i_1-k}\cdots$$
$$+\binom{k-1}{1}AB^{k-2}\sum_{i_{k-3}=2k-2}^{t}\sum_{i_{k-4}=2k-2}^{i_{k-3}}\cdots\sum_{i_1=2k-2}^{i_2}\alpha_{t+3-i_{k-3}}\alpha_{i_{k-3}+3-i_{k-4}}$$
$$\ldots\alpha_{i_2+3-i_1}\alpha_{i_1+5-2k}$$
$$+B^{k-1}\sum_{i_{k-2}=2k-1}^{t}\sum_{i_{k-3}=2k-1}^{i_{k-2}}\cdots\sum_{i_1=2k-1}^{i_2}\alpha_{t+3-i_{k-2}}\alpha_{i_{k-2}+3-i_{k-3}}\cdots$$
$$\alpha_{i_2+3-i_1}\alpha_{i_1+4-2k}\Bigg)\alpha'_k\Bigg), \text{ where } 4 \le t \le n.$$

$$\theta' = \frac{1}{A^{n-1}}\Bigg(A\theta + B\alpha_n - \sum_{k=3}^{n-1}\Bigg(\binom{k-1}{k-2}A^{k-2}B\alpha_{n+2-k}$$
$$+\binom{k-1}{k-3}A^{k-3}B^2\sum_{i_1=k+2}^{n}\alpha_{n+3-i_1}\alpha_{i_1+1-k}+\binom{k-1}{k-4}A^{k-4}B^3$$
$$\sum_{i_2=k+3}^{n}\sum_{i_1=k+3}^{i_2}\alpha_{n+3-i_2}\,\alpha_{i_2+3-i_1}\,\alpha_{i_1-k}\;+\;\ldots\;+\binom{k-1}{1}AB^{k-2}$$
$$\sum_{i_{k-3}=2k-2}^{n}\sum_{i_{k-4}=2k-2}^{i_{k-3}}\cdots\sum_{i_1=2k-2}^{i_2}\alpha_{n+3-i_{k-3}}\alpha_{i_{k-3}+3-i_{k-4}}\cdots\alpha_{i_2+3-i_1}\alpha_{i_1+5-2k}$$
$$+B^{k-1}\sum_{i_{k-2}=2k-1}^{n}\sum_{i_{k-3}=2k-1}^{i_{k-2}}\cdots\sum_{i_1=2k-1}^{i_2}\alpha_{n+3-i_{k-2}}\alpha_{i_{k-2}+3-i_{k-3}}\cdots$$
$$\alpha_{i_2+3-i_1}\alpha_{i_1+4-2k}\Bigg)\alpha'_k\Bigg),$$

b) *Two algebras* $L(\beta_3, \beta_4, \ldots, \beta_n, \gamma)$ *and* $L'(\beta'_3, \beta'_4, \ldots, \beta'_n, \gamma')$ *are isomorphic if and only if there exist* $A, B, D \in \mathbb{C}$ *such that* $AD \neq 0$ *and the following conditions hold:*

$$\gamma' = \frac{D^2}{A^n}\gamma,$$
$$\beta'_3 = \frac{D}{A^2}\beta_3,$$
$$\beta'_t = \frac{1}{A^{t-1}}\Bigg(D\beta_t - \sum_{k=3}^{t-1}\Bigg(\binom{k-1}{k-2}A^{k-2}B\beta_{t+2-k} + \binom{k-1}{k-3}A^{k-3}B^2$$

$$\sum_{i_1=k+2}^{t} \beta_{t+3-i_1}\beta_{i_1+1-k} + \begin{pmatrix} k-1 \\ k-4 \end{pmatrix} A^{k-4}B^3$$
$$\sum_{i_2=k+3}^{t} \sum_{i_1=k+3}^{i_2} \beta_{t+3-i_2}\beta_{i_2+3-i_1}\beta_{i_1-k} + \dots + \begin{pmatrix} k-1 \\ 1 \end{pmatrix} AB^{k-2}$$
$$\sum_{i_{k-3}=2k-2}^{t} \sum_{i_{k-4}=2k-2}^{i_{k-3}} \cdots \sum_{i_1=2k-2}^{i_2} \beta_{t+3-i_{k-3}}\beta_{i_{k-3}+3-i_{k-4}} \cdots \beta_{i_2+3-i_1}\beta_{i_1+5-2k} + B^{k-1}$$
$$\left.\left.\sum_{i_{k-2}=2k-1}^{t} \sum_{i_{k-3}=2k-1}^{i_{k-2}} \cdots \sum_{i_1=2k-1}^{i_2} \beta_{t+3-i_{k-2}}\beta_{i_{k-2}+3-i_{k-3}} \cdots \beta_{i_2+3-i_1}\beta_{i_1+4-2k}\right)\beta'_k\right)$$

where $4 \leq t < n$,

$$\beta'_n = \frac{BD\gamma}{A^n} + \frac{1}{A^{n-1}}\Bigg(D\beta_n - \sum_{k=3}^{n-1}\Bigg(\begin{pmatrix} k-1 \\ k-2 \end{pmatrix} A^{k-2}B\beta_{n+2-k} + \begin{pmatrix} k-1 \\ k-3 \end{pmatrix}$$
$$\cdot A^{k-3}B^2 \sum_{i_1=k+2}^{n} \beta_{n+3-i_1} \cdot \beta_{i_1+1-k} + \begin{pmatrix} k-1 \\ k-4 \end{pmatrix} A^{k-4}B^3$$
$$\cdot \sum_{i_2=k+3}^{n} \sum_{i_1=k+3}^{i_2} \beta_{n+3-i_2}\beta_{i_2+3-i_1}\beta_{i_1-k} + \dots + \begin{pmatrix} k-1 \\ 1 \end{pmatrix} AB^{k-2}$$
$$\cdot \sum_{i_{k-3}=2k-2}^{n} \sum_{i_{k-4}=2k-2}^{i_{k-3}} \cdots \sum_{i_1=2k-2}^{i_2} \beta_{n+3-i_{k-3}}\beta_{i_{k-3}+3-i_{k-4}} \cdots \beta_{i_2+3-i_1}\beta_{i_1+5-2k}$$
$$+B^{k-1} \sum_{i_{k-2}=2k-1}^{n} \sum_{i_{k-3}=2k-1}^{i_{k-2}} \cdots \sum_{i_1=2k-1}^{i_2} \beta_{n+3-i_{k-2}}\beta_{i_{k-2}+3-i_{k-3}} \cdots$$
$$\left.\left.\cdot\beta_{i_2+3-i_1}\beta_{i_1+4-2k}\right)\beta'_k\right).$$

Proof. Consider the class FLb_{n+1}. Let $\{e_0, e_1, \dots, e_n\}$ be a basis of $L(\alpha_3, \alpha_4, \dots, \alpha_n, \theta)$, and let $\{e'_0, e'_1, \dots, e'_n\}$ be a basis of $L'(\alpha'_3, \alpha'_4, \dots, \alpha'_n, \theta')$.

It is easy to see that in $L(\alpha_3, \alpha_4, \dots, \alpha_n, \theta)$ the following holds

$$[[e_0, e_1], e_1] = [[e_1, e_1], e_1].$$

Lemma 5.1 and the identity (5.3) imply

$$R^m_{e_1}(e_{k-m}) = \sum_{i_m=k+m}^{n} \sum_{i_{m-1}=k+m}^{i_m} \cdots \sum_{i_1=2m+1}^{i_2} \alpha_{i_m+3-i_{m-1}} \cdots \alpha_{i_2+3-i_1}\alpha_{i_1+3-(k+m)}e_{i_m}, \tag{5.6}$$

where $m \leq n-k$ and $m \leq k \leq n$.

Now we substitute (5.6) in (5.5) and use the equalities (5.1), (5.3) to get

$$e_k' = (A+B)\left(\sum_{i=0}^{k-2}\binom{k-1}{k-1-i}A^{k-1-i}B^iR_{e_1}^i(e_{k-1}) + B^{k-1}R_{e_1}^{k-1}(e_0)\right)$$

$$= (A+B)(A^{k-1}e_k + \binom{k-1}{k-2}A^{k-2}B\sum_{i=k+1}^{n}\alpha_{i+2-k}e_i + \binom{k-1}{k-3}A^{k-3}B^2$$

$$\cdot\sum_{i=k+2}^{n}\sum_{i_1=k+2}^{i}\alpha_{i+3-i_1}\cdot\alpha_{i_1+1-k}e_i + \binom{k-1}{k-4}A^{k-4}B^3$$

$$\cdot\sum_{i=k+3}^{n}\sum_{i_2=k+3}^{i}\sum_{i_1=k+3}^{i_2}\alpha_{i+3-i_2}\,\alpha_{i_2+3-i_1}\,\alpha_{i_1-k}e_i \;+\; \cdots + \binom{k-1}{1}AB^{k-2}$$

$$\cdot\sum_{i=2k-2}^{n}\sum_{i_{k-3}=2k-2}^{i}\cdots\sum_{i_1=2k-2}^{i_2}\alpha_{i+3-i_{k-3}}\cdots\alpha_{i_2+3-i_1}\cdot\alpha_{i_1+5-2k}e_1$$

$$+B^{k-1}\sum_{i=2k-1}^{n}\sum_{i_{k-2}=2k-1}^{i}\cdots\sum_{i_1=2k-1}^{i_2}\alpha_{i+3-i_{k-2}}\cdots\alpha_{i_2+3-i_1}\alpha_{i_1+4-2k}e_i)$$

$$= (A+B)(\,A^{k-1}e_k + \binom{k-1}{k-2}A^{k-2}B\alpha_3e_{k+1} + \left(\binom{k-1}{k-2}A^{k-2}B\alpha_4\right.$$

$$\left.+\binom{k-1}{k-3}A^{k-3}B^2\sum_{i_1=k+2}^{k+2}\alpha_{k+5-i_1}\cdot\alpha_{i_1+1-k}\right)e_{k+2} + \cdots$$

$$+\left(\binom{k-1}{k-2}A^{k-2}B\alpha_{t+2-k} + \binom{k-1}{k-3}A^{k-3}B^2\sum_{i_1=k+1}^{t}\alpha_{t+3-i_1}\alpha_{i_1+1-k}\right.$$

$$+\binom{k-1}{k-4}A^{k-4}B^3\sum_{i_2=k+3}^{t}\sum_{i_1=k+3}^{i_2}\alpha_{t+3-i_2}\alpha_{i_2+3-i_1}\,\alpha_{i_1-k} + \cdots + \binom{k-1}{1}$$

$$\cdot AB^{k-2}\sum_{i_{k-3}=2k-2}^{t}\sum_{i_{k-4}=2k-2}^{i_{k-3}}\cdots\sum_{i_1=2k-2}^{i_2}\alpha_{t+3-i_{k-3}}\alpha_{i_{k-3}+3-i_{k-4}}\cdots\alpha_{i_2+3-i_1}\alpha_{i_1+5-2k}$$

$$\left.+B^{k-1}\sum_{i_{k-2}=2k-1}^{t}\sum_{i_{k-3}=2k-1}^{i_{k-2}}\cdots\sum_{i_1=2k-1}^{i_2}\alpha_{t+3-i_{k-2}}\alpha_{i_{k-2}+3-i_{k-3}}\cdots\alpha_{i_2+3-i_1}\alpha_{i_1+4-2k}\right)e_t$$

$$+\;\cdots\;+\left(\binom{k-1}{k-2}A^{k-2}B\alpha_{n+2-k} + \binom{k-1}{k-3}A^{k-3}B^2\sum_{i_1=k+1}^{n}\alpha_{n+3-i_1}\alpha_{i_1+1-k}\right.$$

$$+\binom{k-1}{k-4}A^{k-4}B^3\sum_{i_2=k+3}^{n}\sum_{i_1=k+3}^{i_2}\alpha_{n+3-i_2}\,\alpha_{i_2+3-i_1}\,\alpha_{i_1-k} + \;\cdots\; + \binom{k-1}{1}$$

$$\cdot AB^{k-2}\sum_{i_{k-3}=2k-2}^{n}\sum_{i_{k-4}=2k-2}^{i_{k-3}}\cdots\sum_{i_1=2k-2}^{i_2}\alpha_{n+3-i_{k-3}}\alpha_{i_{k-3}+3-i_{k-4}}\cdots\alpha_{i_2+3-i_1}\alpha_{i_1+5-2k}$$

$$\left.+B^{k-1}\sum_{i_{k-2}=2k-1}^{n}\sum_{i_{k-3}=2k-1}^{i_{k-2}}\cdots\sum_{i_1=2k-1}^{i_2}\alpha_{n+3-i_{k-2}}\alpha_{i_{k-2}+3-i_{k-3}}\cdots\alpha_{i_2+3-i_1}\alpha_{i_1+4-2k}\right)e_n)$$

$$= (A+B)\Bigg(A^{k-1}e_k + \sum_{t=k+1}^{n}\Bigg(\Bigg(\binom{k-1}{k-2}A^{k-2}B\alpha_{t+2-k} + \binom{k-1}{k-3}A^{k-3}B^2$$

$$\cdot \sum_{i_1=k+2}^{t} \alpha_{t+3-i_1}\alpha_{i_1+1-k} + \binom{k-1}{k-4}A^{k-4}B^3$$

$$\sum_{i_2=k+3}^{t} \cdot \sum_{i_1=k+3}^{i_2} \alpha_{t+3-i_2}\alpha_{i_2+3-i_1}\ \alpha_{i_1-k} + \quad \dots \quad + \binom{k-1}{1}AB^{k-2}$$

$$\cdot \sum_{i_{k-3}=2k-2}^{t}\ \sum_{i_{k-4}=2k-2}^{i_{k-3}} \cdots \sum_{i_1=2k-2}^{i_2} \alpha_{t+3-i_{k-3}}\alpha_{i_{k-3}+3-i_{k-4}}\dots\alpha_{i_2+3-i_1}\alpha_{i_1+5-2k} + B^{k-1}$$

$$\cdot \sum_{i_{k-2}=2k-1}^{t}\ \sum_{i_{k-3}=2k-1}^{i_{k-2}} \cdots \sum_{i_1=2k-1}^{i_2} \alpha_{t+3-i_{k-2}}\ \alpha_{i_{k-2}+3-i_{k-3}}\ \dots\ \alpha_{i_2+3-i_1}\alpha_{i_1+4-2k}\Bigg)e_t\Bigg).$$

Consider the following products in the algebra $L'(\alpha'_3, \alpha'_4, \dots, \alpha'_n, \theta')$:

$$[e'_0, e'_1] = \sum_{k=3}^{n-1} \alpha'_k e'_k + \theta' e'_n, \qquad [e'_1, e'_1] = \sum_{k=3}^{n} \alpha'_k e'_k.$$

Substituting the expression for e'_k, obtained above, into the product $[e'_0, e'_1]$ and using the equality (5.1) with $p = 1$, we derive:

$$[e'_0, e'_1]$$

$$= \sum_{k=3}^{n-1} \alpha'_k (A+B)\Bigg(A^{k-1}e_k + \sum_{t=k+1}^{n}\Bigg(\Bigg(\binom{k-1}{k-2}A^{k-2}B\alpha_{t+2-k} + \binom{k-1}{k-3}$$

$$\cdot A^{k-3}B^2 \sum_{i_1=k+2}^{t} \alpha_{t+3-i_1}\alpha_{i_1+1-k} + \binom{k-1}{k-4}A^{k-4}B^3$$

$$\cdot \sum_{i_2=k+3}^{t} \sum_{i_1=k+3}^{i_2} \alpha_{t+3-i_2}\alpha_{i_2+3-i_1}\ \alpha_{i_1-k} \quad + \quad \dots + \binom{k-1}{1}AB^{k-2}$$

$$\cdot \sum_{i_{k-3}=2k-2}^{t}\ \sum_{i_{k-4}=2k-2}^{i_{k-3}} \cdots \sum_{i_1=2k-2}^{i_2} \alpha_{t+3-i_{k-3}}\alpha_{i_{k-3}+3-i_{k-4}}\dots\alpha_{i_2+3-i_1}\alpha_{i_1+5-2k} + B^{k-1}$$

$$\cdot \sum_{i_{k-2}=2k-1}^{t}\ \sum_{i_{k-3}=2k-1}^{i_{k-2}} \quad \dots \quad \sum_{i_1=2k-1}^{i_2} \alpha_{t+3-i_{k-2}}\alpha_{i_{k-2}+3-i_{k-3}}\dots\alpha_{i_2+3-i_1}\alpha_{i_1+4-2k}\Bigg)e_t\Bigg)$$

$$+\theta' A^{n-1}(A+B)e_n$$

$$= (A+B)\Bigg(\sum_{k=3}^{n-1} A^{k-1}\alpha'_k e_k + \sum_{k=3}^{n-1}\sum_{t=k+1}^{n}\Bigg(\Bigg(\binom{k-1}{k-2}A^{k-2}B\alpha_{t+2-k}$$

$$+\binom{k-1}{k-3}A^{k-3}B^2 \sum_{i_1=k+2}^{t} \alpha_{t+3-i_1}\alpha_{i_1+1-k} + \binom{k-1}{k-4}A^{k-4}B^3$$

$$\sum_{i_2=k+3}^{t} \sum_{i_1=k+3}^{i_2} \alpha_{t+3-i_2}\alpha_{i_2+3-i_1}\alpha_{i_1-k} + \dots + \binom{k-1}{1}AB^{k-2}$$

$$\sum_{i_{k-3}=2k-2}^{t}\sum_{i_{k-4}=2k-2}^{i_{k-3}}\cdots\sum_{i_1=2k-2}^{i_2}\alpha_{t+3-i_{k-3}}\alpha_{i_{k-3}+3-i_{k-4}}\cdots\alpha_{i_2+3-i_1}\alpha_{i_1+5-2k}+B^{k-1}$$

$$\cdot\sum_{i_{k-2}=2k-1}^{t}\sum_{i_{k-3}=2k-1}^{i_{k-2}}\cdots\sum_{i_1=2k-1}^{i_2}\alpha_{t+3-i_{k-2}}\alpha_{i_{k-2}+3-i_{k-3}}\cdots\alpha_{i_2+3-i_1}\alpha_{i_1+4-2k}\Bigg)$$

$$\cdot\alpha'_k e_t+\theta' A^{n-1}e_n\Bigg)$$

$$=(A+B)\Bigg(A^2\alpha'_3e_3+\sum_{t=3}^{n-1}A^{t-1}\alpha'_te_t+\sum_{t=3}^{n-1}\sum_{k=3}^{t-1}\Bigg(\binom{k-1}{k-2}A^{k-2}B\alpha_{t+2-k}$$

$$+\binom{k-1}{k-3}A^{k-3}B^2\sum_{i_1=k+2}^{t}\alpha_{t+3-i_1}\alpha_{i_1+1-k}+\binom{k-1}{k-4}A^{k-4}B^3$$

$$\cdot\sum_{i_2=k+3}^{t}\sum_{i_1=k+3}^{i_2}\alpha_{t+3-i_2}\alpha_{i_2+3-i_1}\alpha_{i_1-k}+\cdots+\binom{k-1}{1}AB^{k-2}$$

$$\cdot\sum_{i_{k-3}=2k-2}^{t}\sum_{i_{k-4}=2k-2}^{i_{k-3}}\cdots\sum_{i_1=2k-2}^{i_2}\alpha_{t+3-i_{k-3}}\alpha_{i_{k-3}+3-i_{k-4}}\cdots\alpha_{i_2+3-i_1}\alpha_{i_1+5-2k}+B^{k-1}$$

$$\cdot\sum_{i_{k-2}=2k-1}^{t}\sum_{i_{k-3}=2k-1}^{i_{k-2}}\cdots\sum_{i_1=2k-1}^{i_2}\alpha_{t+3-i_{k-2}}\alpha_{i_{k-2}+3-i_{k-3}}\cdots\alpha_{i_2+3-i_1}\alpha_{i_1+4-2k}\Bigg)\alpha'_ke_t\Bigg)$$

$$+(A+B)\Bigg(\theta'A^{n-1}+\sum_{k=3}^{n-1}\Bigg(\binom{k-1}{k-2}A^{k-2}B\alpha_{n+2-k}+\binom{k-1}{k-3}A^{k-3}B^2$$

$$\cdot\sum_{i_1=k+2}^{n}\alpha_{n+3-i_1}\alpha_{i_1+1-k}+\binom{k-1}{k-4}A^{k-4}B^3$$

$$\cdot\sum_{i_2=k+3}^{n}\sum_{i_1=k+3}^{i_2}\alpha_{n+3-i_2}\,\alpha_{i_2+3-i_1}\,\alpha_{i_1-k}\cdots+\binom{k-1}{1}AB^{k-2}$$

$$\cdot\sum_{i_{k-3}=2k-2}^{n}\sum_{i_{k-4}=2k-2}^{i_{k-3}}\cdots\sum_{i_1=2k-2}^{i_2}\alpha_{n+3-i_{k-3}}\alpha_{i_{k-3}+3-i_{k-4}}\cdots\alpha_{i_2+3-i_1}\alpha_{i_1+5-2k}+B^{k-1}$$

$$\cdot\sum_{i_{k-2}=2k-1}^{n}\sum_{i_{k-3}=2k-1}^{i_{k-2}}\cdots\sum_{i_1=2k-1}^{i_2}\alpha_{n+3-i_{k-2}}\alpha_{i_{k-2}+3-i_{k-3}}\cdots\alpha_{i_2+3-i_1}\alpha_{i_1+4-2k}\Bigg)\alpha'_k\Bigg)e_n$$

$$=(A+B)\Bigg(A^2\alpha'_3e_3+\sum_{t=3}^{n-1}\Bigg(A^{t-1}\alpha'_te_t+\sum_{k=3}^{t-1}\Bigg(\binom{k-1}{k-2}A^{k-2}B\alpha_{t+2-k}$$

$$+\binom{k-1}{k-3}A^{k-3}B^2\sum_{i_1=k+2}^{t}\alpha_{t+3-i_1}\alpha_{i_1+1-k}+\binom{k-1}{k-4}A^{k-4}B^3$$

$$\cdot\sum_{i_2=k+3}^{t}\sum_{i_1=k+3}^{i_2}\alpha_{t+3-i_2}\,\alpha_{i_2+3-i_1}\,\alpha_{i_1-k}\quad+\quad\cdots+\binom{k-1}{1}AB^{k-2}$$

$$\cdot\sum_{i_{k-3}=2k-2}^{t}\sum_{i_{k-4}=2k-2}^{i_{k-3}}\cdots\sum_{i_1=2k-2}^{i_2}\alpha_{t+3-i_{k-3}}\alpha_{i_{k-3}+3-i_{k-4}}\cdots\alpha_{i_2+3-i_1}\alpha_{i_1+5-2k}+B^{k-1}$$

$$\cdot\sum_{i_{k-2}=2k-1}^{t}\sum_{i_{k-3}=2k-1}^{i_{k-2}}\cdots\sum_{i_1=2k-1}^{i_2}\alpha_{t+3-i_{k-2}}\alpha_{i_{k-2}+3-i_{k-3}}\cdots\alpha_{i_2+3-i_1}\alpha_{i_1+4-2k}\Bigg)\alpha'_k\Bigg)e_t\Bigg)$$

$$+\Bigg(\theta' A^{n-1} + \sum_{k=3}^{n-1}\Bigg(\binom{k-1}{k-2}A^{k-2}B\alpha_{n+2-k}$$

$$+\binom{k-1}{k-3}B^2\sum_{i_1=k+2}^{n}\alpha_{n+3-i_1}\alpha_{i_1+1-k} + \binom{k-1}{k-4}A^{k-4}B^3$$

$$\cdot\sum_{i_2=k+3}^{n}\sum_{i_1=k+3}^{i_2}\alpha_{n+3-i_2}\,\alpha_{i_2+3-i_1}\,\alpha_{i_1-k} \;+\;\cdots+\binom{k-1}{1}AB^{k-2}$$

$$\cdot\sum_{i_{k-3}=2k-2}^{n}\sum_{i_{k-4}=2k-2}^{i_{k-3}}\cdots\sum_{i_1=2k-2}^{i_2}\alpha_{n+3-i_{k-3}}\alpha_{i_{k-3}+3-i_{k-4}}\cdots\alpha_{i_2+3-i_1}\alpha_{i_1+5-2k}$$

$$+B^{k-1}\sum_{i_{k-2}=2k-1}^{n}\sum_{i_{k-3}=2k-1}^{i_{k-2}}\cdots\sum_{i_1=2k-1}^{i_2}\alpha_{n+3-i_{k-2}}\alpha_{i_{k-2}+3-i_{k-3}}\cdots$$

$$\alpha_{i_2+3-i_1}\alpha_{i_1+4-2k}\Big)\alpha'_k\Big)e_n\Big).$$

The similar expression for $[e'_1, e'_1]$ can be easily obtained by substitution in $[e'_0, e'_1]$ the coefficient α'_n instead of θ':

$$[e'_1, e'_1] \;=(A\;+\;B)\Bigg(A^2\alpha'_3e_3\;+\;\sum_{t=4}^{n}\Bigg(A^{t-1}\alpha'_t\;+\;\sum_{k=3}^{t-1}\Bigg(\binom{k-1}{k-2}$$

$$\cdot A^{k-2}B\alpha_{t+2-k}+\binom{k-1}{k-3}A^{k-3}B^2\sum_{i_1=k+2}^{t}\alpha_{t+3-i_1}\alpha_{i_1+1-k}$$

$$+\binom{k-1}{k-4}A^{k-4}B^3\sum_{i_2=k+3}^{t}\sum_{i_1=k+3}^{i_2}\alpha_{t+3-i_2}\alpha_{i_2+3-i_1}\alpha_{i_1-k}+\binom{k-1}{1}AB^{k-2}$$

$$\cdot\sum_{i_{k-3}=2k-2}^{t}\sum_{i_{k-4}=2k-2}^{i_{k-3}}\cdots\sum_{i_1=2k-2}^{i_2}\alpha_{t+3-i_{k-3}}\alpha_{i_{k-3}+3-i_{k-4}}\cdots\alpha_{i_2+3-i_1}\alpha_{i_1+5-2k}$$

$$+B^{k-1}\sum_{i_{k-2}=2k-1}^{t}\sum_{i_{k-3}=2k-1}^{i_{k-2}}\cdots\sum_{i_1=2k-1}^{i_2}\alpha_{t+3-i_{k-2}}\alpha_{i_{k-2}+3-i_{k-3}}\cdots$$

$$\alpha_{i_2+3-i_1}\alpha_{i_1+4-2k}\Big)\alpha'_k\Big)e_t\Big).$$

On the other hand, we have

$$[e'_0, e'_1] = [Ae_0 + Be_1, (A+B)e_1 + B(\theta-\alpha_n)e_{n-1}]$$

$$= (A+B)^2\sum_{t=3}^{n-1}\alpha_t e_t + (A+B)(A\theta + B\alpha_n)e_n,$$

$$[e'_1, e'_1] = [(A+B)e_1 + B(\theta-\alpha_n)e_{n-1}, (A+B)e_1 + B(\theta-\alpha_n)e_{n-1}]$$

$$= (A+B)\sum_{t=3}^{n}\alpha_t e_t.$$

Comparing the coefficients of the basis elements e_t and keeping in mind that the coefficient of $A+B$ should not be zero, we get the restrictions, that were outlined in the first assertion of the theorem.

Using Corollary 5.2, the assertion b) of the theorem is proved by applying similar arguments. □

Remark. From Theorem 5.1, we have that α'_k is a polynomial of the form $P_k(A, B, \alpha_3, \alpha_4, \ldots, \alpha_k, \alpha'_3, \alpha'_4, \ldots, \alpha'_{k-1})$, where the parameters $\alpha_3, \alpha_4, \ldots, \alpha_k, \alpha'_3, \alpha'_4, \ldots, \alpha'_{k-1}$ are given and coefficients A, B are unknown, but satisfy the condition $A(A+B) \neq 0$. And β'_k is also a polynomial of the form $Q_k(A, B, D, \beta_3, \beta_4, \ldots, \beta_k, \beta'_3, \beta'_4, \ldots, \beta'_{k-1})$, where the parameters $\beta_3, \beta_4, \ldots, \beta_k, \beta'_3, \beta'_4, \ldots, \beta'_{k-1}$ are given and coefficients A, B, D are unknown, but satisfy the condition $AD \neq 0$. Therefore, the calculation of parameters α'_k and β'_k is a recursive procedure. Consequently, we conclude that in any given dimension the problem of the classification (up to an isomorphism) of complex filiform Leibniz algebras, which are obtained from the naturally graded filiform non-Lie algebras, is an algorithmically solvable task.

5.2.1 Isomorphism criteria

In this section we give a few versions of the isomorphism criteria discussed in the previous section. We write them in an invariant form adapted for application to the classification problem of filiform Leibniz algebras in low dimensions.

Let us first treat the classes FLb_{n+1} and SLb_{n+1}. The action of $GL_{ad}(V)$ on LB_n induces the actions on FLb_{n+1} and SLb_{n+1}. Due to Proposition 5.3 the action can be written as a composition of elementary transformations of the second and the fifth types, respectively.

The next two theorems are reformulations of the corresponding results of [88] on isomorphism criteria for filiform Leibniz algebras appearing from the naturally graded non Lie filiform Leibniz algebras.

Introduce the following series of functions:

$$\begin{aligned}
&\varphi_t(y; z) = \varphi_t(y; z_3, z_4, \ldots, z_n, z_{n+1}) \\
&= (1+y)\, z_t - \sum_{k=3}^{t-1} \left(\binom{k-1}{k-2} y z_{t+2-k} + \binom{k-1}{k-3} y^2 \sum_{i_1=k+2}^{t} z_{t+3-i_1}\, z_{i_1+1-k} + \right. \\
&+ \binom{k-1}{k-4} y^3 \sum_{i_2=k+3}^{t} \sum_{i_1=k+3}^{i_2} z_{t+3-i_2}\, z_{i_2+3-i_1}\, z_{i_1-k} + \ldots + \binom{k-1}{1} y^{k-2} \\
&\cdot \sum_{i_{k-3}=2k-2}^{t} \sum_{i_{k-4}=2k-2}^{i_{k-3}} \cdots \sum_{i_1=2k-2}^{i_2} z_{t+3-i_{k-3}}\, z_{i_{k-3}+3-i_{k-4}} \cdots z_{i_2+3-i_1}\, z_{i_1+5-2k} + y^{k-1} \\
&\cdot \left. \sum_{i_{k-2}=2k-1}^{t} \sum_{i_{k-3}=2k-1}^{i_{k-2}} \cdots \sum_{i_1=2k-1}^{i_2} z_{t+3-i_{k-2}}\, z_{i_{k-2}+3-i_{k-3}} \cdots z_{i_2+3-i_1}\, z_{i_1+4-2k} \right) \varphi_k(y; z), \\
&\hspace{20em} \text{for } 3 \leq t \leq n
\end{aligned}$$

and

$$\varphi_{n+1}(y; z) = \left[(z_{n+1} - z_n) + (1+y)\varphi_n(y; z) \right].$$

Theorem 5.2. *Two algebras $L(\alpha)$ and $L(\alpha')$ from* FLb_{n+1}, *where* $\alpha = (\alpha_3, \alpha_4, \ldots, \alpha_n, \theta)$, *and* $\alpha' = (\alpha'_3, \alpha'_4, \ldots, \alpha'_n, \theta')$, *are isomorphic if and only if there exist complex numbers A and B such that $A(A+B) \neq 0$ and the following conditions hold:*

$$\alpha'_t = \frac{1}{A^{t-2}}\varphi_t\left(\frac{B}{A};\alpha\right), \quad 3 \le t \le n,$$

$$\theta' = \frac{1}{A^{n-2}}\varphi_{n+1}\left(\frac{B}{A};\alpha\right),$$

(see Theorem 5.1 a)).

Let

$$\psi_t(y;z) = \psi_t(y;z_3,z_4,\ldots,z_n,z_{n+1})$$

$$= z_t - \sum_{k=3}^{t-1}\left(\binom{k-1}{k-2} y z_{t+2-k} + \binom{k-1}{k-3} y^2 \sum_{i_1=k+2}^{t} z_{t+3-i_1} z_{i_1+1-k}\right.$$

$$+\binom{k-1}{k-4} y^3 \sum_{i_2=k+3}^{t}\sum_{i_1=k+3}^{i_2} z_{t+3-i_2} z_{i_2+3-i_1} z_{i_1-k} + \cdots + \binom{k-1}{1} y^{k-2}$$

$$\cdot \sum_{i_{k-3}=2k-2}^{t}\sum_{i_{k-4}=2k-2}^{i_{k-3}}\cdots\sum_{i_1=2k-2}^{i_2} z_{t+3-i_{k-3}} z_{i_{k-3}+3-i_{k-4}} \cdots z_{i_2+3-i_1} z_{i_1+5-2k} + y^{k-1}$$

$$\left.\cdot \sum_{i_{k-2}=2k-1}^{t}\sum_{i_{k-3}=2k-1}^{i_{k-2}}\cdots\sum_{i_1=2k-1}^{i_2} z_{t+3-i_{k-2}} z_{i_{k-2}+3-i_{k-3}} \cdots z_{i_2+3-i_1} z_{i_1+4-2k}\right)\psi_k(y;z),$$

where $3 \le t \le n$,

and

$$\psi_{n+1}(y;z) = z_{n+1}.$$

Theorem 5.3. *Two algebras $L(\beta)$ and $L(\beta')$ from* SLb_{n+1}, *where* $\beta = (\beta_3, \beta_4, \ldots, \beta_n, \gamma)$, *and* $\beta' = (\beta'_3, \beta'_4, \ldots, \beta'_n, \gamma')$, *are isomorphic if and only if there exist complex numbers A, B and D such that $AD \neq 0$ and the following conditions hold:*

$$\beta'_t = \frac{1}{A^{t-2}}\frac{D}{A}\psi_t\left(\frac{B}{A};\beta\right), \quad 3 \le t \le n-1,$$

$$\beta'_n = \frac{1}{A^{n-2}}\frac{D}{A}\frac{B}{A}\gamma + \psi_n\left(\frac{B}{A};\beta\right),$$

and

$$\gamma' = \frac{1}{A^{n-2}}\left(\frac{D}{A}\right)^2 \psi_{n+1}\left(\frac{B}{A};\beta\right),$$

(see Theorem 5.1 b)).

Now we shall present the classification procedure for Lb_{n+1}.

To simplify notation let us agree that in the above case for transition from $L(\alpha)$ to $L(\alpha')$ and from $L(\beta)$ to $L(\beta')$ we write as $\alpha' = \rho\left(\frac{1}{A}, \frac{B}{A};\alpha\right)$ and $\beta' = \nu\left(\frac{1}{A}, \frac{B}{A}, \frac{D}{A};\beta\right)$, respectively, where

$$\rho\left(\frac{1}{A}, \frac{B}{A};\alpha\right) = \left(\rho_1\left(\frac{1}{A}, \frac{B}{A};\alpha\right), \rho_2\left(\frac{1}{A}, \frac{B}{A};\alpha\right), \ldots, \rho_{n-1}\left(\frac{1}{A}, \frac{B}{A};\alpha\right)\right),$$

with $\rho_t(x,y;z) = x^t\varphi_{t+2}(y;z)$ for $1 \le t \le n-2$, $\rho_{n-1}(x,y;z) = x^{n-2}\varphi_{n+1}(y;z)$

and

$$\nu\left(\tfrac{1}{A},\tfrac{B}{A},\tfrac{D}{A};\beta\right) = \left(\nu_1\left(\tfrac{1}{A},\tfrac{B}{A},\tfrac{D}{A};\beta\right),\nu_2\left(\tfrac{1}{A},\tfrac{B}{A},\tfrac{D}{A};\beta\right),\dots,\nu_{n-1}\left(\tfrac{1}{A},\tfrac{B}{A},\tfrac{D}{A};\beta\right)\right),$$

with

$$\begin{aligned}\nu_t(x,y,v;z) &= x^t v\psi_{t+2}(y;z) \text{ for } 1 \le t \le n-3,\\ \nu_{n-2}(x,y,v;z) &= x^{n-2}v(yz_{n+1}+\psi_n(y;z)),\\ \nu_{n-1}(x,y,v;z) &= x^{n-2}\psi_{n+1}(y;z),\end{aligned}$$

Here are main properties of the operators ρ and ν used in what follows.

Proposition 5.4.

1. $\rho(1,0;\cdot)$ *is the identity operator.*
2. $\rho\left(\frac{1}{A_2},\frac{B_2}{A_2};\rho\left(\frac{1}{A_1},\frac{B_1}{A_1};\alpha\right)\right) = \rho\left(\frac{1}{A_1A_2},\frac{A_1B_2+A_2B_1+B_1B_2}{A_1A_2};\alpha\right).$
3. *If* $\alpha' = \rho\left(\frac{1}{A},\frac{B}{A};\alpha\right)$ *then* $\alpha = \rho\left(A,-\frac{B}{A+B};\alpha'\right).$
4. $\nu(1,0,1;\cdot)$ *is the identity operator.*
5. $\nu\left(\frac{1}{A_2},\frac{B_2}{A_2},\frac{D_2}{A_2};\nu\left(\frac{1}{A_1},\frac{B_1}{A_1},\frac{D_1}{A_1};\beta\right)\right) = \nu\left(\frac{1}{A_1A_2},\frac{B_1A_2+B_2D_1}{A_1A_2},\frac{D_1D_2}{A_1A_2};\beta\right).$
6. *If* $\beta' = \nu\left(\frac{1}{A},\frac{B}{A},\frac{D}{A};\beta\right)$ *then* $\beta = \nu\left(A,-\frac{B}{D},\frac{A}{D};\beta'\right).$

From now on we assume that $n \ge 4$ since there are complete classifications of complex nilpotent Leibniz algebras of dimension at most four in Albeverio et al. [13].

5.2.2 Classification procedure

We remind the reader of the following definition of the action of an algebraic group on an algebraic variety.

Definition 5.4. *An action of an algebraic group G on a variety X is a morphism $\sigma : G \times X \longrightarrow X$ with*

(*i*) $\sigma(e,x) = x$, *where* e *is an identity element of* G *and* $x \in X$,
(*ii*) $\sigma(g,\sigma(h,x)) = \sigma(gh,x)$, *for all* $g,h \in G$ *and* $x \in X$.

We write gx for $\sigma(g,x)$, and call X a G-variety.

Definition 5.5. *A function $f : X \longrightarrow \mathbb{F}$ is said to be invariant if $f(gx) = f(x)$ for any $g \in G$ and $x \in X$.*

We consider the case when $G = G_{ad}$ and $X = Lb_{n+1}$. Then the orbits with respect to the action of $G = G_{ad}$ on $X = Lb_{n+1}$ consist of mutually isomorphic algebras.

5.2.2.1 *Classification algorithm and invariants for* FLb_{n+1}

Consider the following representation of $\mathrm{FLb}_{n+1} = U \cup F$, where

$$U = \{L(\alpha) \in \mathrm{FLb}_{n+1} : \alpha_3 \neq 0\} \text{ and } F = \{L(\alpha) \in \mathrm{FLb}_{n+1} : \alpha_3 = 0\}.$$

Then U can be represented as a disjoint union of the subsets

$$U_1 = \{L(\alpha) \in U : \alpha_4 \neq -2\alpha_3^2\} \text{ and } F_1 = \{L(\alpha) \in U : \alpha_4 = -2\alpha_3^2\}.$$

Theorem 5.4.

i) *Two algebras $L(\alpha)$ and $L(\alpha')$ from U_1 are isomorphic if and only if*

$$\rho_i\left(\frac{2\alpha_3}{\alpha_4 + 2\alpha_3^2}, \frac{\alpha_4}{2\alpha_3^2}; \alpha\right) = \rho_i\left(\frac{2\alpha_3'}{\alpha_4' + 2\alpha_3'^2}, \frac{\alpha_4'}{2\alpha_3'^2}; \alpha'\right),$$

whenever $i = 3, 4, \ldots, n-1.$

ii) *For any $(a_3, a_4, \ldots, a_{n-1}) \in \mathbb{C}^{n-3}$ there is an algebra $L(\alpha)$ from U_1 such that*

$$\rho_i\left(\frac{2\alpha_3}{\alpha_4 + 2\alpha_3^2}, \frac{\alpha_4}{2\alpha_3^2}; \alpha\right) = a_i, \quad i = 3, 4, \ldots, n-1.$$

Proof. i). "If" part. Let two algebras $L(\alpha)$ and $L(\alpha')$ be isomorphic. Then there exist $A, B \in \mathbb{C}$ such that $A(A+B) \neq 0$ and $\alpha' = \rho\left(\frac{1}{A}, \frac{B}{A}; \alpha\right)$. Hence, making use of Properties 5.4 we have

$$\alpha = \rho\left(A, \frac{-B}{A+B}; \alpha'\right).$$

Consider the algebra $L(\alpha^0)$, where

$$\alpha^0 = \rho\left(\frac{1}{A_0}, \frac{B_0}{A_0}; \alpha\right) \text{ and } A_0 = \frac{\alpha_4 + 2\alpha_3^2}{2\alpha_3}, \ B_0 = \frac{\alpha_4(\alpha_4 + 2\alpha_3^2)}{4\alpha_3^3}.$$

Then again due to Properties 5.4 we get

$$\begin{aligned} \alpha^0 &= \rho\left(\tfrac{2\alpha_3}{\alpha_4+2\alpha_3^2}, \tfrac{\alpha_4}{2\alpha_3^2}; \alpha\right) \\ &= \rho\left(\tfrac{1}{A_0}, \tfrac{B_0}{A_0}; \rho\left(A, \tfrac{-B}{A+B}; \alpha'\right)\right) \\ &= \rho\left(\tfrac{A}{A_0}, \tfrac{B_0 A - A_0 B}{A(A+B)}; \alpha'\right). \end{aligned}$$

It is easy to check that

$$\frac{A}{A_0} = \frac{2\alpha'_3}{\alpha'_4 + 2\alpha'^2_3} \text{ and } \frac{B_0A - A_0B}{A(A+B)} = \frac{\alpha'_4}{2\alpha'^2_3}.$$

Therefore,

$$\rho\left(\frac{2\alpha_3}{\alpha_4 + 2\alpha_3^2}, \frac{\alpha_4}{2\alpha_3^2}; \alpha\right) = \rho\left(\frac{2\alpha'_3}{\alpha'_4 + 2\alpha'^2_3}, \frac{\alpha'_4}{2\alpha'^2_3}; \alpha'\right)$$

and, hence,

$$\rho_i\left(\frac{2\alpha_3}{\alpha_4 + 2\alpha_3^2}, \frac{\alpha_4}{2\alpha_3^2}; \alpha\right) = \rho_i\left(\frac{2\alpha'_3}{\alpha'_4 + 2\alpha'^2_3}, \frac{\alpha'_4}{2\alpha'^2_3}; \alpha'\right),$$

for all $i = 3, 4, \dots, n-1$.

This procedure can be shown schematically as follows:

$$\begin{array}{ccc} \alpha & \overset{(A_0,B_0)}{\longrightarrow} & \alpha^0 \\ (A,B) \searrow & & \nearrow \left(A_0A^{-1}, \frac{B_0A - A_0B}{A_0(A+B)}\right) \\ & \alpha' & \end{array}$$

"Only if" part. Let the equalities $\rho_i\left(\frac{2\alpha_3}{\alpha_4+2\alpha_3^2}, \frac{\alpha_4}{2\alpha_3^2}; \alpha\right) = \rho_i\left(\frac{2\alpha'_3}{\alpha'_4+2\alpha'^2_3}, \frac{\alpha'_4}{2\alpha'^2_3}; \alpha'\right)$, $i = 3, 4, \dots, n-1$ hold. Then it is easy to see that

$$\rho_i\left(\frac{2\alpha_3}{\alpha_4 + 2\alpha_3^2}, \frac{\alpha_4}{2\alpha_3^2}; \alpha\right) = \rho_i\left(\frac{2\alpha'_3}{\alpha'_4 + 2\alpha'^2_3}, \frac{\alpha'_4}{2\alpha'^2_3}; \alpha'\right) \text{ for } i = 1, 2$$

as well and therefore,

$$\rho\left(\frac{2\alpha_3}{\alpha_4 + 2\alpha_3^2}, \frac{\alpha_4}{2\alpha_3^2}; \alpha\right) = \rho\left(\frac{2\alpha'_3}{\alpha'_4 + 2\alpha'^2_3}, \frac{\alpha'_4}{2\alpha'^2_3}; \alpha'\right)$$

that means the algebras $L(\alpha)$ and $L(\alpha')$ are isomorphic.

ii). The system of equations

$$\rho_i\left(\frac{2\alpha_3}{\alpha_4 + 2\alpha_3^2}, \frac{\alpha_4}{2\alpha_3^2}; \alpha\right) = a_i, \quad 3 \le i \le n-1, \tag{5.7}$$

where $(a_3, a_4, \dots, a_{n-1})$ is given and $\alpha = (\alpha_3, \alpha_4, \dots, \alpha_{n-1}, \theta)$ is unknown, has a solution as far as for any $3 \le i \le n-1$ in $\rho_i\left(\frac{2\alpha_3}{\alpha_4+2\alpha_3^2}, \frac{\alpha_4}{2\alpha_3^2}; \alpha\right)$

only variables $\alpha_3, \alpha_4, \ldots, \alpha_i$ occur and each of these equations is a linear equation with respect to its last variable. Hence, making each of α_i the subject of (5.7) for $i = 3, \ldots, n-1$, one can find the required algebra $L(\alpha)$. □

Here are the corresponding invariants for some low-dimensional cases.

Case of $n = 4$ i.e., **dim L=5:**

$$\rho_3\left(\frac{2\alpha_3}{\alpha_4+2\alpha_3^2}, \frac{\alpha_4}{2\alpha_3^2}; \alpha\right) = \left(\frac{2\alpha_3}{\alpha_4+2\alpha_3^2}\right)^2 (\theta - \alpha_4).$$

Case of $n = 5$ i.e., **dim L=6:**

$$\rho_3\left(\frac{2\alpha_3}{\alpha_4+2\alpha_3^2}, \frac{\alpha_4}{2\alpha_3^2}; \alpha\right) = \frac{4\alpha_3\alpha_5-5\alpha_4^2}{(\alpha_4+2\alpha_3^2)^2}$$

$$\rho_4\left(\frac{2\alpha_3}{\alpha_4+2\alpha_3^2}, \frac{\alpha_4}{2\alpha_3^2}; \alpha\right) = \left(\frac{2\alpha_3}{\alpha_4+2\alpha_3^2}\right)^3 (\theta - \alpha_5) + \rho_3\left(\frac{2\alpha_3}{\alpha_4+2\alpha_3^2}, \frac{\alpha_4}{2\alpha_3^2}; \alpha\right).$$

Let us now consider the isomorphism criterion for F_1. This set in its turn can be written as a disjoint union of the subsets

$$V_1 = \{L(\alpha) \in F_1 : \alpha_5 \neq 5\alpha_3^3\} \text{ and } G_1 = \{L(\alpha) \in F_1 : \alpha_5 = 5\alpha_3^3\}.$$

Further, V_1 can be represented as a disjoint union of the subsets

$$U_2 = \{L(\alpha) \in V_1 : \alpha_6 + 6\alpha_3\alpha_5 - 16\alpha_3^4 \neq 0\}$$

and

$$G_2 = \{L(\alpha) \in V_1 : \alpha_6 + 6\alpha_3\alpha_5 - 16\alpha_3^4 = 0\}.$$

Then the isomorphism criterion for U_2 can be spelled out as follows.

Theorem 5.5. i) *Two algebras $L(\alpha)$ and $L(\alpha')$ from U_2 are isomorphic if and only if*

$$\rho_i\left(\frac{5\alpha_3^3-\alpha_5}{\alpha_6+6\alpha_3\alpha_5-16\alpha_3^4}, \frac{\alpha_6+7\alpha_3\alpha_5-21\alpha_3^4}{\alpha_3(5\alpha_3^3-\alpha_5)}; \alpha\right)$$
$$= \rho_i\left(\frac{5\alpha_3'^3-\alpha_5'}{\alpha_6'+6\alpha_3'\alpha_5'-16\alpha_3'^4}, \frac{\alpha_6'+7\alpha_3'\alpha_5'-21\alpha_3'^4}{\alpha_3'(5\alpha_3'^3-\alpha_5')}; \alpha'\right)$$

for $i = 4, \ldots, n-1.$

ii) *For any* $(a_4, \ldots, a_{n-1}) \in \mathbb{C}^{n-4}$ *there is an algebra* $L(\alpha)$ *from* U_2 *such that*

$$\rho_i\left(\frac{5\alpha_3^3 - \alpha_5}{\alpha_6 + 6\alpha_3\alpha_5 - 16\alpha_3^4}, \frac{\alpha_6 + 7\alpha_3\alpha_5 - 21\alpha_3^4}{\alpha_3\left(5\alpha_3^3 - \alpha_5\right)}; \alpha\right) = a_i$$

for all $i = 4, 5, \ldots, n-1$.

The proof can be carried out with minor changes in the proof of Theorem 5.4.

Here are the corresponding invariants for $n = 7$ case:

$$\begin{cases} \rho_4\left(\frac{5\alpha_3^3-\alpha_5}{\alpha_6+6\alpha_3\alpha_5-16\alpha_3^4}, \frac{\alpha_6+7\alpha_3\alpha_5-21\alpha_3^4}{\alpha_3(5\alpha_3^3-\alpha_5)}; \alpha\right) = \frac{\left(5\alpha_3^3-\alpha_5\right)^3}{\alpha_3\left(\alpha_6+6\alpha_3\alpha_5-16\alpha_3^4\right)^2} \\ \\ \rho_5\left(\frac{5\alpha_3^3-\alpha_5}{\alpha_6+6\alpha_3\alpha_5-16\alpha_3^4}, \frac{\alpha_6+7\alpha_3\alpha_5-21\alpha_3^4}{\alpha_3\left(5\alpha_3^3-\alpha_5\right)}; \alpha\right) = \left(\frac{5\alpha_3^3-\alpha_5}{\alpha_6+6\alpha_3\alpha_5-16\alpha_3^4}\right)^4 \frac{\alpha_7+7\alpha_3\alpha_6-14\alpha_3^5}{\alpha_3} \\ \quad -\left(\frac{5\alpha_3^3-\alpha_5}{\alpha_6+6\alpha_3\alpha_5-16\alpha_3^4}\right)^3 \frac{42\alpha_3(\alpha_5-5\alpha_3^3)+7(\alpha_6+14\alpha_3^4}{\alpha_3} + \left(\frac{5\alpha_3^3-\alpha_5}{\alpha_6+6\alpha_3\alpha_5-16\alpha_3^4}\right)^2 \frac{28(\alpha_5-5\alpha_3^3)}{\alpha_3} \\ \\ \rho_6\left(\frac{5\alpha_3^3-\alpha_5}{\alpha_6+6\alpha_3\alpha_5-16\alpha_3^4}, \frac{\alpha_6+7\alpha_3\alpha_5-21\alpha_3^4}{\alpha_3(5\alpha_3^3-\alpha_5)}; \alpha\right) = \left(\frac{5\alpha_3^3-\alpha_5}{\alpha_6+6\alpha_3\alpha_5-16\alpha_3^4}\right)^5(\theta-\alpha_7) \\ \quad +\left(\frac{5\alpha_3^3-\alpha_5}{\alpha_6+6\alpha_3\alpha_5-16\alpha_3^4}\right)^4 \frac{\alpha_7+7\alpha_3\alpha_6-14\alpha_3^5}{\alpha_3} - \left(\frac{5\alpha_3^3-\alpha_5}{\alpha_6+6\alpha_3\alpha_5-16\alpha_3^4}\right)^3 \frac{42\alpha_3(\alpha_5-5\alpha_3^3)+7(\alpha_6+14\alpha_3^4}{\alpha_3} \\ \qquad + \left(\frac{5\alpha_3^3-\alpha_5}{\alpha_6+6\alpha_3\alpha_5-16\alpha_3^4}\right)^2 \frac{28(\alpha_5-5\alpha_3^3)}{\alpha_3}. \end{cases}$$

As for the subsets F, G_1 and G_2 the isomorphism criteria for them can be treated likewise.

5.2.2.2 *Classification algorithm and invariants for* SLb_{n+1}

In this section we treat SLb_{n+1}. The classification algorithm in this case works effectively as well. However, in this case instead of the representation ρ we have to use the representation ν (see Proposition 5.4).

Introduce the following notations. Let $x = (x_3, x_4, x_5, x_6)$ be a vector variable.

$\chi_1(x) = 4x_3^2x_6 - 12x_3x_4x_6 + x_4^3$;
$\chi_2(x) = 4x_3x_5 - 5x_4^2$;
$\chi_3(x) = x_3^2x_6 - 3x_3x_4x_5 + 2x_4^3$;
$\chi_4(x) = 4x_5x_3^2 - 5x_4^2x_3 + 2x_4\gamma$;
$\chi_5(x) = 2x_3^2x_6 - 6x_3x_4x_5 + x_4\gamma + 4x_4^3$

Consider the following presentation of SLb_{n+1}:

$$\mathrm{SLb}_{n+1} = U \cup F,$$

where

$$U = \{L(\beta) : \beta_3\chi_1(\beta)\chi_2(\beta) \neq 0\}, \quad \text{and} \quad F = \{L(\beta) : \beta_3\chi_1(\beta)\chi_2(\beta) = 0\}.$$

Theorem 5.6.

i) *Two algebras $L(\beta)$ and $L(\beta')$ from U are isomorphic if and only if*

$$\nu_i\left(\frac{\beta_3\chi_2(\beta)}{4\chi_3(\beta)}, \frac{\beta_4}{2\beta_3^2}, \frac{4\chi_3(\beta)}{\beta_3^2\chi_2(\beta)}; \beta\right) = \nu_i\left(\frac{\beta_3'\chi_2(\beta')}{4\chi_3(\beta')}, \frac{\beta_4'}{2\beta_3'^2}, \frac{4\chi_3(\beta')}{\beta_3'^2\chi_2(\beta')}; \beta'\right),$$

whenever $i = 3, 4, \ldots, n-1$.

ii) *For any $(\lambda_3, \lambda_4, \ldots, \lambda_{n-1}) \in \mathbb{C}^{n-3}$ there is an algebra $L(\beta)$ from U such that*

$$\nu_i\left(\frac{\beta_3\chi_2(\beta)}{4\chi_3(\beta)}, \frac{\beta_4}{2\beta_3^2}, \frac{4\chi_3(\beta)}{\beta_3^2\chi_2(\beta)}; \beta\right) = \lambda_i \ \textit{for all} \ i = 3, 4, \ldots, n-1.$$

Proof. i). Let $L(\beta)$ and $L(\beta')$ be isomorphic. Then there exist $A, B, D \in \mathbb{C}$ such that $AD \neq 0$ and $\beta' = \nu(\frac{1}{A}, \frac{B}{A}, \frac{D}{A}; \beta)$. Consider the algebra $L(\beta^0)$, where

$$\beta^0 = \nu\left(\frac{1}{A_0}, \frac{B_0}{A_0}, \frac{D_0}{A_0}; \beta\right)$$

with

$$A_0 = \frac{4\chi_3(\beta)}{\beta_3\chi_2(\beta)}, \quad B_0 = \frac{2\beta_4\chi_3(\beta)}{\beta_3^3\chi_2(\beta)} \quad \text{and} \quad D_0 = \frac{4\chi_3(\beta)}{\beta_3^2\chi_2(\beta)}.$$

Since $\beta = \nu\left(A, \frac{-B}{D}, \frac{A}{D}; \beta'\right)$ and

$$\begin{aligned}
\beta^0 &= \nu\left(\tfrac{1}{A_0}, \tfrac{B_0}{A_0}, \tfrac{D_0}{A_0}; \beta\right) \\
&= \nu\left(\tfrac{1}{A_0}, \tfrac{B_0}{A_0}, \tfrac{D_0}{A_0}; \nu\left(A, \tfrac{-B}{D}, \tfrac{A}{D}; \beta'\right)\right) \\
&= \nu\left(\tfrac{A}{A_0}, \tfrac{B_0A - A_0B}{A_0D}, \tfrac{D_0A}{A_0D}; \beta'\right).
\end{aligned}$$

One can easily check that $\frac{A}{A_0} = \frac{\beta'_3\chi_2(\beta')}{4\chi_3(\beta')}$, $\frac{B_0A - A_0B}{A_0D} = \frac{\beta'_4}{2\beta'^2_3}$, and $\frac{D_0A}{A_0D} = \frac{4\chi_3(\beta')}{\beta'^2_3\chi_2(\beta')}$.

Therefore

$$\nu\left(\frac{\beta_3\chi_2(\beta)}{4\chi_3(\beta)}, \frac{\beta_4}{2\beta_3^2}, \frac{4\chi_3(\beta)}{\beta_3^2\chi_2(\beta)}; \beta\right) = \nu\left(\frac{\beta'_3\chi_2(\beta')}{4\chi_3(\beta')}, \frac{\beta'_4}{2\beta'^2_3}, \frac{4\chi_3(\beta')}{\beta'^2_3\chi_2(\beta')}; \beta'\right),$$

and, in particular,

$$\nu_i\left(\frac{\beta_3\chi_2(\beta)}{4\chi_3(\beta)}, \frac{\beta_4}{2\beta_3^2}, \frac{4\chi_3(\beta)}{\beta_3^2\chi_2(\beta)}; \beta\right) = \nu_i\left(\frac{\beta'_3\chi_2(\beta')}{4\chi_3(\beta')}, \frac{\beta'_4}{2\beta'^2_3}, \frac{4\chi_3(\beta')}{\beta'^2_3\chi_2(\beta')}; \beta'\right),$$

for all $i = 3, 4, \ldots, n-1$.

This procedure can be shown schematically as follows

$$\begin{array}{ccc} \beta & \xrightarrow{(\frac{1}{A_0}, \frac{B_0}{A_0}, \frac{D_0}{A_0})} & \beta^0 \\ (\frac{1}{A}, \frac{B}{A}, \frac{D}{A}) \searrow & & \nearrow (\frac{A}{A_0}, \frac{B_0A - A_0B}{A_0D}, \frac{D_0A}{A_0D}) \\ & \beta' & \end{array}$$

Conversely, let the equalities

$$\nu_i\left(\frac{\beta_3\chi_2(\beta)}{4\chi_3(\beta)}, \frac{\beta_4}{2\beta_3^2}, \frac{4\chi_3(\beta)}{\beta_3^2\chi_2(\beta)}; \beta\right) = \nu_i\left(\frac{\beta'_3\chi_2(\beta')}{4\chi_3(\beta')}, \frac{\beta'_4}{2\beta'^2_3}, \frac{4\chi_3(\beta')}{\beta'^2_3\chi_2(\beta')}; \beta'\right),$$

hold for $i = 3, 4, \ldots, n-1$. Then it is easy to see that

$$\nu_i\left(\frac{\beta_3\chi_2(\beta)}{4\chi_3(\beta)}, \frac{\beta_4}{2\beta_3^2}, \frac{4\chi_3(\beta)}{\beta_3^2\chi_2(\beta)}; \beta\right) = \nu_i\left(\frac{\beta'_3\chi_2(\beta')}{4\chi_3(\beta')}, \frac{\beta'_4}{2\beta'^2_3}, \frac{4\chi_3(\beta')}{\beta'^2_3\chi_2(\beta')}; \beta'\right),$$

for $i = 1, 2$ as well and, therefore

$$\nu\left(\frac{\beta_3\chi_2(\beta)}{4\chi_3(\beta)}, \frac{\beta_4}{2\beta_3^2}, \frac{4\chi_3(\beta)}{\beta_3^2\chi_2(\beta)}; \beta\right) = \nu\left(\frac{\beta'_3\chi_2(\beta')}{4\chi_3(\beta')}, \frac{\beta'_4}{2\beta'^2_3}, \frac{4\chi_3(\beta')}{\beta'^2_3\chi_2(\beta')}; \beta'\right),$$

that means that the algebras $L(\beta)$ and $L(\beta')$ are isomorphic.

The proof of Part ii), is similar to that of Theorem 5.4. □

Here are examples of invariants for some low-dimensional cases.
Let $\dim L = 6$:

$$\nu_3\left(\frac{\beta_3^2}{\gamma}, \frac{\beta_4}{2\beta_3^2}, \frac{1}{\beta_3^3}; \beta\right) = \frac{\beta_3\chi_4(\beta)}{4\gamma^2}.$$

Let $\dim L = 7$:

$$\nu_3\left(\frac{\beta_3\chi_2(\beta)}{2\chi_5(\beta)}, \frac{\beta_4}{2\beta_3^2}, \frac{2\chi_5(\beta)}{\beta_3^2\chi_2(\beta)}; \beta\right) = \nu_4\left(\frac{\beta_3\chi_2(\beta)}{2\chi_5(\beta)}, \frac{\beta_4}{2\beta_3^2}, \frac{2\chi_5(\beta)}{\beta_3^2\chi_2(\beta)}; \beta\right) = \frac{\chi_2^3(\beta)}{16\chi_5^2(\beta)}$$

$$\nu_5\left(\frac{\beta_3^2}{\gamma}, \frac{\beta_4}{2\beta_3^2}, \frac{1}{\beta_3^3}; \beta\right) = \frac{\gamma\chi_2^2(\beta)}{4\chi_5^2(\beta)}.$$

Let $\dim L = 8$:

$$\nu_3\left(\frac{\beta_3\chi_2(\beta)}{2\chi_5(\beta)}, \frac{\beta_4}{2\beta_3^2}, \frac{2\chi_5(\beta)}{\beta_3^2\chi_2(\beta)}; \beta\right) = \nu_4\left(\frac{\beta_3\chi_2(\beta)}{2\chi_5(\beta)}, \frac{\beta_4}{2\beta_3^2}, \frac{2\chi_5(\beta)}{\beta_3^2\chi_2(\beta)}; \beta\right) = \frac{\chi_2^3(\beta)}{16\chi_5^2(\beta)}.$$

$$\nu_5\left(\frac{\beta_3\chi_2(\beta)}{2\chi_5(\beta)}, \frac{\beta_4}{2\beta_3^2}, \frac{2\chi_5(\beta)}{\beta_3^2\chi_2(\beta)}; \beta\right) = \frac{\chi_2^4(\beta)\chi_6(\beta)}{(\chi_5(\beta) - \beta_4\gamma)}.$$

$$\nu_6\left(\frac{\beta_3\chi_2(\beta)}{2\chi_5(\beta)}, \frac{\beta_4}{2\beta_3^2}, \frac{2\chi_5(\beta)}{\beta_3^2\chi_2(\beta)}; \beta\right) = \frac{\beta_3\gamma\chi_2^3(\beta)}{(\chi_5(\beta) - \beta_4\gamma))^3}.$$

In regard to the set F, it can be split into its subsets and the algorithm can be applied with ν, instead of ρ, by using the properties of ν (see Proposition 5.4). Applications of the procedure are demonstrated in Chapter 6.

5.2.2.3 *Classification algorithm and invariants for* TLb_{n+1}

In this section we treat filiform Leibniz algebras whose natural gradation is an algebra from NGF_3. This class has been denoted by TLb_{n+1}. Here we investigate the behavior of structure constants under the base change in TLb_{n+1}. Recall that $(n+1)$–dimensional filiform Lie algebras are also in TLb_{n+1}.

Further,

$$L(\bar{a}) = L(b_{0,0}, b_{0,1}, b_{1,1}, a_{i,j}^1, \ldots, a_{i,j}^{n-(i+j+1)}, b_{i,j})$$

or sometimes just

$$\bar{a} = (b_{0,0}, b_{0,1}, b_{1,1}, a_{i,j}^1, \ldots, a_{i,j}^{n-(i+j+1)}, b_{i,j})$$

stand for an algebra from TLb_{n+1}, with the structure constants $b_{0,0}, b_{0,1}, b_{1,1}, a_{i,j}^1, \ldots, a_{i,j}^{n-(i+j+1)}, b_{i,j}$.

Let $L(\bar{a}') = L(b'_{0,0}, b'_{0,1}, b'_{1,1}, a_{i,j}^{1\,\prime}, \ldots, a_{i,j}^{n-(i+j+1)\,\prime}, b'_{i,j})$ be the image of $L(\bar{a})$ under an adapted basis change.

The proof of the following proposition with specified elements of G_{ad}, is straightforward.

Proposition 5.5. *Let $f \in G_{ad}$. Then f can be represented as follows:*

$$\begin{aligned} f(e_0) = e'_0 &= A_0 e_0 + A_1 e_1 + \cdots + A_n e_n, \\ f(e_1) = e'_1 &= \qquad\quad B_1 e_1 + \cdots + B_n e_n, \\ f(e_i) = e'_i &= [f(e_{i-1}), f(e_0)], \qquad 2 \le i \le n, \end{aligned}$$

where A_0, A_i, B_j, $(i, j = 1, \ldots, n)$ are complex numbers and $A_0\, B_1 (A_0 + A_1 b) \neq 0$.

The following elements of G_{ad} are said to be elementary with respect to the structure of $L \in \mathrm{TLb}_{n+1}$:

$$\sigma(b,k) = \begin{cases} f(e_0) = e_0, & \\ f(e_1) = e_1 + b\, e_k, & b \in \mathbb{C}, \quad 2 \le k \le n, \\ f(e_{i+1}) = [f(e_i), f(e_0)], & 1 \le i \le n-1, \end{cases}$$

$$\tau(a,k) = \begin{cases} f(e_0) = e_0 + a\, e_k, & a \in \mathbb{C}, \quad 1 \le k \le n, \\ f(e_1) = e_1, & \\ f(e_{i+1}) = [f(e_i), f(e_0)], & 1 \le i \le n-1, \end{cases}$$

$$\upsilon(a,b) = \begin{cases} f(e_0) = a\, e_0, & \\ f(e_1) = b\, e_1, & a, b \in \mathbb{C}^*, \\ f(e_{i+1}) = [f(e_i), f(e_0)], & 1 \le i \le n-1. \end{cases}$$

Proposition 5.6. *Let f be an adapted transformation of L. Then it can be represented as the following composition:*
$f = \tau(A_n, n) \circ \tau(A_{n-1}, n-1) \circ \cdots \circ \tau(A_2, 2) \circ \sigma(B_n, n) \circ \sigma(B_{n-1}, n-1) \circ \cdots \circ \sigma(B_2, 2) \circ \tau(A_1, 1) \circ \upsilon(A_0, B_1)$.

Proof. The proof is straightforward. □

Proposition 5.7. *The transformations*

$$\begin{aligned} g = \tau(A_n, n) \circ \tau(A_{n-1}, n-1) \circ \cdots \circ \tau(A_{n-4}, n-4) \\ \circ \sigma(B_n, n) \circ \sigma(B_{n-1}, n-1) \circ \cdots \circ \sigma(B_{n-3}, n-3) \end{aligned}$$

if n is even, and

$$g = \tau(A_n, n) \circ \tau(A_{n-1}, n-1) \circ \cdots \circ \tau(A_{n-3}, n-3)$$

$$\circ\sigma(B_n, n) \circ \sigma(B_{n-1}, n-1) \circ \cdots \circ \sigma(B_{n-2}, n-2)$$

for odd n, do not change the structure constants of these cases.

Lemma 5.2. *For $b'_{0,0}$, $b'_{0,1}$ and $b'_{1,1}$ the following hold:*

$$\begin{aligned} b'_{0,0} &= \frac{A_0^2 b_{0,0} + A_0 A_1 b_{0,1} + A_1^2 b_{1,1}}{A_0^{n-2} B_1 (A_0 + A_1\, b)} \\ b'_{0,1} &= \frac{A_0 b_{0,1} + 2A_1 b_{1,1}}{A_0^{n-2}(A_0 + A_1\, b)}, \\ b'_{1,1} &= \frac{B_1 b_{1,1}}{A_0^{n-2}(A_0 + A_1\, b)}. \end{aligned}$$

Proof. Consider the product $[f(e_0), f(e_0)] = b'_{0,0} f(e_n)$. Equating the coefficients of e_n in it we get

$$A_0^2 b_{0,0} + A_0 A_1 b_{0,1} + A_1^2 b_{1,1} = b'_{0,0} A_0^{n-2} B_1 (A_0 + A_1 b).$$

Then

$$b'_{0,0} = \frac{A_0^2 b_{0,0} + A_0 A_1 b_{0,1} + A_1^2 b_{1,1}}{A_0^{n-2} B_1 (A_0 + A_1 b)}.$$

The product $[f(e_1), f(e_1)] = b'_{1,1} f(e_n)$ yields

$$b'_{1,1} = \frac{B_1 b_{1,1}}{A_0^{n-2}(A_0 + A_1 b)}.$$

Consider the equality

$$b'_{0,1} f(e_n) = [f(e_1), f(e_0)] + [f(e_0), f(e_1)].$$

Then

$$b'_{0,1} A_0^{n-2} B_1 (A_0 + A_1 b) = A_0 B_1 b_{0,1} + 2A_1 B_1 b_{1,1}$$

and this implies that

$$b'_{0,1} = \frac{A_0 b_{0,1} + 2A_1 b_{1,1}}{A_0^{n-2}(A_0 + A_1 b)}.$$

□

The complete implementation of the procedure for some low-dimensional cases is given in the next chapters.

[illegible]

[illegible] do not change the structure constants [illegible]

Lemma 8.2. *For* [illegible] *the following* [illegible]

[illegible]

Proof. Consider the product [illegible] Equating the coefficients of [illegible] in both [illegible] we get

[illegible]

[illegible]

Consider the equality

[illegible]

Then

[illegible]

and this implies

[illegible]

∎

[illegible] complete implementation of the procedure [illegible] is given in the next chapter.

CHAPTER 6

CLASSIFICATION OF FILIFORM LEIBNIZ ALGEBRAS IN LOW DIMENSIONS

6.1 ISOMORPHISM CRITERIA FOR FIRST CLASS

We are now ready to obtain a complete classification of filiform Leibniz algebras in low dimensions. In this chapter we start with the class FLb_n. Considering the natural gradation on algebras from FLb_n we obtain the algebra NGF_1 (see [148]).

In accordance with results of [88] the class FLb_n is represented by the following table of multiplication in the so-called adapted basis $\{e_0, e_1, \ldots, e_n\}$:

$$\mathrm{FLb}_{n+1} := \begin{cases} [e_0, e_0] = e_2, \\ [e_i, e_0] = e_{i+1}, \quad 1 \leq i \leq n-1, \\ [e_0, e_1] = \alpha_3 e_3 + \alpha_4 e_4 + \cdots + \alpha_{n-1} e_{n-1} + \theta e_n, \\ [e_j, e_1] = \alpha_3 e_{j+2} + \alpha_4 e_{j+3} + \cdots + \alpha_{n+1-j} e_n, \\ \text{ where } 1 \leq j \leq n-2 \text{ and } \alpha_3, \alpha_4, \ldots, \alpha_n, \theta \in \mathbb{C}. \end{cases}$$

Recall the isomorphism criteria for FLb_{n+1} (see Theorem 5.5).

Criterion 1.

Introduce the following series of functions:

$$\varphi_t(y;z) = \varphi_t(y;z_3,z_4,\dots,z_n,z_{n+1})$$
$$= z_t - \sum_{k=3}^{t-1}\Bigg(\binom{k-1}{k-2} y z_{t+2-k} + \binom{k-1}{k-3} y^2 \sum_{i_1=k+2}^{t} z_{t+3-i_1} z_{i_1+1-k}$$
$$+ \binom{k-1}{k-4} y^3 \sum_{i_2=k+3}^{t} \sum_{i_1=k+3}^{i_2} z_{t+3-i_2} z_{i_2+3-i_1} z_{i_1-k} + \dots + \binom{k-1}{1} y^{k-2}$$
$$\cdot \sum_{i_{k-3}=2k-2}^{t} \sum_{i_{k-4}=2k-2}^{i_{k-3}} \dots \sum_{i_1=2k-2}^{i_2} z_{t+3-i_{k-3}} z_{i_{k-3}+3-i_{k-4}} \cdots z_{i_2+3-i_1} z_{i_1+5-2k} + y^{k-1}$$
$$\cdot \sum_{i_{k-2}=2k-1}^{t} \sum_{i_{k-3}=2k-1}^{i_{k-2}} \dots \sum_{i_1=2k-1}^{i_2} z_{t+3-i_{k-2}} z_{i_{k-2}+3-i_{k-3}} \cdots$$
$$z_{i_2+3-i_1} z_{i_1+4-2k}\Bigg) \varphi_k(y;z),$$
$$\text{for } 3 \le t \le n,$$

and

$$\varphi_{n+1}(y;z) = \left[(z_{n+1} - z_n) + (1+y)\varphi_n(y;z)\right].$$

Theorem 6.1. *Two algebras $L(\alpha)$ and $L(\alpha')$ from* FLb_{n+1}, *where* $\alpha = (\alpha_3,\alpha_4,\dots,\alpha_n,\theta)$, *and* $\alpha' = (\alpha'_3,\alpha'_4,\dots,\alpha'_n,\theta')$, *are isomorphic if and only if there exist complex numbers A and B such that $A(A+B) \neq 0$ and the following conditions hold:*

$$\alpha'_t = \frac{1}{A^{t-2}}\left(1+\frac{B}{A}\right)\varphi_t\left(\frac{B}{A};\alpha\right), \quad 3 \le t \le n, \quad \theta' = \frac{1}{A^{n-2}}\varphi_{n+1}\left(\frac{B}{A};\alpha\right).$$

We now demonstrate the systems of equalities for the cases of FLb_5 and FLb_6.

Case of $n=4$ i.e., $\dim L = 5$:

$$\alpha'_3 = \tfrac{1}{A}\left(1+\tfrac{B}{A}\right)\alpha_3 = \tfrac{1}{A}\left(1+\tfrac{B}{A}\right)\varphi_3\left(\tfrac{B}{A};\alpha\right),$$
$$\alpha'_4 = \tfrac{1}{A^2}\left(1+\tfrac{B}{A}\right)\left(\alpha_4 - 2\tfrac{B}{A}\alpha_3^2\right) = \tfrac{1}{A^2}\left(1+\tfrac{B}{A}\right)\varphi_4\left(\tfrac{B}{A};\alpha\right),$$
$$\theta' = \tfrac{1}{A^2}\left(\theta - \alpha_4 + \left(1+\tfrac{B}{A}\right)\left(\alpha_4 - 2\tfrac{B}{A}\alpha_3^2\right)\right) = \tfrac{1}{A^2}\varphi_5\left(\tfrac{B}{A};\alpha\right).$$

Case of $n=5$ i.e., $\dim L = 6$:

$$\alpha'_3 = \tfrac{1}{A}(1+\tfrac{B}{A})\alpha_3 = \tfrac{1}{A}\left(1+\tfrac{B}{A}\right)\varphi_3\left(\tfrac{B}{A};\alpha\right),$$
$$\alpha'_4 = \tfrac{1}{A^2}(1+\tfrac{B}{A})(\alpha_4 - 2\tfrac{B}{A}\alpha_3^2) = \tfrac{1}{A^2}\left(1+\tfrac{B}{A}\right)\varphi_4\left(\tfrac{B}{A};\alpha\right),$$

$$\alpha'_5 = \tfrac{1}{A^3}(1+\tfrac{B}{A})(\alpha_5 - 5\tfrac{B}{A}(\alpha_4 - \tfrac{B}{A}\alpha_3^2)\alpha_3 = \tfrac{1}{A^3}\left(1+\tfrac{B}{A}\right)\varphi_5\left(\tfrac{B}{A};\alpha\right),$$

$$\theta' = \tfrac{1}{A^3}[\theta - \alpha_5 + (1+\tfrac{B}{A})(\alpha_5 - 5\tfrac{B}{A}(\alpha_4 - \tfrac{B}{A}\alpha_3^2)\alpha_3)] = \tfrac{1}{A^2}\varphi_6\left(\tfrac{B}{A};\alpha\right).$$

Based on this isomorphism criterion we create another two equivalent criteria (Theorem 3.9 and Theorem 6.3) for FLb_{n+1}.

Criterion 2.

We introduce the notations:

$$\begin{array}{lll}
\Delta_3 = \alpha_3, & \Delta'_3 = \alpha'_3, & \Delta_4 = \alpha_4 + 2\alpha_3^2, \\
\Delta'_4 = \alpha'_4 + 2\alpha_3'^2, & \Delta_5 = \alpha_5 - 5\alpha_3^3, & \Delta'_5 = \alpha'_5 - 5\alpha_3'^3, \\
\Delta_6 = \alpha_6 + 14\alpha_3^4, & \Delta'_6 = \alpha'_6 + 14\alpha_3'^4, & \Delta_7 = \alpha_7 - 42\alpha_3^5, \\
\Delta'_7 = \alpha'_7 - 42\alpha_3'^5, & \Delta_8 = \alpha_8 + 132\alpha_3^6, & \Delta'_8 = \alpha'_8 + 132\alpha_3'^6, \\
\Theta_8 = \theta - \alpha_8, & \Theta'_8 = \theta' - \alpha'_8. &
\end{array} \tag{6.1}$$

In general,

$$\Delta_s = \alpha_s + (-1)^s C_{s-2}\alpha_3^{s-2}, \text{ and } \Delta'_s = \alpha'_s + (-1)^{s-2} C_{s-2}\alpha_3'^{s-2},$$

$$\Theta_s = \theta - \alpha_s, \text{ and } \Theta'_s = \theta' - \alpha'_s,$$

where $s = 4, 5, \ldots$. and C_k are Catalan numbers, i.e., $C_k = \frac{(2k)!}{(k+1)!k!}$, $k = 0, 1, 2, \ldots$.

Remark. Under these notations the structure constants $\alpha_3, \alpha_4, \ldots, \alpha_n, \theta$ are transformed into $\Delta_3, \Delta_4, \Delta_5, \ldots, \Delta_n, \Theta_n$ and this gives rise to a transition from the adapted basis $\{e_0, e_1, \ldots, e_n\}$ into another adapted basis. We shall keep the notation $L(\Delta_3, \Delta_4, \Delta_5, \ldots, \Delta_n, \Theta_n)$ for algebras from FLb_{n+1}.

Let us define the following functions

$$f_t(y;z) = f_t(y; z_3, z_4, \ldots, z_n, z_{n+1})$$

$$= z_t - \sum_{k=3}^{t-1}\left(\binom{k-1}{k-2} y(z_{t+2-k} + (-1)^{t+1-k}C_{t-k}z_3^{t-k}) + \binom{k-1}{k-3} y^2\right.$$

$$\cdot \sum_{i_1=k+3}^{t} (z_{t+3-i_1} + (-1)^{t-i_1}C_{t+1-i_1}z_3^{t+1-i_1})\,(z_{i_1+1-k} + (-1)^{i_1-k}C_{i_1-1-k}z_3^{i_1-1-k})$$

$$+ \binom{k-1}{k-4} y^3 \sum_{i_2=k+3}^{t}\ \sum_{i_1=k+3}^{i_2} (z_{t+3-i_2} + (-1)^{t-i_2}C_{t+1-i_2}z_3^{t+1-i_2})(z_{i_2+3-i_1}$$

$$+(-1)^{i_2-i_1}C_{i_2+1-i_1}z_3^{i_2+1-i_1})(z_{i_1-k}+(-1)^{i_1-k-1}C_{i_1-k-2}z_3^{i_1-k-2})\dots+\binom{k-1}{1}$$

$$\cdot y^{k-2}\sum_{i_{k-3}=2k-2}^{t}\sum_{i_{k-4}=2k-2}^{i_{k-3}}\dots\sum_{i_1=2k-2}^{i_2}(z_{t+3-i_{k-3}}+(-1)^{t-i_{k-3}}C_{t+1-i_{k-3}}z_3^{t+1-i_{k-3}})$$

$$\cdot(z_{i_{k-3}+3-i_{k-4}}+(-1)^{i_{k-3}-i_{k-4}}C_{i_{k-3}+1-i_{k-4}}z_3^{i_{k-3}+1-i_{k-4}})\dots(z_{i_2+3-i_1}$$

$$+(-1)^{i_2-i_1}C_{i_2+1-i_1}z_3^{i_2+1-i_1})(z_{i_1+5-2k}+(-1)^{i_2-i_1}C_{i_2+3-i_1}z_3^{i_2+3-i_1})$$

$$+y^{k-1}\sum_{i_{k-2}=2k-1}^{t}\sum_{i_{k-3}=2k-1}^{i_{k-2}}\dots\sum_{i_1=2k-1}^{i_2}(z_{t+3-i_{k-2}}+(-1)^{t-i_{k-2}}C_{t+1-i_{k-2}}z_3^{t+1-i_{k-2}})$$

$$\cdot(z_{i_{k-2}+3-i_{k-3}}+(-1)^{i_{k-2}-i_{k-3}}C_{i_{k-2}+1-i_{k-3}}z_3^{i_{k-2}+1-i_{k-3}})\dots$$

$$\cdot(z_{i_2+3-i_1}+(-1)^{i_2-i_1}C_{i_2+1-i_1}z_3^{i_2+1-i_1})(z_{i_1+4-2k}+(-1)^{i_1+1-2k}C_{i_1+2-2k}z_3^{i_1+2-2k})\Big)$$

$$\cdot f_k(y;z)+(-1)^t(1+y)^{t-3}z_3^{t-2}-(-1)^{t-2}C_{t-2}z_3^{t-2},$$

$$\text{for } 3\le t\le n,$$

and

$$f_{n+1}(y;z)=z_{n+1}.$$

Then one has the following theorem.

Theorem 6.2. *Two algebras $L(\Delta)$ and $L(\Delta')$ from* FLb_{n+1}, *where* $\Delta=(\Delta_3,\Delta_4,\dots,\Delta_n,\Theta_n)$, *and* $\Delta'=(\Delta'_3,\Delta'_4,\dots,\Delta'_n,\Theta'_n)$, *are isomorphic if and only if there exist complex numbers A and B such that $A(A+B)\neq 0$ and the following conditions hold:*

$$\Delta'_t=\frac{1}{A^{t-2}}\left(1+\frac{B}{A}\right)f_t\left(\frac{B}{A};\Delta\right),\quad 3\le t\le n,\quad \Theta'_n=\frac{1}{A^{n-2}}\Theta_n. \tag{6.2}$$

Criterion 3.

In order to state the **Criterion 3** let us consider the following functions

$$f_t(z)=f_t(z_3,z_4,\dots,z_n,z_{n+1})=f_t(-1;z)=f_t(-1;z_3,z_4,\dots,z_n,z_{n+1})$$

$$=z_t+\sum_{k=3}^{t-1}\left(\binom{k-1}{k-2}(z_{t+2-k}+(-1)^{t+1-k}C_{t-k}z_3^{t-k})+\binom{k-1}{k-3}\right.$$

$$\cdot\sum_{i_1=k+2}^{t}(z_{t+3-i_1}+(-1)^{t-i_1}C_{t+1-i_1}z_3^{t+1-i_1})(z_{i_1+1-k}+(-1)^{i_1-k}C_{i_1-1-k}z_3^{i_1-1-k})$$

$$+\binom{k-1}{k-4}\sum_{i_2=k+3}^{t}\sum_{i_1=k+3}^{i_2}(z_{t+3-i_2}+(-1)^{t-i_2}C_{t+1-i_2}z_3^{t+1-i_2})(z_{i_2+3-i_1}+(-1)^{i_2-i_1}$$

$$\cdot C_{i_2+1-i_1}z_3^{i_2+1-i_1})(z_{i_1-k}+(-1)^{i_1-k-1}C_{i_1-k-2}z_3^{i_1-k-2})+\ldots+(-1)^{k-2}\binom{k-1}{1}$$

$$\cdot\sum_{i_{k-3}=2k-2}^{t}\sum_{i_{k-4}=2k-2}^{i_{k-3}}\cdots\sum_{i_1=2k-2}^{i_2}(z_{t+3-i_{k-3}}+(-1)^{t-i_{k-3}}C_{t+1-i_{k-3}}z_3^{t+1-i_{k-3}})$$

$$\cdot\ (z_{i_{k-3}+3-i_{k-4}}+(-1)^{i_{k-3}-i_{k-4}}C_{i_{k-3}+1-i_{k-4}}z_3^{i_{k-3}+1-i_{k-4}})\ldots(z_{i_2+3-i_1}$$

$$+(-1)^{i_2-i_1}C_{i_2+1-i_1}z_3^{i_2+1-i_1})(z_{i_1+5-2k}+(-1)^{i_2-i_1}C_{i_2+3-i_1}z_3^{i_2+3-i_1})$$

$$+(-1)^{k-1}\sum_{i_{k-2}=2k-1}^{t}\sum_{i_{k-3}=2k-1}^{i_{k-2}}\cdots\sum_{i_1=2k-1}^{i_2}(z_{t+3-i_{k-2}}+(-1)^{t-i_{k-2}}C_{t+1-i_{k-2}}z_3^{t+1-i_{k-2}})$$

$$\cdot\ (z_{i_{k-2}+3-i_{k-3}}+(-1)^{i_{k-2}-i_{k-3}}C_{i_{k-2}+1-i_{k-3}}z_3^{i_{k-2}+1-i_{k-3}})\ldots$$

$$\cdot\ (z_{i_2+3-i_1}+(-1)^{i_2-i_1}C_{i_2+1-i_1}z_3^{i_2+1-i_1})(z_{i_1+4-2k}+(-1)^{i_1+1-2k}C_{i_1+2-2k}z_3^{i_1+2-2k})\Bigg)$$

$$\cdot f_k(z)+(-1)^{t-1}C_{t-2}z_3^{t-2},$$

for $3 \le t \le n$,

and

$$f_{n+1}(z) = z_{n+1}.$$

Criterion 3 is now stated as follows.

Theorem 6.3. *Two algebras $L(\Delta)$ and $L(\Delta')$ from* FLb_{n+1}, *where $\Delta = (\Delta_3, \Delta_4, \ldots, \Delta_n, \Theta_n)$, and $\Delta' = (\Delta'_3, \Delta'_4, \ldots, \Delta'_n, \Theta'_n)$, are isomorphic if and only if there exist complex numbers A and B such that $A(A+B) \neq 0$ and the following conditions hold:*

$$f_t(\Delta') = \frac{1}{A^{t-2}}\left(1+\frac{B}{A}\right)f_t(\Delta),\ \ 3 \le t \le n \ \text{ and } \ \Theta'_n = \frac{1}{A^2}\Theta_n.$$

The equivalence of the criteria 1 and 2 can be shown by substitutions. The equivalence of the criteria 2 and 3 is based on the following properties of the Catalan numbers, which are new to the authors' best knowledge. These properties are of interest in their own right.

- $A_1(t) = \sum\limits_{k=3}^{t-1}\binom{k-1}{k-2}C_{t-k}C_{k-2} = \binom{t-3}{1}C_{t-2};$

- $$A_2(t, j_1) = \sum_{k=3}^{t-1} \binom{k-1}{k-2} C_{t-k} \sum_{l=3}^{k-1} \binom{l-1}{l-2} C_{k-l} C_{l-2} - \sum_{k=3}^{t-1} \binom{k-1}{k-3} \sum_{i=k+2}^{t} C_{t+1-i} C_{i-1-k} C_{k-2} = \binom{t-3}{2} C_{t-2};$$

- $$A_3(t, j_1, j_2) = \sum_{k=3}^{t-1} \binom{k-1}{k-2} \binom{k-3}{2} C_{t-k} C_{k-2} - \sum_{k=3}^{t-1} \binom{k-1}{k-3} \binom{k-3}{1} \sum_{i_1=k+2}^{t} C_{t+1-i_1} C_{i_1-1-k} C_{k-2} + \sum_{k=3}^{t-1} \binom{k-1}{k-4} \sum_{i_2=k+3}^{t} \sum_{i_1=k+3}^{i_2} C_{t+1-i_2} C_{i_2+1-i_1} C_{i_1-k-2} C_{k-2} = \binom{t-3}{3} C_{t-2};$$

- $$A_4(t, j_1, j_2, j_3) = \sum_{k=3}^{t-1} \binom{k-1}{k-2} \binom{k-3}{3} C_{t-k} C_{k-2} - \sum_{k=3}^{t-1} \binom{k-1}{k-3} \binom{k-3}{2} \sum_{i_1=k+2}^{t} C_{t+1-i_1} C_{i_1-1-k} C_{k-2} + \sum_{k=3}^{t-1} \binom{k-1}{k-4} \binom{k-3}{1} \sum_{i_2=k+3}^{t} \sum_{i_1=k+3}^{i_2} C_{t+1-i_2} C_{i_2+1-i_1} C_{i_1-k-2} C_{k-2} - \sum_{k=3}^{t-1} \binom{k-1}{k-5} \sum_{i_3=k+4}^{t} \sum_{i_2=k+4}^{i_3} \sum_{i_1=k+4}^{i_2} C_{t+1-i_3} C_{i_3+1-i_2} C_{i_2+1-i_1} C_{i_1-k-2} C_{k-2} = \binom{t-3}{4} C_{t-2}.$$

In general,

- $$A_k(t, j_1, j_2, \dots, j_{k-1}) = \sum_{j_1=3}^{t-1} \binom{j_1-1}{j_1-2} C_{t-j_1} A_{k-1}(t, j_1, j_2, \dots, j_{k-2}) - \sum_{j_1=3}^{t-1} \binom{j_1-1}{j_1-3} \sum_{i_1=j_1+2}^{t} C_{t+1-i_1} C_{i_1-1-j_1} A_{k-2}(t, j_1, j_2, \dots, j_{k-3}) + \sum_{j_1=3}^{t-1} \binom{j_1-1}{j_1-4} \sum_{i_2=j_1+3}^{t} \sum_{i_1=j_1+3}^{i_2} C_{t+1-i_2} C_{i_2+1-i_1} C_{i_1-j_1-2}$$

$$\cdot A_{k-3}(t, j_1, j_2, \ldots, j_{k-4}) + \cdots + (-1)^k \sum_{j_1=3}^{t-1} \binom{j_1 - 1}{j_1 - k} \sum_{i_{k-2}=j_1+k-1}^{t} \sum_{i_{k-3}=j_1+k-1}^{i_{k-2}}$$

$$\ldots \sum_{i_1=j_1+(k-1)}^{i_2} C_{t+1-i_{k-2}} C_{i_{k-1}+1-i_{k-3}} \ldots C_{i_2+1-i_1} C_{i_1-j_1-2} A_1(j_1)$$

$$+(-1)^{k+1} \sum_{j_1=3}^{t-1} \binom{j_1 - 1}{j_1 - (k+1)} \sum_{i_{k-2}=j_1+k}^{t} \sum_{i_{k-3}=j_1+k}^{i_{k-2}}$$

$$\ldots \sum_{i_1=j_1+k}^{i_2} C_{t+1-i_{k-2}} C_{i_{k-1}+1-i_{k-3}} \ldots C_{i_2+1-i_1} C_{i_1-j_1-2} C_{j_1-2}$$

$$= \binom{t-3}{k} C_{t-2}.$$

The details in dimensions ≤ 9 can be found in [174].

Remark. Further we equally use all the three criteria in our research.

From now on we assume that $n \geq 4$ is a positive integer, since there are complete classifications of complex nilpotent Leibniz algebras of dimension at most four [13], [27], [50], [57], [160] (see previous Chapters as well).

Remark. The same notation $\Delta' = \rho\left(\frac{1}{A}, \frac{B}{A}; \Delta\right)$ will be used for the transition from an $(n+1)$-dimensional filiform Leibniz algebra $L(\Delta)$ to $(n+1)$-dimensional filiform Leibniz algebra $L(\Delta')$.

Introduce the notation:

$$\Phi(z) = \Phi(z_3, z_4, z_5) = z_5 + 5z_3z_4.$$

6.2 CLASSIFICATION OF FIRST CLASS IN LOW DIMENSIONS

6.2.1 Classification in dimension five

According to (5.2) the isomorphism criterion for FLb_5 stands as follows.

Theorem 6.4. *Two algebras $L(\Delta)$ and $L(\Delta')$ from* FLb_5, *where $\Delta = (\Delta_3, \Delta_4, \Theta_4)$ and $\Delta' = (\Delta'_3, \Delta'_4, \Theta'_4)$, are isomorphic if and only if there exist complex numbers A and B such that $A(A+B) \neq 0$ and the following conditions hold:*

$$\Delta'_3 = \frac{1}{A}\left(1 + \frac{B}{A}\right)\Delta_3, \quad \Delta'_4 = \frac{1}{A^2}\left(1 + \frac{B}{A}\right)\Delta_4, \quad \Theta'_4 = \frac{1}{A^2}\Theta_4.$$

In order to describe the orbits of the adapted base change on FLb_5 we split FLb_5 into the following subsets:

$$
\begin{aligned}
U_1 &= \{L(\Delta) \in FLb_5 : \Delta_3 \neq 0, \Delta_4 \neq 0\},\\
U_2 &= \{L(\Delta) \in FLb_5 : \Delta_3 \neq 0, \Delta_4 = 0, \Theta_4 \neq 0\},\\
U_3 &= \{L(\Delta) \in FLb_5 : \Delta_3 \neq 0, \Delta_4 = 0, \Theta_4 = 0\},\\
U_4 &= \{L(\Delta) \in FLb_5 : \Delta_3 = 0, \Delta_4 \neq 0, \Theta_4 \neq 0\},\\
U_5 &= \{L(\Delta) \in FLb_5 : \Delta_3 = 0, \Delta_4 \neq 0, \Theta_4 = 0\},\\
U_6 &= \{L(\Delta) \in FLb_5 : \Delta_3 = 0, \Delta_4 = 0, \Theta_4 \neq 0\},\\
U_7 &= \{L(\Delta) \in FLb_5 : \Delta_3 = 0, \Delta_4 = 0, \Theta_4 = 0\}.
\end{aligned}
$$

It is obvious that $\{U_i\}, i = 1, \dots, 7$, is a partition of FLb_5.

The following proposition shows that U_1 is a union of infinitely many orbits and these orbits can be parameterized by $\mathbb{C}$.

Proposition 6.1.

i) *Two algebras $L(\Delta)$ and $L(\Delta')$ from U_1 are isomorphic if and only if*

$$\left(\frac{\Delta_3}{\Delta_4}\right)^2 \Theta_4 = \left(\frac{\Delta'_3}{\Delta'_4}\right)^2 \Theta'_4.$$

ii) *Orbits in U_1 can be parameterized by $\mathbb{C}$ and $L(1, 1, \lambda)$, $\lambda \in \mathbb{C}$, are representatives of the orbits.*

The following proposition is a description of the subsets U_i, $i = 2, \dots, 7$.

Proposition 6.2. *The subsets U_2, U_3, U_4, U_5, U_6 and U_7 are single orbits with representatives $L(1, 0, 1), L(1, 0, 0), L(0, 1, 1), L(0, 1, 0)$, $L(0, 0, 1)$ and $L(0, 0, 0)$, respectively.*

We summarize the above observations in the following classification theorem.

Theorem 6.5. *Let L be a non-Lie complex filiform Leibniz algebra in* FLb_5. *Then it is isomorphic to one of the following pairwise non-isomorphic Leibniz algebras:*

1. $L(0,0,0) := \ L_5^s = \{[e_0, e_0] = e_2, \ [e_i, e_0] = e_{i+1}, \ 1 \le i \le 3\}.$
2. $L(0,0,1) := L_5^s, \ [e_0, e_1] = e_4.$
3. $L(0,1,0) := L_5^s, \ [e_0, e_1] = e_4, \ [e_1, e_1] = e_4.$
4. $L(0,1,1) := L_5^s, \ [e_0, e_1] = 2e_4, \ [e_1, e_1] = e_4.$
5. $L(1,0,0) := L_5^s, \ [e_0, e_1] = e_3 - 2e_4,$
$$[e_1, e_1] = e_3 - 2e_4, \ [e_2, e_1] = e_4.$$
6. $L(1,0,1) := L_5^s, \ [e_0, e_1] = e_3 - e_4,$
$$[e_1, e_1] = e_3 - 2e_4, \ [e_2, e_1] = e_4.$$
7. $L(1,1,\lambda) := L_5^s, \ [e_0, e_1] = e_3 + (\lambda - 1)e_4, \ [e_1, e_1] = e_3 - e_4,$
$$[e_2, e_1] = e_4, \ \lambda \in \mathbb{C}.$$

Conclusion. The class FLb_5 consists of one parametric family and six isolated isomorphism classes of filiform Leibniz algebras.

6.2.2 Classification in dimension six

The isomorphism criterion for FLb_6 in terms of Δ-notations is written as follows.

Theorem 6.6. *Two algebras $L(\Delta)$ and $L(\Delta')$ from* FLb_6*, where $\Delta = (\Delta_3, \Delta_4, \Delta_5, \Theta_5)$ and $\Delta' = (\Delta'_3, \Delta'_4, \Delta'_5, \Theta'_5)$, are isomorphic if and only if there exist complex numbers A and B such that $A(A+B) \neq 0$ and the following conditions hold:*

$$\begin{array}{llll} \Delta'_3 & = \frac{1}{A}\left(1 + \frac{B}{A}\right)\Delta_3, & \Delta'_4 & = \frac{1}{A^2}\left(1 + \frac{B}{A}\right)\Delta_4, \\ \Phi(\Delta') & = \frac{1}{A^3}\left(1 + \frac{B}{A}\right)\Phi(\Delta), & \Theta'_5 & = \frac{1}{A^3}\Theta_5. \end{array}$$

We represent FLb_6 as a disjoint union of the following subsets:

$$\begin{array}{lll} U_1 & = & \{L(\Delta) \in \mathrm{FLb}_6 : \Delta_3 \neq 0, \Delta_4 \neq 0\}, \\ U_2 & = & \{L(\Delta) \in \mathrm{FLb}_6 : \Delta_3 \neq 0, \Delta_4 = 0, \Delta_5 \neq 0, \Theta_5 \neq 0\}, \\ U_3 & = & \{L(\Delta) \in \mathrm{FLb}_6 : \Delta_3 = 0, \Delta_4 \neq 0, \Delta_5 \neq 0\}, \\ U_6 & = & \{L(\Delta) \in \mathrm{FLb}_6 : \Delta_3 \neq 0, \Delta_4 = 0, \Delta_5 \neq 0, \Theta_5 = 0\}, \\ U_5 & = & \{L(\Delta) \in \mathrm{FLb}_6 : \Delta_3 \neq 0, \Delta_4 = 0, \Delta_5 = 0, \Theta_5 \neq 0\}, \\ U_6 & = & \{L(\Delta) \in \mathrm{FLb}_6 : \Delta_3 \neq 0, \Delta_4 = 0, \Delta_5 = 0, \Theta_5 = 0\}, \\ U_7 & = & \{L(\Delta) \in \mathrm{FLb}_6 : \Delta_3 = 0, \Delta_4 \neq 0, \Delta_5 = 0, \Theta_5 \neq 0\}, \\ U_8 & = & \{L(\Delta) \in \mathrm{FLb}_6 : \Delta_3 = 0, \Delta_4 \neq 0, \Delta_5 = 0, \Theta_5 = 0\}, \end{array}$$

$$\begin{aligned}
U_9 &= \{L(\Delta) \in \mathrm{FLb}_6 : \Delta_3 = 0, \Delta_4 = 0, \Delta_5 \neq 0, \Theta_5 \neq 0\},\\
U_{10} &= \{L(\Delta) \in \mathrm{FLb}_6 : \Delta_3 = 0, \Delta_4 = 0, \Delta_5 \neq 0, \Theta_5 = 0\},\\
U_{11} &= \{L(\Delta) \in \mathrm{FLb}_6 : \Delta_3 = 0, \Delta_4 = 0, \Delta_5 = 0, \Theta_5 \neq 0\},\\
U_{12} &= \{L(\Delta) \in \mathrm{FLb}_6 : \Delta_3 = 0, \Delta_4 = 0, \Delta_5 = 0, \Theta_5 = 0\}.
\end{aligned}$$

The following propositions show that each of these subsets is either a single orbit or a union of infinitely many orbits of the group G_{ad}. Indeed, the subsets U_1, U_2 and U_3 turn out to be a union of infinitely many orbits and they can be described as follows:

Proposition 6.3.

i) *Two algebras $L(\Delta)$ and $L(\Delta')$ from U_1 are isomorphic if and only if*

$$\frac{\Delta_3\Phi(\Delta)}{\Delta_4^2} = \frac{\Delta_3'\Phi(\Delta')}{\Delta_4'^2}, \quad \frac{\Delta_3^3\Theta_5}{\Delta_4^3} = \frac{\Delta_3'^3\Theta_5'}{\Delta_4'^3}.$$

ii) *Orbits in U_1 can be parameterized as $L(1, 1, \lambda_1, \lambda_2)$, $\lambda_1, \lambda_2 \in \mathbb{C}$.*

Proposition 6.4.

i) *Two algebras $L(\Delta)$ and $L(\Delta')$ from U_2 are isomorphic if and only if*

$$\frac{\Delta_5^3}{\Delta_3^3\Theta_5^2} = \frac{\Delta_5'^3}{\Delta_3'^3\Theta_5'^2}.$$

ii) *Orbits in U_2 can be parameterized as $L(1, 0, \lambda, \lambda)$, $\lambda \in \mathbb{C}^*$.*

Proposition 6.5.

i) *Two algebras $L(\Delta)$ and $L(\Delta')$ from U_3 are isomorphic if and only if*

$$\frac{\Delta_4^3\Theta_5}{\Delta_5^3} = \frac{\Delta_4'^3\Theta_5'}{\Delta_5'^3}.$$

ii) *Orbits in U_3 can be parameterized as $L(0, 1, 1, \lambda)$, $\lambda \in \mathbb{C}$.*

Each of the sets $U_4 - U_{12}$ is a single orbit and here is their description.

Proposition 6.6. *The subsets $U_4, U_5, U_6, U_7, U_8, U_9, U_{10}, U_{11}$ and U_{12} are single orbits with representatives $L(1, 0, 1, 0)$, $L(1, 0, 0, 1)$, $L(1, 0, 0, 0)$, $L(0, 1, 0, 1)$, $L(0, 1, 0, 0)$, $L(0, 0, 1, 1)$, $L(0, 0, 1, 0)$, $L(0, 0, 0, 1)$ and $L(0, 0, 0, 0)$, respectively.*

The result of all the observation above can be written down as follows:

Theorem 6.7. *Let L be a non-Lie complex filiform Leibniz algebra in* FLb_6*. Then it is isomorphic to one of the following pairwise non-isomorphic Leibniz algebras:*

1. $L(0,0,0,0) := \ L_6^s = \{[e_0,e_0] = e_2,\ \ [e_i,e_0] = e_{i+1},\ \ 1 \le i \le 4\}.$
2. $L(0,0,0,1) := \ L_6^s,\ \ [e_0,e_1] = e_5.$
3. $L(0,0,1,0) := \ L_6^s,\ \ [e_0,e_1] = e_5,\ \ [e_1,e_1] = e_5.$
4. $L(0,0,1,1) := \ L_6^s,\ \ [e_0,e_1] = 2e_5,\ \ [e_1,e_1] = e_5.$
5. $L(0,1,0,0) := \ L_6^s,\ \ [e_0,e_1] = e_4,\ \ [e_1,e_1] = e_4,\ \ [e_2,e_1] = e_5.$
6. $L(0,1,0,1) := \ L_6^s,\ \ [e_0,e_1] = e_4 + e_5,$ $[e_1,e_1] = e_4,\ \ [e_2,e_1] = e_5.$
7. $L(1,0,0,1) := \ L_6^s,\ \ [e_0,e_1] = e_3 - 2e_4 + 6e_5,$ $[e_1,e_1] = e_3 - 2e_4 + 5e_5, [e_2,e_1] = e_4 - 2e_5,\ \ [e_3,e_1] = e_5.$
8. $L(1,0,1,0) := \ L_6^s,\ \ [e_0,e_1] = e_3 - 2e_4 + 6e_5,$ $[e_1,e_1] = e_3 - 2e_4 + 6e_5, [e_2,e_1] = e_4 - 2e_5,\ \ [e_3,e_1] = e_5.$
9. $L(0,1,1,\lambda) := \ L_6^s,\ \ [e_0,e_1] = e_4 + (\lambda+1)e_5,$ $[e_1,e_1] = e_4 + e_5,\ \ [e_2,e_1] = e_5, \lambda \in \mathbb{C}.$
10. $L(1,0,\lambda,\lambda) := \ L_6^s,\ \ [e_0,e_1] = e_3 - 2e_4 + (2\lambda+5)e_5,$ $[e_1,e_1] = e_3 - 2e_4 + (\lambda+5)e_5,\ \ [e_2,e_1] = e_4 - 2e_5,\ \ [e_3,e_1] = e_5,$ $\lambda \in \mathbb{C}.$
11. $L(1,1,\lambda_1,\lambda_2) := L_6^s, [e_0,e_1] = e_3 - e_4 + (\lambda_1+\lambda_2+5)e_5,\ \ [e_1,e_1] = e_3 - e_4 + (\lambda_1+5)e_5,\ \ [e_2,e_1] = e_4 - e_5, [e_3,e_1] = e_5,\ \ \lambda_1,\lambda_2 \in \mathbb{C}.$

Note 6.1. *The algebra* $L(1,0,0,0)$ *from* U_6 *can be included into the parametric family* $L(1,0,\lambda,\lambda)$ *at* $\lambda = 0$.

There are 11 isomorphism classes in FLb_6, three of them are parametric families and the others are isolated orbits.

6.3 ISOMORPHISM CRITERIA FOR SECOND CLASS

According to Theorem 4.3 there exists an adapted basis $\{e_0, e_1, \dots, e_n\}$ for algebras from the class SLb_{n+1} such that the multiplication table can be represented as

$$\mathrm{SLb}_{n+1} := \begin{cases} [e_0, e_0] = e_2, \\ [e_i, e_0] = e_{i+1}, \ \ 2 \leq i \leq n-1 \\ [e_0, e_1] = \beta_3 e_3 + \beta_4 e_4 + \cdots + \beta_n e_n, \\ [e_1, e_1] = \gamma e_n, \\ [e_j, e_1] = \beta_3 e_{j+2} + \beta_4 e_{j+3} + \cdots + \beta_{n+1-j} e_n, \\ \qquad\qquad 2 \leq j \leq n-2 \end{cases}$$

Further an algebra from SLb_{n+1} will be denoted as $L(\beta)$, where $\beta = (\beta_3, \beta_4, \dots, \beta_n, \gamma)$ are structure constants defined above.

Recall that the isomorphism criterion for SLb_{n+1} is written as follows

Theorem 6.8. *Two algebras $L(\beta)$ and $L(\beta')$ from SLb_{n+1}, where $\beta = (\beta_3, \beta_4, \dots, \beta_n, \gamma)$, and $\beta' = (\beta'_3, \beta'_4, \dots, \beta'_n, \gamma')$, are isomorphic if and only if there exist complex numbers A, B and D such that $AD \neq 0$ and the following conditions hold:*

$$\begin{aligned} \beta'_t &= \tfrac{1}{A^{t-2}} \tfrac{D}{A} \psi_t \left(\tfrac{B}{A}; \beta\right), \ \ 3 \leq t \leq n-1, \\ \beta'_n &= \tfrac{1}{A^{n-2}} \tfrac{D}{A} \left(\tfrac{B}{A} \gamma + \psi_n \left(\tfrac{B}{A}; \beta\right)\right), \end{aligned}$$

and

$$\gamma' = \tfrac{1}{A^{n-2}} \left(\tfrac{D}{A}\right)^2 \psi_{n+1} \left(\tfrac{B}{A}; \beta\right).$$

6.4 CLASSIFICATION OF SECOND CLASS IN LOW DIMENSIONS

This section deals with finding the isomorphism classes of SLb_n for $n = 5, 6$. Every filiform Leibniz algebra admits an adapted basis with respect to which the table of multiplication can be represented as in Theorem 6.8. Recall that a filiform Leibniz algebra is two-generated, therefore it is sufficient to give nonsingular transformation for the elements e_0 and e_1 of the adapted basis. Note that in each specified dimension we split the class SLb_n into its subsets. Meanwhile some of the subsets turn out to be a union of infinitely many orbits. In this case we give a proposition on structure of these subsets. The propositions consist of two parts; the first part gives orbit functions (invariants), another part guaranties that the algebras with the different invariants are not isomorphic.

For the simplification purpose, we use the following notations:

$$\Lambda_1 = 4\beta_3\beta_5 - 5\beta_4^2, \text{ and } \Lambda'_1 = 4\beta'_3\beta'_5 - 5\beta'^2_4.$$

6.4.1 Classification in dimension five

As we have seen above, the table of multiplication of SLb_5 with respect to the adapted basis is as follows

$$\begin{cases} \beta'_3 = \frac{1}{A}\frac{D}{A}\beta_3, \\ \beta'_4 = \frac{1}{A^2}\frac{D}{A}\left(\frac{B}{A}\gamma + \beta_4 - 2\frac{B}{A}\beta_3^2\right), \\ \gamma' = \frac{1}{A^2}\left(\frac{D}{A}\right)^2\gamma. \end{cases}$$

The class SLb_5 can be represented as a disjoint union of its subsets as follows:

$$SLb_5 = U_1 \bigcup U_2 \bigcup U_3 \bigcup U_4 \bigcup U_5 \bigcup F,$$

where

$U_1 = \{L(\beta) \in SLb_5 : \beta_3 \neq 0 \text{ and } \gamma - 2\beta_3^2 \neq 0\}$,
$U_2 = \{L(\beta) \in SLb_5 : \beta_3 \neq 0, \ \gamma - 2\beta_3^2 = 0 \text{ and } \beta_4 \neq 0\}$,
$U_3 = \{L(\beta) \in SLb_5 : \beta_3 \neq 0, \ \gamma - 2\beta_3^2 = 0 \text{ and } \beta_4 = 0\}$,
$U_4 = \{L(\beta) \in SLb_5 : \beta_3 = 0, \ \gamma \neq 0\}$,
$U_5 = \{L(\beta) \in SLb_5 : \beta_3 = 0, \ \gamma = 0 \text{ and } \beta_4 \neq 0\}$,
$F = \{L(\beta) \in SLb_5 : \beta_3 = 0, \ \gamma = 0 \text{ and } \beta_4 = 0\}$.

Proposition 6.7.

i) *Two algebras $L(\beta)$ and $L(\beta')$ from U_1 are isomorphic if and only if*

$$\frac{\gamma}{\beta_3^2} = \frac{\gamma'}{\beta'^2_3}.$$

ii) *For any $\lambda \in \mathbb{C}$ there is an algebra $L(\beta)$ from U_1 such that $\frac{\gamma}{\beta_3^2} = \lambda$.*

Proof. If this system of equalities holds true, then the base change

$$e'_0 = e_0 - \frac{\beta_4}{\gamma - 2\beta_3^2}e_1, \ e'_1 = \frac{1}{\beta_3}e_1 - \frac{\beta_4}{\beta_3(\gamma - 2\beta_3^2)}e_3$$

transforms the algebra $L(\beta)$ to the algebra $L(\beta')$.

Conversely, according to Theorem 5.6 there is a base change that brings $L(\beta) = L(\beta_3, \beta_4, \gamma)$ and $L(\beta') = L(\beta'_3, \beta'_4, \gamma')$ into $L(1, 0, \frac{\gamma}{\beta_3^2})$ and $L(1, 0, \frac{\gamma'}{\beta'^2_3})$, respectively. But due to the condition the last two algebras are the same.

The second part is obvious if we take into account that the equation $\frac{\gamma}{\beta_3^2} = \lambda$ is solvable with respect to β_3 and γ in U_1. □

Proposition 6.8. *The subsets U_2, U_3, U_4, U_5 and F are single orbits with representatives $L(1, 1, 2)$, $L(1, 0, 2)$, $L(0, 0, 1)$, $L(0, 1, 0)$ and $L(0, 0, 0)$, respectively.*

Proof. We show base changes transforming algebras from U_2 to the algebra $L(1, 1, 2)$. The others can be shown similarly. Indeed, it can be easily verified that $e'_0 = e_0$ and $e'_1 = \frac{1}{\beta_3} e_1$ is the required base change. □

Theorem 6.9.

Any five-dimensional complex filiform Leibniz algebra from SLb_5 *is isomorphic to one of the following pairwise nonisomorphic non-Lie filiform complex Leibniz algebras, whose tables of multiplication with respect to the adapted basis $\{e_0, e_1, e_2, e_3, e_4\}$ are as follows:*

1. $L(1, 0, \lambda) :=$ $[e_0, e_0] = e_3, [e_2, e_0] = e_4, [e_3, e_0] = e_5,$ $[e_0, e_1] = e_3, [e_1, e_1] = \lambda e_4, [e_2, e_1] = e_5, \ \lambda \in \mathbb{C}.$

2. $L(1, 1, 2) :=$ $[e_0, e_0] = e_3, [e_2, e_0] = e_4, [e_3, e_0] = e_5,$ $[e_0, e_1] = e_3 + e_4, [e_1, e_1] = 2e_4, [e_2, e_1] = e_5.$

3. $L(0, 0, 1) :=$ $[e_0, e_0] = e_3, [e_2, e_0] = e_4, [e_3, e_0] = e_5, [e_1, e_1] = e_4.$

4. $L(0, 1, 0) :=$ $[e_0, e_0] = e_3, [e_2, e_0] = e_4, [e_3, e_0] = e_5, [e_0, e_1] = e_4.$

5. $L(0, 0, 0) :=$ $[e_0, e_0] = e_3, [e_2, e_0] = e_4, [e_3, e_0] = e_5.$

6.4.2 Classification in dimension six

This section concerns the six-dimensional case of SLb_n. Let $\{e_0, e_1, e_2, e_3, e_4, e_5\}$ be an adapted basis. Then the multiplication table of the class SLb_6 with respect to this basis is as follows:

$$SLb_6 := \begin{cases} [e_0, e_0] = e_2, & \\ [e_i, e_0] = e_{i+1}, & i = 2, 3, 4, \\ [e_0, e_1] = \beta_3 e_3 + \beta_4 e_4 + \beta_5 e_5, & \\ [e_1, e_1] = \gamma e_5, & \\ [e_j, e_1] = \beta_3 e_{j+2} + \beta_4 e_{j+3} + \cdots + \beta_{6-j} e_5, & j = 2, 3. \end{cases}$$

The representative of the class SLb_6 is written as $L(\beta_3, \beta_4, \beta_5, \gamma)$. Here is the isomorphism criterion for algebras from SLb_6:

$$\begin{cases} \beta_3' = \frac{1}{A}\frac{D}{A}\beta_3, \\ \beta_4' = \frac{1}{A^2}\frac{D}{A}\left(\beta_4 - 2\frac{B}{A}\beta_3^2\right), \\ \beta_5' = \frac{1}{A^3}\frac{D}{A}\left(\frac{B}{A}\gamma + \beta_5 - 5\frac{B}{A}\beta_3\beta_4 + 5\left(\frac{B}{A}\right)^2 \beta_3^3\right), \\ \gamma' = \frac{1}{A^3}\left(\frac{D}{A}\right)^2 \gamma. \end{cases}$$

The set SLb_6 can be represented as a disjoint union of its subsets as follows:

$$SLb_6 = \bigcup_{i=1}^{9} U_i,$$

where

$$\begin{aligned}
U_1 &= \{L(\beta) \in SLb_6 : \beta_3 \neq 0, \gamma \neq 0\}, \\
U_2 &= \{L(\beta) \in SLb_6 : \beta_3 \neq 0, \gamma = 0, \Lambda_1 \neq 0\}, \\
U_3 &= \{L(\beta) \in SLb_6 : \beta_3 \neq 0, \gamma = 0, \Lambda_1 = 0\}, \\
U_4 &= \{L(\beta) \in SLb_6 : \beta_3 = 0, \beta_4 \neq 0, \gamma \neq 0\}, \\
U_5 &= \{L(\beta) \in SLb_6 : \beta_3 = 0, \beta_4 \neq 0, \gamma = 0, \beta_5 \neq 0\}, \\
U_6 &= \{L(\beta) \in SLb_6 : \beta_3 = 0, \beta_4 \neq 0, \gamma = 0, \beta_5 = 0\}, \\
U_7 &= \{L(\beta) \in SLb_6 : \beta_3 = 0, \beta_4 = 0, \gamma \neq 0\}, \\
U_8 &= \{L(\alpha) \in SLb_6 : \beta_3 = 0, \beta_4 = 0, \gamma = 0, \beta_5 \neq 0\}, \\
U_9 &= \{L(\alpha) \in SLb_6 : \beta_3 = 0, \beta_4 = 0, \gamma = 0, \beta_5 = 0\}.
\end{aligned}$$

"If" part of the following proposition is just a substitution and "Only if" part is a particular case of Theorem 5.6.

Proposition 6.9.

i) *Two algebras* $L(\beta)$ *and* $L(\beta')$ *from* U_1 *are isomorphic, if and only if*

$$\frac{2\beta_3\beta_4\gamma+\beta_3^2\Lambda_1}{\gamma^2}=\frac{2\beta_3'\beta_4'\gamma'+\beta_3'^2\Lambda_1'}{\gamma'^2}.$$

ii) *Orbits in* U_1 *can be parameterized as* $L(1,0,\lambda,1),\ \lambda\in\mathbb{C}$.

The proof of the proposition below can carried out by showing the base change leading to the specified representatives.

Proposition 6.10.

The subsets U_2, U_3, U_4, U_5, U_6, U_7, U_8, *and* U_9 *are single orbits under the action of* G_{ad} *with the representatives* $L(1,0,1,0)$, $L(1,0,0,0)$, $L(0,1,0,1)$, $L(0,1,1,0)$, $L(0,1,0,0)$, $L(0,0,0,1)$, $L(0,0,1,0)$, *and* $L(0,0,0,0)$, *respectively.*

Theorem 6.10. *Let L be algebra from* SLb_6. *Then, it is isomorphic to one of the following pairwise non-isomorphic Leibniz algebras:*

1. $L(0,0,0,0) := G_6^s = \{[e_0,e_0]=e_2,\ [e_i,e_0]=e_{i+1},\ 2\le i\le 4\}.$

2. $L(0,0,1,0) := G_6^s,\ [e_0,e_1]=e_5.$

3. $L(0,0,0,1) := G_6^s,\ [e_1,e_1]=e_5.$

4. $L(0,1,0,0) := G_6^s,\ [e_0,e_1]=e_4,\ [e_2,e_1]=e_5.$

5. $L(0,1,1,0) := G_6^s,\ [e_0,e_1]=e_4+e_5,\ [e_2,e_1]=e_5.$

6. $L(0,1,0,1) := G_6^s,\ [e_0,e_1]=e_4,\ [e_1,e_1]=e_5,\ [e_2,e_1]=e_5.$

7. $L(1,0,0,0) := G_6^s,\ [e_0,e_1]=e_3,\ [e_2,e_1]=e_4,\ [e_3,e_1]=e_5.$

8. $L(1,0,1,0) := G_6^s,\ [e_0,e_1]=e_3+e_5,$

   $[e_2,e_1]=e_4,\ [e_3,e_1]=e_5.$

9. $L(1,0,\lambda,1) := \ G_6^s, \ [e_0,e_1] = e_3 + \lambda e_5, \ [e_1,e_1] = e_5,$
$[e_2,e_1] = e_4, \ [e_3,e_1] = e_5, \ \lambda \in \mathbb{C}.$

6.5 SIMPLIFICATIONS AND NOTATIONS IN THIRD CLASS

In this section we treat filiform Leibniz algebras whose natural gradation is an algebra from NGF_3. This class has been denoted by TLb_{n+1} in dimension $n+1$. We recall that $(n+1)$–dimensional filiform Lie algebras are in TLb_{n+1}. The study of TLb_{n+1} has been initiated in Omirov et al. [140]. The multiplication table of TLb_{n+1} on an adapted basis is as follows.

$$\mathrm{TLb}_{n+1} := \begin{cases} [e_i,e_0] = e_{i+1}, & 1 \le i \le n-1, \\ [e_0,e_i] = -e_{i+1}, & 2 \le i \le n-1, \\ [e_0,e_0] = b_{0,0}e_n, & \\ [e_0,e_1] = -e_2 + b_{0,1}e_n, & \\ [e_1,e_1] = b_{1,1}e_n, & \\ [e_i,e_j] = a^1_{i,j}e_{i+j+1} + \cdots + a^{n-(i+j+1)}_{i,j}e_{n-1} + b_{i,j}e_n, & \\ 1 \le i < j \le n-1, & \\ [e_i,e_j] = -[e_j,e_i], & \\ 1 \le i < j \le n-1, & \\ [e_i,e_{n-i}] = -[e_{n-i},e_i] = (-1)^i\, b_{i,n-i}e_n, & \end{cases}$$

where $a^k_{i,j}, b_{i,j} \in \mathbb{C}$ and $b_{i,n-i} = b$ whenever $1 \le i \le n-1$, and $b = 0$ for even n.

An element of $\mathrm{TLb}(n+1)$ is denoted by

$$L(\alpha) = L(b_{00}, b_{01}, b_{11}, a^1_{ij}, \ldots, a^{n-(i+j+1)}_{ij}, b_{ij}).$$

Recall that, in general, the structure constants $b_{00}, b_{01}, b_{11}, a^1_{ij}, \ldots, a^{n-(i+j+1)}_{ij}, b_{ij}$ are not free and in each fixed dimensional case we re-arrange them taking into account the relations between them from Lemma 4.6. Mainly, we keep lexicographical order with respect to the indices. Moreover, we express the a^k_{ij} via b_{st}, $s,t = 0,1,\ldots$.

The complete implementation of the classification procedure for TLb_{n+1} in low dimensional cases will be given in the next sections.

For the purpose of simplification, we introduce the following notations:

$$\begin{array}{llll} \Pi_1(Z) & = 4z_{00}z_{11} - z_{01}^2; & \Pi_2(Z) & = 2z_{11} - z_{01}z_{23}; \\ \Pi_3(Z) & = z_{01} - z_{00}z_{23}; & \Pi_4(Z) & = 4z_{00}z_{12}^4 - 2z_{01}z_{12}^2z_{13} + z_{11}z_{13}^2; \\ \Pi_5(Z) & = z_{01}z_{12}^2 - z_{11}z_{13}; & \Pi_6(Z) & = 2z_{00}z_{12}^2 - z_{01}z_{13}; \end{array}$$

Introduce functions:

$$\chi_0(X;Z) = 1 + x_1 z_{23};$$

$$\chi_1(X;Z) = z_{00} + x_1 z_{01} + x_1^2 z_{11};$$

$$\chi_2(X;Z) = z_{01} + 2x_1 z_{11};$$

$$\chi_3(X;Y;Z) = z_{13} + (y_2^2 - 2y_3)z_{23} + x_1^2 z_{12}^2 z_{23};$$

$$\chi_4(X;Z) = z_{13} + 2x_1 z_{12}^2,$$

where $X = (x_1, x_2, \ldots, x_n)$,
$Y = (y_2, \ldots, y_n)$ and
$Z = (z_{00}, z_{01}, z_{02}, \ldots, z_{11}, z_{12}, \ldots)$

Let

$$\mathbf{A} = \left(\frac{A_1}{A_0}, \frac{A_2}{A_0}, \ldots, \frac{A_n}{A_0}\right),$$

$$\mathbf{B} = \left(\frac{B_2}{B_1}, \frac{B_3}{B_1}, \ldots, \frac{B_n}{B_1}\right),$$

and

$$\alpha = (b_{00}, b_{01}, b_{11}, b_{12}, b_{13}, \ldots).$$

6.6 CLASSIFICATION IN DIMENSION FIVE

In this section we deal with the class TLb_5. By virtue of Theorem 4.3 we can represent TLb_5 as follows:

$$TLb_5 := \begin{cases} [e_i, e_0] = e_{i+1}, & 1 \le i \le 3, \\ [e_0, e_i] = -e_{i+1}, & 2 \le i \le 3, \\ [e_0, e_0] = b_{00}e_4, \\ [e_0, e_1] = -e_2 + b_{01}e_4, \\ [e_1, e_1] = b_{11}e_4, \\ [e_1, e_2] = -[e_2, e_1] = b_{12}e_4, \\ b_{00}, b_{01}, b_{11}, b_{12} \in \mathbb{C}. \end{cases}$$

Further, the elements of TLb_5 will be denoted by $L(\alpha) = L(b_{00}, b_{01}, b_{11}, b_{12})$.

Theorem 6.11. *(Isomorphism criterion for* TLb_5*) Two algebras* $L(\alpha)$ *and* $L(\alpha')$ *from* TLb_5 *are isomorphic, if and only if there exist complex numbers* A_0, A_1 *and* B_1 *such that* $A_0 B_1 \neq 0$ *and the following conditions hold:*

$$\begin{aligned} b'_{00} &= \tfrac{1}{A_0 B_1}\chi_1(\mathbf{A};\alpha);\\ b'_{01} &= \tfrac{1}{A_0^2}\chi_2(\mathbf{A};\alpha);\\ b'_{11} &= \tfrac{B_1}{A_0^3} b_{11};\\ b'_{12} &= \tfrac{B_1}{A_0^2} b_{12}. \end{aligned}$$

Proof. **Necessity**. Let L_1 and L_2 from TLb_5 be isomorphic: $f : L_1 \cong L_2$. We choose the corresponding adapted bases $\{e_0, e_1, e_2, e_3, e_4\}$ and $\{e'_0, e'_1, e'_2, e'_3, e'_4\}$ in L_1 and L_2, respectively. Then, in these bases the algebras will be presented as $L(\alpha)$ and $L(\alpha')$, where $\alpha = (b_{00}, b_{01}, b_{11}, b_{12})$, and $\alpha' = (b'_{00}, b'_{01}, b'_{11}, b'_{12})$.

The change of the generators e_0 and e_2 is as follows:

$$\begin{aligned} e'_0 = f(e_0) &= A_0e_0 + A_1e_1 + A_2e_2 + A_3e_3 + A_4e_4, \\ e'_1 = f(e_1) &= B_1e_1 + B_2e_2 + B_3e_3 + B_4e_4. \end{aligned} \tag{6.3}$$

Then we obtain

$$\begin{aligned} e'_2 = f(e_2) &= [f(e_1), f(e_0)] = A_0B_1e_2 + A_0B_2e_3 \\ &+ (A_0B_3 + A_1B_1b_{11} + (A_2B_1 - A_1B_2)b_{12})\, e_4, \\ e'_3 = f(e_3) &= [f(e_2), f(e_0)] = A_0^2B_1e_3 + (A_0^2B_2 - A_0A_1B_1b_{12})e_4, \\ e'_4 = f(e_4) &= [f(e_3), f(e_0)] = A_0^3B_1e_4. \end{aligned} \tag{6.4}$$

By using the adapted bases $\{e_0, e_1, e_2, e_3, e_4\}$ and $\{e'_0, e'_1, e'_2, e'_3, e'_4\}$ one finds the relations between the structure constants $b_{00}, b_{01}, b_{11}, b_{12}$ and $b'_{00}, b'_{01}, b'_{11}, b'_{1,2}$. First, we consider the equality $[f(e_0), f(e_0)] = b'_{00}f(e_4)$, and get the equation (7.1) and from the equality $[f(e_1), f(e_0)]+ [f(e_0), f(e_1)] = b'_{01}f(e_4)$ we obtain (7.2), and $[f(e_1), f(e_1)] = b'_{11}\, f(e_4)$ gives (7.3). Finally, the equality (7.4) comes out from $[f(e_1), f(e_2)] = b'_{12}f(e_4)$.

Sufficiency. Let the equalities of Theorem 6.11 hold. Then, the base change (6.3) above is adapted and it transforms $L(\alpha)$ into $L(\alpha')$.

Indeed,

$$\begin{aligned}
[e'_0, e'_0] &= \left[\sum_{i=0}^{4} A_i e_i, \sum_{i=0}^{4} A_i e_i\right] \\
&= A_0^2[e_0, e_0] + A_0 A_1[e_0, e_1] + A_0 A_1[e_1, e_0] + A_1^2[e_1, e_1] \\
&= \left(A_0^2 b_{00} + A_0 A_1 b_{01} + A_1^2 b_{11}\right) e_4 = b'_{00} A_0^3 B_1 e_4 = b'_{00} e'_4.
\end{aligned}$$

$$\begin{aligned}
[e'_0, e'_1] &= \left[\sum_{i=0}^{4} A_i e_i, \sum_{i=1}^{4} B_i e_i\right] \\
&= -(A_0 B_1 e_2 + A_0 B_2 e_3 + (A_1 B_1 b_{11} + A_2 B_1 b_{12} \\
&\quad -A_1 B_2 b_{1,2} + A_0 B_3) e_4) + B_1 \left(b_{01} A_0 + 2 A_1 b_{11}\right) e_4 \\
&= -e'_2 + A_0^3 B_1 b'_{01} e_4 = -e'_2 + b'_{01} e'_4.
\end{aligned}$$

In the same way one can prove that $[e'_1, e'_1] = b'_{11} e'_4$, $[e'_1, e'_2] = b'_{12} e'_4$ and the other products to be zero.

□

6.6.1 Isomorphism classes in TLb$_5$

Now, we list the isomorphism classes of algebras from TLb$_5$. First we represent TLb$_5$ as a disjoint union of its subsets as follows: TLb$_5$ = $\bigcup_{i=1}^{9} U_5^i$, where

$U_5^1 = \{L(\alpha) \in \text{TLb}_5 \ : b_{11} \neq 0,\ b_{12} \neq 0\};$
$U_5^2 = \{L(\alpha) \in \text{TLb}_5 \ : b_{11} \neq 0,\ b_{12} = 0,\ \Pi_1(\alpha) \neq 0\};$
$U_5^3 = \{L(\alpha) \in \text{TLb}_5 \ : b_{11} \neq 0,\ b_{12} = \Pi_1(\alpha) = 0\};$
$U_5^4 = \{L(\alpha) \in \text{TLb}_5 \ : b_{11} = 0,\ b_{01} \neq 0,\ b_{12} \neq 0\};$
$U_5^5 = \{L(\alpha) \in \text{TLb}_5 \ : b_{11} = 0,\ b_{01} \neq 0,\ b_{12} = 0\};$
$U_5^6 = \{L(\alpha) \in \text{TLb}_5 \ : b_{11} = b_{01} = 0,\ b_{00} \neq 0,\ b_{12} \neq 0\};$
$U_5^7 = \{L(\alpha) \in \text{TLb}_5 \ : b_{11} = b_{01} = 0,\ b_{00} \neq 0,\ b_{12} = 0\};$
$U_5^8 = \{L(\alpha) \in \text{TLb}_5 \ : b_{11} = b_{01} = b_{00} = 0,\ b_{12} \neq 0\};$
$U_5^9 = \{L(\alpha) \in \text{TLb}_5 \ : b_{11} = b_{01} = b_{00} = b_{12} = 0\}.$

For each of the subsets we state an isomorphism criterion.

Proposition 6.11.

1. *Two algebras $L(\alpha)$ and $L(\alpha')$ from U_5^1 are isomorphic, if and only if*
$$\left(\frac{b'_{12}}{b'_{11}}\right)^4 \Pi_1(\alpha') = \left(\frac{b_{12}}{b_{11}}\right)^4 \Pi_1(\alpha).$$

2. *For any λ from $\mathbb{C}$, there exists $L(\alpha) \in U_5^1$ such that $\left(\frac{b_{12}}{b_{11}}\right)^4 \Pi_1(\alpha) = \lambda$.*

Proof. 1. **Necessity**. Let $L(\alpha)$ and $L(\alpha')$ be isomorphic. Then by a substitution it is easy to see that $\left(\frac{b'_{12}}{b'_{11}}\right)^4 \Pi_1(\alpha') = \left(\frac{b_{12}}{b_{11}}\right)^4 \Pi_1(\alpha)$.

Sufficiency. Let the equality $\left(\frac{b'_{12}}{b'_{11}}\right)^4 \Pi_1(\alpha') = \left(\frac{b_{12}}{b_{11}}\right)^4 \Pi_1(\alpha)$ hold. Consider the base change (6.3) above with $A_0 = \frac{b_{11}}{b_{12}}$, $A_1 = -\frac{b_{01}}{2b_{12}}$ and $B_1 = \frac{b_{11}^2}{b_{12}^3}$. This base change leads $L(\alpha)$ to $L\left(\left(\frac{b_{12}}{b_{11}}\right)^4 \Pi_1(\alpha), 0, 1, 1\right)$. A similar base change with $A_0 = \frac{b'_{11}}{b'_{12}}$, $A_1 = -\frac{b'_{01}}{2b'_{12}}$ and $B_1 = \frac{b'^2_{11}}{b'^3_{12}}$ transforms $L(\alpha')$ to $L\left(\left(\frac{b'_{12}}{b'_{11}}\right)^4 \Pi_1(\alpha'), 0, 1, 1\right)$.

Since $\left(\frac{b'_{12}}{b'_{11}}\right)^4 \Pi_1(\alpha') = \left(\frac{b_{12}}{b_{11}}\right)^4 \Pi_1(\alpha)$, then $L(\alpha)$ is isomorphic to $L(\alpha')$.

2. Obvious.

□

Proposition 6.12. *The subsets U_5^2, U_5^3, U_5^4, U_5^5, U_5^6, U_5^7, U_5^8 and U_5^9 are single orbits with representatives $L(1,0,1,0)$, $L(0,0,1,0)$, $L(0,1,0,1)$, $L(0,1,0,0)$, $L(1,0,0,1)$, $L(1,0,0,0)$, $L(0,0,0,1)$ and $L(0,0,0,0)$, respectively.*

Proof. To prove this, we give the appropriate values of A_0, A_1 and B_1 in the base change (the other A_i, B_j, $i, j = 2, 3, 4$ are arbitrary, except where specified otherwise).

For U_5^2 :
$e_0' = A_0e_0 + A_1e_1 + A_2e_2 + A_3e_3 + A_4e_4,$
$e_1' = B_1e_1 + B_2e_2 + B_3e_3 + B_4e_4,$
$e_2' = A_0B_1e_2 + A_0B_2e_3 + (A_1B_1b_{11} + A_0B_3)\, e_4,$
$e_3' = A_0^2B_1e_3 + A_0^2B_2e_4,$
$e_4' = {A_0}^3B_1e_4,$
where
$A_0^4 = \frac{\Pi_1(\alpha)}{4}$, $A_1^4 = \frac{b_{01}^4\,\Pi_1(\alpha)}{64b_{11}^4}$ and $B_1^4 = \frac{\Pi_1(\alpha)^3}{64b_{11}^4}$.
For U_5^3 :
$e_0' = A_0e_0 + A_1e_1 + A_2e_2 + A_3e_3 + A_4e_4,$
$e_1' = B_1e_1 + B_2e_2 + B_3e_3 + B_4e_4,$
$e_2' = A_0B_1e_2 + A_0B_2e_3 + (A_1B_1b_{11} + A_0B_3)\, e_4,$
$e_3' = A_0^2B_1e_3 + {A_0}^2B_2e_4,$
$e_4' = {A_0}^3B_1e_4,$
where $A_0 \in \mathbb{C}^*$, $A_1 = -\frac{A_0b_{01}}{2b_{11}}$ and $B_1 = \frac{A_0^3}{b_{11}}$.
For U_5^4 :
$e_0' = A_0e_0 + A_1e_1 + A_2e_2 + A_3e_3 + A_4e_4,$
$e_1' = B_1e_1 + B_2e_2 + B_3e_3 + B_4e_4,$
$e_2' = A_0B_1e_2 + A_0B_2e_3 + (A_0B_3 + (A_2B_1 - A_1B_2)b_{12})e_4,$
$e_3' = A_0^2B_1e_3 + (A_0^2B_2 - A_1A_0B_1b_{12})e_4,$
$e_4' = {A_0}^3B_1e_4,$
where $A_0^2 = b_{01}$, $A_1 = -\frac{A_0b_{00}}{b_{01}}$ and $B_1 = \frac{A_0^2}{b_{12}}$.
For U_5^5 :
$e_0' = A_0e_0 + A_1e_1 + A_2e_2 + A_3e_3 + A_4e_4,$
$e_1' = B_1e_1 + B_2e_2 + B_3e_3 + B_4e_4,$
$e_2' = A_0B_1e_2 + A_0B_2e_3 + A_0B_3e_4,$
$e_3' = A_0^2B_1e_3 + A_0^2B_2e_4,$
$e_4' = A_0^3B_1e_4,$
where
$A_0^2 = b_{01}$, $A_1^2 = \frac{b_{00}^2}{b_{01}}$ and $B_1 \in \mathbb{C}^*$.
For U_5^6 :
$e_0' = A_0e_0 + A_1e_1 + A_2e_2 + A_3e_3 + A_4e_4,$
$e_1' = B_1e_1 + B_2e_2 + B_3e_3 + B_4e_4,$
$e_2' = A_0B_1e_2 + A_0B_2e_3 + (A_0B_3 + (A_2B_1 - A_1B_2)b_{12})e_4,$
$e_3' = A_0^2B_1e_3 + (A_0^2B_2 - A_1A_0B_1b_{12})e_4,$
$e_4' = {A_0}^3B_1e_4,$
where $A_0^3 = b_{00}b_{12}$, $A_1 \in \mathbb{C}$ and $B_1^3 = \frac{{b^2}_{0,0}}{b_{1,2}}$.

$A_0^3 = b_{00}b_{12}$, $A_1 \in \mathbb{C}$ and $B_1^3 = \frac{b_{0,0}^2}{b_{12}}$.

For U_5^7 :

$e_0' = A_0e_0 + A_1e_1 + A_2e_2 + A_3e_3 + A_4e_4,$
$e_1' = B_1e_1 + B_2e_2 + B_3e_3 + B_4e_4,$
$e_2' = A_0B_1e_2 + A_0B_2e_3 + A_0B_3e_4,$
$e_3' = A_0^2B_1e_3 + A_0^2B_2e_4,$
$e_4' = A_0^3B_1e_4,$

where $A_0 \in \mathbb{C}^*$, $A_1 \in \mathbb{C}$ and $B_1 = \frac{b_{00}}{A_0}$.

For U_5^8 :

$e_0' = A_0e_0 + A_1e_1 + A_2e_2 + A_3e_3 + A_4e_4,$
$e_1' = B_1e_1 + B_2e_2 + B_3e_3 + B_4e_4,$
$e_2' = A_0B_1e_2 + A_0B_2e_3 + (A_0B_3 + (A_2B_1 - A_1B_2)b_{12})e_4,$
$e_3' = A_0^2B_1e_3 + (A_0^2B_2 - A_1A_0B_1b_{12})e_4,$
$e_4' = {A_0}^3B_1e_4,$

where $A_0 \in \mathbb{C}^*$, $A_1 \in \mathbb{C}$ and $B_1 = \frac{A_0^2}{b_{12}}$.

For U_5^9 :

$e_0' = A_0e_0 + A_1e_1 + A_2e_2 + A_3e_3 + A_4e_4,$
$e_1' = B_1e_1 + B_2e_2 + B_3e_3 + B_4e_4,$
$e_2' = A_0B_1e_2 + A_0B_2e_3 + A_0B_3e_4,$
$e_3' = A_0^2B_1e_3 + A_0^2B_2e_4,$
$e_4' = A_0^3B_1e_4,$

where $A_0, B_1 \in \mathbb{C}^*$ and $A_1 \in \mathbb{C}$. □

6.7 CLASSIFICATION IN DIMENSION SIX

This section is devoted to the description of isomorphism classes in TLb_6. The class TLb_6 is represented by the following table of multiplication:

$$\mathrm{TLb}_6 := \begin{cases} [e_i, e_0] = e_{i+1}, & 1 \le i \le 4, \\ [e_0, e_i] = -e_{i+1}, & 2 \le i \le 4, \\ [e_0, e_0] = b_{00}e_5, \\ [e_0, e_1] = -e_2 + b_{01}e_5, \\ [e_1, e_1] = b_{11}e_5, \\ [e_1, e_2] = -[e_2, e_1] = b_{12}e_4 + b_{13}e_5, \\ [e_1, e_3] = -[e_3, e_1] = b_{12}e_5, \\ [e_1, e_4] = -[e_4, e_1] = -[e_2, e_3] = [e_3, e_2] = -b_{23}e_5. \end{cases}$$

Further elements of TLb_6 will be denoted by $L(\alpha)$, where $\alpha = (b_{01}, b_{11}, b_{12}, b_{13}, b_{23})$.

Theorem 6.12. (***Isomorphism criterion for*** TLb$_6$) *Two filiform Leibniz algebras $L(\alpha')$ and $L(\alpha)$ from* TLb$_6$ *are isomorphic, if and only if there exist $A_0, A_1, B_1, B_2, B_3 \in \mathbb{C}$ such that $A_0B_1(A_0 + A_1 b_{23}) \neq 0$ and the following equalities hold:*

$$\begin{aligned}
b'_{00} &= \frac{1}{A_0^2B_1}\frac{\chi_1(A;\alpha)}{\chi_0(A;\alpha)};\\
b'_{01} &= \frac{1}{A_0^3}\frac{\chi_2(A;\alpha)}{\chi_0(A;\alpha)}; \alpha);\\
b'_{11} &= \frac{B_1}{A_0^4}\frac{1}{\chi_0(A;\alpha)}b_{11};\\
b'_{12} &= \frac{B_1}{A_0^2}b_{12};\\
b'_{13} &= \frac{B_1}{A_0^3}\frac{\chi_3(A;B;\alpha)}{\chi_0(A;\alpha)};\\
b'_{23} &= \frac{B_1}{A_0}\frac{1}{\chi_0(A;\alpha)}b_{23}.
\end{aligned}$$

The proof is the similar to that of TLb$_5$.

6.7.1 Isomorphism classes in TLb$_6$

Represent TLb$_6$ as a union of the following subsets:

$U_6^1 = \{L(\alpha) \in \text{TLb}_6 : b_{23} \neq 0,\ b_{11} \neq 0\};$
$U_6^2 = \{L(\alpha) \in \text{TLb}_6 : b_{23} \neq 0,\ b_{11} = 0,\ b_{01} \neq 0\};$
$U_6^3 = \{L(\alpha) \in \text{TLb}_6 : b_{23} \neq 0,\ b_{11} = b_{01} = 0,\ b_{12} \neq 0,\ b_{00} \neq 0\};$
$U_6^4 = \{L(\alpha) \in \text{TLb}_6 : b_{23} \neq 0,\ b_{11} = b_{01} = 0,\ b_{12} \neq 0,\ b_{00} = 0\};$
$U_6^5 = \{L(\alpha) \in \text{TLb}_6 : b_{23} \neq 0,\ b_{11} = b_{01} = b_{12} = 0,\ b_{00} \neq 0\};$
$U_6^6 = \{L(\alpha) \in \text{TLb}_6 : b_{23} \neq 0,\ b_{11} = b_{01} = b_{12} = b_{00} = 0\};$
$U_6^7 = \{L(\alpha) \in \text{TLb}_6 : b_{23} = 0,\ b_{12} \neq 0,\ b_{11} \neq 0\};$
$U_6^8 = \{L(\alpha) \in \text{TLb}_6 : b_{23} = 0,\ b_{12} \neq 0,\ b_{11} = 0,\ b_{01} \neq 0\};$
$U_6^9 = \{L(\alpha) \in \text{TLb}_6 : b_{23} = 0,\ b_{12} \neq 0,\ b_{11} = b_{01} = 0,\ b_{00} \neq 0\};$
$U_6^{10} = \{L(\alpha) \in \text{TLb}_6 : b_{23} = 0,\ b_{12} \neq 0,\ b_{11} = b_{01} = b_{00} = 0\};$
$U_6^{11} = \{L(\alpha) \in \text{TLb}_6 : b_{23} = b_{12} = 0,\ b_{11} \neq 0,\ b_{13} \neq 0\};$
$U_6^{12} = \{L(\alpha) \in \text{TLb}_6 : b_{23} = b_{12} = 0,\ b_{11} \neq 0,\ b_{13} = 0,$
$\Pi_1(\alpha) \neq 0\};$
$U_6^{13} = \{L(\alpha) \in \text{TLb}_6 : b_{23} = b_{12} = 0,\ b_{11} \neq 0,\ b_{13} = \Pi_1(\alpha) = 0\};$
$U_6^{14} = \{L(\alpha) \in \text{TLb}_6 : b_{23} = b_{12} = b_{11} = 0,\ b_{01} \neq 0,\ b_{13} \neq 0\};$
$U_6^{15} = \{L(\alpha) \in \text{TLb}_6 : b_{23} = b_{12} = b_{11} = 0,\ b_{01} \neq 0,\ b_{13} = 0\};$
$U_6^{16} = \{L(\alpha) \in \text{TLb}_6 : b_{23} = b_{12} = b_{11} = b_{01} = 0,\ b_{00} \neq 0,$
$b_{13} \neq 0\};$

$$U_6^{17} = \{L(\alpha) \in \mathrm{TLb}_6 \ : b_{23} = b_{12} = b_{11} = b_{01} = 0, \ b_{00} \neq 0, \ b_{13} = 0\};$$
$$U_6^{18} = \{L(\alpha) \in \mathrm{TLb}_6 \ : b_{23} = b_{12} = b_{11} = b_{01} = b_{00} = 0, \ b_{13} \neq 0\};$$
$$U_6^{19} = \{L(\alpha) \in \mathrm{TLb}_6 \ : b_{23} = b_{12} = b_{11} = b_{01} = b_{00} = b_{13} = 0\}.$$

Proposition 6.13.

1. *Two algebras $L(\alpha')$ and $L(\alpha)$ from U_6^1 are isomorphic, if and only if*

$$\left(\frac{b'_{23}}{\Pi_2(\alpha')}\right)^2 \Pi_1(\alpha') = \left(\frac{b_{23}}{\Pi_2(\alpha)}\right)^2 \Pi_1(\alpha),$$

$$\frac{(\Pi_2(\alpha'))^3\, b'^3_{12}}{b'^2_{23} b'^4_{11}} = \frac{(\Pi_2(\alpha))^3\, b^3_{1,2}}{b^2_{23} b^4_{11}}.$$

2. *For any $\lambda_1, \lambda_2 \in \mathbb{C}$, there exists $L(\alpha) \in U_6^1$ such that*

$$\left(\frac{b_{23}}{\Pi_2(\alpha)}\right)^2 \Pi_1(\alpha) = \lambda_1, \quad \frac{(\Pi_2(\alpha))^3\, b^3_{1,2}}{b^2_{23} b^4_{11}} = \lambda_2.$$

Then orbits from the set U_6^1 can be parameterized as

$$L(\lambda_1, 0, 1, \lambda_2, 0, 1) \ \lambda_1, \lambda_2 \in \mathbb{C}.$$

Proposition 6.14.

1. *Two algebras $L(\alpha')$ and $L(\alpha)$ from U_6^2 are isomorphic, if and only if*

$$\frac{\Pi_3(\alpha')^4 b'^3_{12}}{b'^3_{23} b'^5_{01}} = \frac{\Pi_3(\alpha)^4 b^3_{1,2}}{b^3_{23} b^5_{01}}.$$

2. *For any $\lambda \in \mathbb{C}$, there exists $L(\alpha) \in U_6^2$ such that $\frac{\Pi_3(\alpha)^4 b^3_{12}}{b^3_{23} b^5_{01}} = \lambda$.*

Therefore orbits from U_6^2 can be parameterized as $L(0, 1, 0, \lambda, 0, 1)$, $\lambda \in \mathbb{C}$.

Proposition 6.15.

1. *Two algebras $L(\alpha')$ and $L(\alpha)$ from U_6^7 are isomorphic, if and only if*
$$\frac{\Pi_4(\alpha')}{b'_{12}b'^2_{11}} = \frac{\Pi_4(\alpha)}{b_{12}b^2_{11}},$$
$$\frac{(\Pi_5(\alpha'))^2}{b'_{12}b'^3_{11}} = \frac{(\Pi_5(\alpha))^2}{b_{12}b^3_{11}}.$$
2. *For any $\lambda_1, \lambda_2 \in \mathbb{C}$, there exists $L(\alpha) \in U_6^7$ such that*
$$\frac{\Pi_4(\alpha)}{b_{12}b^2_{11}} = \lambda_1\,, \quad \frac{(\Pi_5(\alpha))^2}{b_{12}b^3_{11}} = \lambda_2.$$

The orbits from U_6^7 are parameterized as $L(\lambda_1, \lambda_2, 1, 1, 0, 0)$, $\lambda_1, \lambda_2 \in \mathbb{C}$.

Proposition 6.16.

1. *Two algebras $L(\alpha')$ and $L(\alpha)$ from U_6^8 are isomorphic, if and only if*
$$\frac{(\Pi_6(\alpha'))^3}{b'^3_{12}b'^4_{01}} = \frac{(\Pi_6(\alpha))^3}{b^3_{12}b^4_{01}}.$$
2. *For any $\lambda \in \mathbb{C}$, there exists $L(\alpha) \in U_6^8$ such that* $\frac{(\Pi_6(\alpha))^3}{b^3_{12}b^4_{01}} = \lambda$.

The orbits from the set U_6^8 can be parameterized as $L(\lambda, 1, 0, 1, 0, 0)$, $\lambda \in \mathbb{C}$.

Proposition 6.17.

1. *Two algebras $L(\alpha')$ and $L(\alpha)$ from U_6^{11} are isomorphic, if and only if*
$$\left(\frac{b'_{13}}{b'_{11}}\right)^6 \Pi_1(\alpha') = \left(\frac{b_{13}}{b_{11}}\right)^6 \Pi_1(\alpha).$$
2. *For any $\lambda \in \mathbb{C}$, there exists $L(\alpha) \in U_6^{11}$ such that* $\left(\frac{b_{13}}{b_{11}}\right)^6 \Pi_1(\alpha) = \lambda$.

The orbits from U_6^{11} can be parameterized as $L(\lambda, 0, 1, 0, 1, 0)$, $\lambda \in \mathbb{C}$.

Proposition 6.18.

The subsets $U_6^3, U_6^4, U_6^5, U_6^6, U_6^9, U_6^{10}, U_6^{12}, U_6^{13}, U_6^{14}, U_6^{15}, U_6^{16}, U_6^{17}, U_6^{18}$ *and* U_6^{19} *are single orbits with representatives* $L(1,0,0,1,0,1)$, $L(0,0,0,1,0,1)$, $L(1,0,0,0,0,1)$, $L(0,0,0,0,0,1)$, $L(1,0,0,1,0,0)$, $L(0,0,0,1,0,0)$, $L(1,0,1,0,0,0)$, $L(0,0,1,0,0,0)$, $L(0,1,0,0,1,0)$, $L(0,1,0,0,0,0)$, $L(1,0,0,0,1,0)$, $L(1,0,0,0,0,0)$, $L(0,0,0,0,1,0)$ *and* $L(0,0,0,0,0,0)$, *respectively.*

APPENDIX A

Linear Algebra

A.1 VECTOR SPACES AND SUBSPACES

Definition A.1. *A vector space over* $\mathbb{F}$ *is a set* V *together with a function* $V \times V \longrightarrow V$ *called addition, denoted* $(x, y) \longrightarrow x + y$*, and a function* $\mathbb{F} \times V \longrightarrow V$ *called scalar multiplication and denoted* $(c, x) \longrightarrow cx$*, which satisfy the following axioms:*

1. *For each* $x, y \in V$, $x + y = y + x$.
2. *For each* $x, y, z \in V$, $(x + y) + z = x + (y + z)$.
3. *There exists a zero vector in* V*, denoted* 0*, such that* $0 + x = x$ *for each* $x \in V$.
4. *For each* $x \in V$*, there exists* $-x \in V$ *such that* $(-x) + x = 0$.
5. *For each* $a \in \mathbb{F}$ *and* $x, y \in V$, $a(x + y) = ax + ay$.
6. *For each* $a, b \in \mathbb{F}$ *and* $x \in V$, $(a + b)x = ax + bx$.
7. *For each* $a, b \in \mathbb{F}$ *and* $x \in V$, $(ab)x = a(bx)$.
8. *For each* $x \in V$, $1x = x$.

The elements of V are called vectors, whereas the elements of $\mathbb{F}$ are scalars.

Facts:

Let V be a vector space over a field $\mathbb{F}$.

- The vector 0 is the only additive identity in V.

- For each $x \in V$, the element $-x$ is the only additive inverse for x in V.
- For each $x \in V$, $-x = (-1)x$.
- If $a \in \mathbb{F}$ and $x \in V$, then $ax = 0$ if and only if $a = 0$ or $x = 0$.
- If $x, y, z \in V$ and $x + y = x + z$, then $y = z$.

Example A.1.
- *Any field is a vector space over itself and over its subfield. Particularly, the number fields $\mathbb{Q}$, $\mathbb{R}$, $\mathbb{C}$ are examples of vector spaces.*
- *The set $\mathbb{F}^n$ $(n \geq 1)$ is a vector space over $\mathbb{F}$.*
- *The set of polynomials $\mathbb{F}[x_1, x_2, \ldots, x_n]$ at variables $x_1, x_2, \ldots, x_n$ with coefficients from a field $\mathbb{F}$ is a vector space over $\mathbb{F}$.*
- *The set of $n \times n$ matrices with entries in a field $\mathbb{F}$ is a vector space over $\mathbb{F}$.*
- *Let X be a non-empty set and $F(X) = \{f : X \longrightarrow \mathbb{F}\}$ be the set of functions on X with values in a field $\mathbb{F}$. Define*

$$(f \star g)(x) = f(x) + g(x) \text{ and } (\lambda \cdot f)(x) = \lambda f(x), \text{ for } f, g \in F(X),\ \lambda \in \mathbb{F}.$$

The set $(F(X), \star, \cdot)$ is a vector space over $\mathbb{F}$.

Let $x_1, x_2, \ldots, x_k \in V$ and $a_1, a_2, \ldots, a_k \in \mathbb{F}$. The vector $x = a_1x_1 + a_2x_2 + \cdots + a_kx_k \in V$ is called a linear combination of vectors $x_1, x_2, \ldots, x_k$ with the coefficients $a_1, a_2, \ldots, a_k$.

Definition A.2. *A subset $W \subset V$ of a vector space V is said to be a subspace if it is a vector space with respect to operations "$+$" and "$\cdot$" defined in V.*

Note that this is equivalent to

- For any $x, y \in W$ one has $x + y \in W$;
- For any $a \in \mathbb{F}$ and $x \in W$ we have $ax \in W$.

Example A.2.

1. *For any vector space V the subset $\{0\}$ and V itself are subspaces of V.*

2. *The intersection of any nonempty collection of subspaces of V is a subspace of V.*

3. *Let W_1 and W_2 be subspaces of V and $W_1 + W_2 = \{x \in V | \, x_1 \in W_1$ and $x_2 \in W_2\}$. Then $W_1 + W_2$ is a subspace of V containing W_1 and W_2. It is the smallest subspace that contains them in the sense that any subspace that contains both W_1 and W_2 must contain $W_1 + W_2$.*

4. *Let W_1 and W_2 be subspaces of V such that $W_1 \cap W_2 = \{0\}$. Then the subspace $W_1 + W_2$ is denoted by $W_1 \oplus W_2$ and it is called the direct sum of the subspaces W_1 and W_2. If $V = W_1 \oplus W_2$ then V is the direct sum of its subspaces W_1 and W_2. Note that the sum and the direct sum of subspaces $W_1, W_2, \ldots, W_s$ are subspaces and defined by induction. They are denoted by $W_1 + W_2 + \cdots + W_s$ and $W_1 \oplus W_2 \oplus \cdots \oplus W_s$, respectively.*

5. *For a set of vectors $x_1, x_2, \ldots, x_k \in V$ one defines*

$$\mathrm{Span}_{\mathbb{F}}\{x_1, x_2, \ldots, x_k\} = \{x \in V \mid x = \sum_{i=1}^{k} a_i x_i, a_1, a_2, \ldots, a_k \in \mathbb{F}\}.$$

The subset $\mathrm{Span}_{\mathbb{F}}$ *is a subspace of V.*

Definition A.3. *A set of vectors $x_1, x_2, \ldots, x_k \in V$ is said to be linearly independent if*

$$a_1 x_1 + a_2 x_2 + \cdots + a_k x_k = 0 \text{ if and only if } a_1 = a_2 = \cdots = a_k = 0.$$

Definition A.4. *A set of vectors $x_1, x_2, \ldots, x_k \in V$ is said to be a basis of V if*

- *The set $\{x_1, x_2, \ldots, x_k\}$ is linearly independent;*
- $V = \mathrm{Span}_{\mathbb{F}}\{x_1, x_2, \ldots, x_k\}$.

Definition A.5. *If such a set of vectors $x_1, x_2, \ldots, x_k \in V$ exists then the number of elements k of the set $x_1, x_2, \ldots, x_k$ is said to be the dimension of the vector space V. The dimension of the vector space V is denoted by* $\dim_{\mathbb{F}} V = k$ *(sometimes just* $\dim V = k$ *if the main field is predefined).*

Remark. Note that

- $\dim(W_1 + W_2) = \dim W_1 + \dim W_2 - \dim W_1 \cap W_2$;
- $\dim(W_1 \oplus W_2) = \dim W_1 + \dim W_2$.

For any subspace W_1 of a finite-dimensional space V there exists a subspace W_2 such that $V = W_1 \oplus W_2$. The subspaces W_1 and W_2 are called the direct complements to each other.

A.2 LINEAR TRANSFORMATIONS

Definition A.6. *Let V_1 and V_2 be vector spaces over a field $\mathbb{F}$. A function $f : V_1 \longrightarrow V_2$ is said to be a linear transformation if*

$$f(a_1x_1 + a_2x_2) = a_1f(x_1) + a_2f(x_2), \textit{ for all } x_1, x_2 \in V_1, \textit{ and } a_1, a_2 \in \mathbb{F}.$$

Let $\mathrm{Hom}(V_1, V_2)$ denote the set of linear transformations from V_1 to V_2. For $f, g \in \mathrm{Hom}(V_1, V_2)$ and $a \in \mathbb{F}$ we define $f \oplus g$ and $a \odot f$ by the formulas

$$(f \oplus g)(x) = f(x) + g(x) \text{ and } (a \odot f)(x) = af(x), \text{ for all } x \in V_1.$$

The set $\mathrm{Hom}(V_1, V_2)$ is a vector space over $\mathbb{F}$ with respect to the operations $\oplus$ and $\odot$.

Definition A.7. *Vector spaces V_1 and V_2 are called isomorphic if there exists a bijective linear transformation $f : V_1 \longrightarrow V_2$.*

Theorem A.1. *Vector spaces V_1 and V_2 are isomorphic if and only if they have the same dimension.*

A.3 DUAL VECTOR SPACE

Let V be a vector space over a field $\mathbb{F}$ and $F(V)$ be the vector space of functions defined on V. Consider a subset V^* of $F(V)$ satisfying the conditions:

- $f(x_1 + x_2) = f(x_1) + f(x_2)$ for all $x_1, x_2 \in V$;
- $f(\lambda x) = \lambda f(x)$ for all $x \in V$, $\lambda \in \mathbb{F}$.

The subset V^* is a subspace of the vector space $F(V)$ and it is said to be a dual vector space to V.

Let V be a finite-dimensional vector space. The dual space has the following properties:

- $\dim V = \dim V^*$;
- The finite-dimensional vector spaces V, V^* and V^{**} are isomorphic;
- Let $\{e_1, e_2, \ldots, e_n\}$ be a basis of V. Introduce the function $e^i(x) = x_i$, where x_i is the coordinate of the vector in the basis $\{e_1, e_2, \ldots, e_n\}$. An equivalent description of $\{e^i\}$ is as follows: $e^i(e_j) = \delta_{ij}$ (the Kronecker delta symbol: 1 for $i = j$ and 0 for $i \neq j$). The set of vectors $\{e^1, e^2, \ldots, e^n\}$ form a basis of V^* and it is said to be dual to $\{e_1, e_2, \ldots, e_n\}$ basis.

A.4 TENSOR PRODUCTS

In this subsection we recall the notion of tensor product of vector spaces.

Definition A.8. *The tensor product $V \otimes W$ of vector spaces V and W over a field $\mathbb{F}$ is the quotient of the space $V * W$ whose basis is given by formal symbols $v \otimes w, \quad v \in V, \quad w \in W$, by the subspace spanned by the elements*

$$(v_1 + v_2) \otimes w - v_1 \otimes w - v_2 \otimes w,$$
$$v \otimes (w_1 + w_2) - v \otimes w_1 - v \otimes w_2,$$
$$av \otimes w - a(v \otimes w), \quad v \otimes aw - a(v \otimes w),$$

where $v \in V, \quad w \in W, \quad a \in \mathbb{F}$.

It is easy to see that $V \otimes W$ can be equivalently defined as the quotient of the free abelian group $v \cdot w$ generated by $v \otimes w, \quad v \in V, \quad w \in W$ by the subgroup generated by

$$(v_1 + v_2) \otimes w - v_1 \otimes w - v_2 \otimes w,$$
$$v \otimes (w_1 + w_2) - v \otimes w_1 - v \otimes w_2,$$
$$av \otimes w - a(v \otimes w), \quad v \otimes aw - a(v \otimes w),$$

where $v \in V, \quad w \in W, \quad a \in \mathbb{F}$.

The elements $v \otimes w \in V \otimes W$ for $v \in V$, $w \in W$ are called pure tensors. Note that in general, there are elements of $V \otimes E$ which are not pure tensors.

This allows one to define the tensor product of any number of vector spaces, $V_1 \otimes \cdots \otimes V_n$. Note that this tensor product is associative, in the sense that $(V_1 \otimes V_2) \otimes V_3$ can be naturally identified with $V_1 \otimes (V_2 \otimes V_3)$.

In particular, people often consider tensor products of the form $V^{\otimes n} = V \otimes \cdots \otimes V$ (n times) for a given vector space V, and, more generally, $E := V^{\otimes n} \otimes (V^*)^{\otimes m}$. This space is called the space of tensors of type (n, m) on V. For instance, tensors of type $(0, 1)$ are vectors, of type $(1, 0)$-linear functionals (covectors), of type $(1, 1)$-linear operators, of type $(2, 0)$-bilinear forms, of type $(2, 1)$-algebra structures, etc.

If V is finite-dimensional with basis e_i, $i = 1, \ldots, N$, and e^i is the dual basis of V^*, then a basis of E is the set of vectors

$$e_{i_1} \otimes \cdots \otimes e_{i_n} \otimes e^{j_1} \otimes \cdots \otimes e^{j_m},$$

and a typical element of E is

$$\sum_{i_1,\ldots,i_n,j_1,\ldots,j_m=1}^{N} T^{i_1,\ldots,i_n}_{j_1,\ldots,j_m} e_{i_1} \otimes \cdots \otimes e_{i_n} \otimes e^{j_1} \otimes \cdots \otimes e^{j_m},$$

where T is a multidimensional table of numbers.

A.5 TENSOR ALGEBRA

The notion of tensor products allows us to give more conceptual (i.e., coordinate free) definitions of the free algebra, polynomial algebra, exterior algebra, and universal enveloping algebra of a Lie algebra.

Namely, given a vector space V, define its *tensor algebra* TV over a field $\mathbb{F}$ to be $TV = \oplus_{n \geq 0} V^{\otimes n}$, with multiplication defined by $a \cdot b := a \otimes b$, $a \in V^{\otimes r}$, $b \in V^{\otimes s}$. Observe that a choice of a basis $\{x_1, \ldots, x_m\}$ in V defines an isomorphism of TV with the free algebra $\mathbb{F}\langle x_1, \ldots, x_m \rangle$.

Also, one can make the following definition.

Definition A.9.

(i) *The symmetric algebra SV of V is the quotient of TV by the ideal generated by $v \otimes w - w \otimes v$, $v, w \in V$.*

(ii) *The exterior algebra $\wedge V$ of V is the quotient of TV by the ideal generated by $v \otimes v$, $v \in V$.*

(iii) *If V is a Lie algebra, the universal enveloping algebra $U(L)$ of V is the quotient of TV by the ideal generated by $v \otimes w - w \otimes v - [v, w]$, $v, w \in V$.*

It is easy to see that a choice of a basis $\{x_1, \ldots, x_m\}$ in V identifies SV with the polynomial algebra $\mathbb{F}[x_1, \ldots, x_m]$, $\wedge V$ with the exterior algebra $\wedge_k(x_1, \ldots, x_m)$, and the universal enveloping algebra $U(V)$ with one defined previously.

Also, it is easy to see that we have decompositions

$$SV = \bigoplus_{n \geq 0} S^n V, \text{ and } \wedge V = \bigoplus_{n \geq 0} \wedge^n V.$$

A.6 MATRIX OF A LINEAR TRANSFORMATION

Let V_1 and V_2 be vector spaces over a field $\mathbb{F}$ with the distinguished bases $\mathrm{e} = \{e_1, e_2, \ldots, e_n\}$ and $\mathrm{e}' = \{e_1', e_2', \ldots, e_m'\}$, respectively. Consider an arbitrary linear transformation $f : V_1 \longrightarrow V_2$ and associate with it an $m \times n$ matrix A_f with entries from $\mathbb{F}$ as follows (note that the sizes of A_f are the same as the dimensions of V_1, V_2 in reverse order). Represent the vector $f(e_k)$ as linear combinations: $f(e_k) = \sum_{i=1}^{m} a_{ik} e_i'$.

The matrix $A_f = (a_{ij})_{i=1,2,\ldots,m,\ j=1,2,\ldots,n}$ is called the matrix of the linear transformation f with respect to bases e and e′. The correspondence $f \longmapsto A_f$ gives the isomorphism between the vector spaces $\mathrm{Hom}(V_1, V_2)$ and $M_{m \times n}(\mathbb{F})$.

Let now $V_1 = V_2 = V$ then for $f : V \longrightarrow V$ the matrix $A_f = (a_{ij})_{i,j=1,2,\ldots,n}$ and it is called the matrix of the linear transformation f of V on the basis $\mathrm{e} = \{e_1, e_2, \ldots, e_n\}$. If we take another basis $\mathrm{e}' = \{e_1', e_2', \ldots, e_n'\}$ with the matrix A_f' of f on e′ and the base change matrix $M : \mathrm{e} = \{e_1, e_2, \ldots, e_n\} \longrightarrow \mathrm{e}' = \{e_1', e_2', \ldots, e_n'\}$ then one has $A_f' = M^{-1} A_f M$. One of the main problems of linear algebra is to choose M

so that $M^{-1}AM$ is as nearly diagonal as possible. Let us consider the case when $M^{-1}AM$ has the diagonal form.

Definition A.10. *A subspace W of V is said to be invariant with respect to a linear transformation $f : V \longrightarrow V$ if $f(W) \subset W$.*

Definition A.11. *A linear transformation $f : V \longrightarrow V$ is diagonalizable if either one of the following two equivalent conditions holds:*

i) The vector space V decomposes into a direct sum of one-dimensional invariant subspaces;

ii) There exists a basis of V with respect to which the matrix of f is diagonal.

Note that if there exists a k-dimensional invariant subspace W of an n-dimensional vector space V with respect to a linear transformation $f : V \longrightarrow V$ then there exists a basis of V such that the matrix of f with respect to the basis has the form

$$\begin{pmatrix} a_{11} & a_{12} & \dots & a_{1k} & a_{1k+1} & \dots & a_{1n} \\ a_{21} & a_{22} & \dots & a_{2k} & a_{2k+1} & \dots & a_{2n} \\ & & \dots & & & \dots & \\ a_{k1} & a_{k2} & \dots & a_{kk} & a_{kk+1} & \dots & a_{kn} \\ 0 & 0 & \dots & 0 & a_{k+1k} & \dots & a_{k+1n} \\ & & \dots & & & & \\ 0 & 0 & \dots & 0 & a_{nk+1} & \dots & a_{nn} \end{pmatrix}.$$

Note that if there exist two such invariant subspaces W_1 ($\dim W_1 = k$) and W_2 with $V = W_1 \oplus W_2$ then there exists a basis so that the matrix of f has the form

$$\begin{pmatrix} a_{11} & a_{12} & \dots & a_{1k} & 0 & \dots & 0 \\ a_{21} & a_{22} & \dots & a_{2k} & 0 & \dots & 0 \\ & & \dots & & & \dots & \\ a_{k1} & a_{k2} & \dots & a_{kk} & 0 & \dots & 0 \\ 0 & 0 & \dots & 0 & a_{k+1k} & \dots & a_{k+1n} \\ & & \dots & & & & \\ 0 & 0 & \dots & 0 & a_{nk+1} & \dots & a_{nn} \end{pmatrix}.$$

Moreover, if there exist k_i-dimensional ($i = 1, 2, \dots, s$) invariant subspaces W_i ($\dim W_i = k_i$) with $V = W_1 \oplus W_2 \oplus \dots \oplus W_s$ then one can find a basis with respect to which the matrix of f has the form

$$\left(\begin{array}{c|c|c|c} A_1 & 0 & 0 & 0 \\ \hline 0 & A_2 & 0 & 0 \\ \hline 0 & 0 & \dots & 0 \\ \hline 0 & 0 & 0 & A_s \end{array}\right).$$

Definition A.12.

i) *A one-dimensional subspace* $V_1 \subset V$ *is said to be a proper subspace for the operator* $f : V \longrightarrow V$ *if it is invariant, i.e.,* $f(V_1) \subset V_1$. *If* V_1 *is such a subspace, then the effect of* f *on it is equivalent to multiplication by a scalar* $\lambda \in \mathbb{F}$. *This scalar is called the eigenvalue of* f *(on* V_1*).*

ii) *The vector* $x \in V$ *is said to be an eigenvector of* f *if* $x \neq 0$ *and the linear span* Span$x$ *is a proper subspace. In other words,* $f(x) = \lambda x$ *for an appropriate* $\lambda \in \mathbb{F}$.

According to the definition above, diagonalizable operators f admit a decomposition of V into a direct sum of its proper subspaces. Let us now determine when f has at least one proper subspace.

Definition A.13. *Let* V *be a finite-dimensional vector space,* $f : V \longrightarrow V$ *be a linear transformation and let* A *be its matrix in some basis. Denote by* $P(t)$ *the polynomial* $\det(tI - A)$ *(*I *is the identity matrix) with coefficients in* $\mathbb{F}$ *and call it the characteristic polynomial of the operator* f *and of the matrix* A.

Theorem A.2.

a) *The characteristic polynomial of* f *does not depend on the choice of basis in which its matrix is represented;*

b) *Any eigenvalue of* f *is a root of* $P(t)$ *and any root of* $P(t)$ *lying in* $\mathbb{F}$ *is an eigenvalue of* f, *corresponding to some (not necessarily the unique) proper subspace of* V.

Since the field $\mathbb{F}$ is algebraically closed any polynomial with leading coefficient equal to one is represented as a product of linear polynomials, therefore, $P(t) = \prod_{i=1}^{n}(t - \lambda_i)^{r_i}$, where $\lambda_i \in \mathbb{F}$, $\lambda_i \neq \lambda_j$ for $i \neq j$. The number r_j is called the multiplicity of the root λ_i of the polynomial $P(t)$. The set of all roots of the characteristic polynomial $P(t)$ is called the spectrum of f. If all multiplicities are equal to one, the spectrum of f is said to be simple.

Let us give the following important theorem on the characteristic polynomial of a linear transformation.

Theorem A.3. *(The Cayley-Hamilton Theorem) If $P(t)$ is the characteristic polynomial for a linear transformation $f : V \longrightarrow V$ with a matrix A on some basis of V then $P(A) = 0$.*

Definition A.14.

a) A matrix A of the form

$$\begin{pmatrix} \lambda & 1 & 0 & \dots & 0 \\ 0 & \lambda & 1 & \dots & 0 \\ \dots & \dots & \dots & \dots & \dots \\ 0 & 0 & 0 & \dots & \lambda \end{pmatrix}$$

is called an $r \times r$ Jordan block $J_r(\lambda)$ with the eigenvalue λ.

b) A Jordan matrix is a matrix consisting of diagonal blocks

$$\left(\begin{array}{c|c|c|c} J_{r_1}(\lambda_1) & 0 & 0 & 0 \\ \hline 0 & J_{r_2}(\lambda_2) & 0 & 0 \\ \hline 0 & 0 & \dots & 0 \\ \hline 0 & 0 & 0 & J_{r_s}(\lambda_s) \end{array}\right)$$

c) A Jordan basis for the operator $f : V \longrightarrow V$ is a basis of the space V with respect to that the matrix of f is a Jordan matrix or, as it is usually said, that the matrix has the Jordan normal form.

d) The solution of a matrix equation of the form $X^{-1}AX = J$, where A is a square matrix, X is an unknown non-singular matrix, and J is an unknown Jordan matrix, is called the reduction of A to Jordan normal form.

Definition A.15. *The vector $x \in V$ is called a root vector of the operator f, corresponding to $\lambda \in \mathbb{F}$, if there exists an r such that $(f - \lambda I)^r x = 0$.*

Note that all eigenvectors are evidently root vectors. The set of all root vectors of f corresponding to λ is denoted by $V(\lambda)$.

Theorem A.4. *a) The set $V(\lambda)$ is an invariant subspace of V and $V(\lambda) \neq \{0\}$ if and only if λ is an eigenvalue of f.*

b) $V = \oplus V(\lambda_i)$, where λ_i runs through all the eigenvalues of the operator f, i.e., $\lambda_i \neq \lambda_j$, $i, j = 1, 2, \dots, s$.

A.7 JORDAN NORMAL FORM

The main goal of this section is to prove the following theorem on the existence and uniqueness of the Jordan normal form for matrices and linear operators.

Theorem A.5. *Let $\mathbb{F}$ be an algebraically closed field, V a finite-dimensional linear space over $\mathbb{F}$, and $f : V \to V$ be a linear operator. Then:*

a) A Jordan basis exists for the operator f, i.e., the matrix of the operator A in the original basis can be reduced by a change X of basis to the Jordan form $X^{-1}AX = J$.

b) The matrix J is unique, apart from a permutation of its constituent Jordan blocks.

The proof of the theorem is divided into a series of intermediate steps. We begin by constructing the direct composition $V = \oplus_{i=1}^{n} V_i$, where the V_i, are invariant subspaces for f, which will later correspond to the *set of Jordan blocks for f with the same number λ along the diagonal*. In order to characterize these subspaces in an invariant manner, we recall that

$$(J_r(\lambda) - \lambda I_r)^n = 0, \text{ where } I_r \text{ is a } r \times r \text{ identity matrix.}$$

If some power of an operator is zero then the operator is said to be *nilpotent*. Thus the operator $f - \lambda$ is nilpotent on the subspace corresponding to the block $J_r(\lambda)$. The same is true for its restriction to the sum of subspaces for fixed λ. This motivates the following definition.

Definition A.16. *The vector $l \in V$ is called a root vector of the operator f, corresponding to $\lambda \in \mathbb{F}$, if there exists an r such that $(f - \lambda)^r l = 0$ (here $(f - \lambda)$ denotes the operator $f - \lambda \mathrm{id}$).*

All eigenvectors are evidently root vectors.

Proposition A.1. *We denote by $V(\lambda)$ the set of root vectors of the operator f in V corresponding to λ. Then $V(\lambda)$ is a linear subspace in V and $V(\lambda) \neq \{0\}$ if and only if λ is an eigenvalue of f.*

Proof. Let $(f - \lambda)^{r_1} l_1 = (f - \lambda)^{r_2} l_2 = 0$. Setting $r = \max(r_1, r_2)$, we find that $(f - \lambda)^r (l_1 + l_2) = 0$ and $(f - \lambda)^{r_1}(a l_1) = 0$. Therefore, $V(\lambda)$ is a subspace.

If λ is an eigenvalue of f, then there exists an eigenvector corresponding to λ such that $V(\lambda) \neq \{0\}$. Conversely, let $l \in V(\lambda)$, $l \neq 0$. We select the *smallest* value of r for which $(f - \lambda)^r l = 0$. Obviously, $r \geq 1$. The vector $l' = (f - \lambda)^{r-1} l$ is an eigenvector of f with eigenvalue $\lambda : l' \neq 0$ according to the choice of r and $(f - \lambda)l' = 0$, whence $f(l') = \lambda l'$. □

Proposition A.2. $V = \oplus V(\lambda_i)$, *where* λ_i *runs through all the eigenvalues of the operator* f, *i.e., the different roots of the characteristic polynomial of* f.

Proof. Let $P(t) = \prod_{i=1}^{s} (t - \lambda_i)^{r_i}$ be the characteristic polynomial of f, $\lambda_i \neq \lambda_j$ for $i \neq j$. Set $F_i(t) = P(t)(t - \lambda_i)^{-r_i}$, $f_i = F_i(f)$, $V_i = \operatorname{im} f_i$. We check the following series of assertions.

a) $f - \lambda_i^{r_i} V_i = \{0\}$, that is, $V_i \subset V(\lambda_i)$. Indeed,

$$(f - \lambda_i)^{r_i} f_i = (f - \lambda_i)^{r_i} F_i(f) = P(f) = 0$$

according to the Cayley-Hamilton theorem.

b) $V = V_1 + \cdots + V_s$. Indeed, since the polynomials $F_i(t)$ in aggregate are relatively prime, there exist polynomials $X_i(t)$ such that $\sum_{i=1}^{s} F_i(t) X_i(t) = 1$. Therefore, substituting f for t, we have

$$\sum_{i=1}^{s} F_i(f) X_i(f) = \text{id}.$$

Applying this identity to any vector $l \in V$, we find

$$l = \sum_{i=1}^{s} F_i(X_i)(f), l \in \sum_{i=1}^{s} V_i.$$

c) $V = V_1 \oplus \cdots \oplus V_s$. Indeed, we choose $1 \leq i \leq s$ and verify that $V_i \cap (\sum_{j \neq i} V_j) = \{0\}$. Let l be a vector from this intersection. Then

$$(f - \lambda_i)^{r_i} l = 0, \quad \text{since} \quad l \in V_i;$$

$$F_i(f) l = \prod_{j \neq i} (f - \lambda_j)^{r_j} l = 0, \quad \text{since} \quad l \in \sum_{j \neq i} V_j.$$

Since $(t - \lambda_i)^{r_i}$ and $F_i(t)$ are relatively prime polynomials, there exist polynomials $X(t)$ and $Y(t)$ such that $X(t)(t - \lambda_i)^{r_i} + Y(t) F_i(t) = 1$.

Substituting here f for t and applying the operator identity obtained to l, we obtain $X(f)(0) + Y(f)(0) = l = 0$.

d) $V_i = V(\lambda_i)$. Indeed, we have already verified that $V_i \subset V(\lambda_i)$. To prove the converse we choose a vector $l \in V(\lambda_i)$ and represent it in the form $l = l' + l''$, $l' \in V_i$, $l'' \in \oplus_{j \neq i} V_i$. There exists a number r' such that $(f - \lambda_i)^{r'} l'' = 0$, because $l'' = l - l' \in V(\lambda_i)$. In addition, $F_i(f)l'' = 0$. Writing the identity $X(t)(t - \lambda_i)^{r'} + Y(t)F_i(t) = 1$ and replacing t by f, we find that $l'' = 0$, so that $l = l' \in V_i$. □

Corollary A.1. *If the spectrum of an operator f is simple, then f is diagonalizable.*

Proof. Indeed, the number of different eigenvalues of f then equals $n = \deg P(t) = \dim V$. Hence, in the decomposition $V = \oplus_{i=1}^{n} V(\lambda_i)$ all spaces $V(\lambda_i)$ are one-dimensional and since each of them contains an eigenvector, the operator f becomes diagonal in a basis consisting of these vectors. □

We now fix one of the eigenvalues λ and prove that the restriction of f to $V(\lambda)$ has a Jordan basis, corresponding to this value of λ. To avoid introducing a new notation we shall assume up to the end of the proof of the proposition A.3 that f has only one eigenvalue λ and $V = V(\lambda)$. Moreover, since any Jordan basis for the operator f is simultaneously a Jordan basis for the operator $f + \mu$, where μ is any constant, we can even assume that $\lambda = 0$. Then, according to the Cayley-Hamilton theorem, the operator f is nilpotent: $P(t) = t^n$, $f^n = 0$. We shall now prove the following proposition.

Proposition A.3. *A nilpotent operator f on a finite-dimensional space V has a Jordan basis; the matrix of f in this basis is a combination of blocks of the form $J_r(0)$.*

Proof. If we already have a Jordan basis in the space V, it is convenient to represent it by a diagram D, similar to the one shown here.

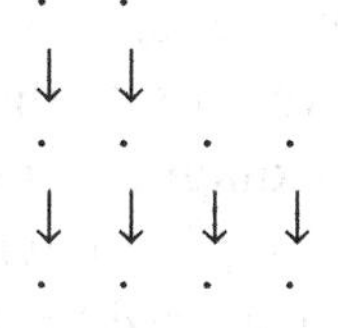

In this diagram, the dots denote elements of the basis and the arrows describe the action of f (in the general case, the action of $f - \lambda$). The operator f transforms to zero the elements in the lowest row, that is, the eigenvectors of f entering into the basis occur in this row. Each column thus stands for a basis of the invariant subspace, corresponding to one Jordan block, whose dimension equals the height of this column (the number of points in it): if

$$f(e_h) = e_{h-1},\ f(e_{h-1}) = e_{h-2}, \dots,\ f(e_1) = 0,$$

then

$$f(e_1, \dots, e_h) = (e_1, \dots, e_h) \begin{pmatrix} 0 & 1 & 0 & \dots & 0 \\ 0 & 0 & 1 & \dots & 0 \\ \dots & \dots & \dots & \dots & \dots \\ 0 & 0 & 0 & \dots & 0 \end{pmatrix}$$

Conversely, if we find a basis of V whose elements are transformed by f into other elements of the basis or into zero, so that the elements of this basis together with the action of f can be described by such a diagram, then it will be the Jordan basis for V.

We shall prove existence by induction on the dimension of V. If $\dim V = 1$, then the nilpotent operator f is a zero operator and any non-zero vector in V forms its Jordan basis. Now, let $\dim V = n > 1$ and assume that the existence of a Jordan basis has already been proved for dimensions less than n. We denote by $V_0 \subset V$ the subspace of eigenvectors for f, that is, $\ker f$. Since $\dim V_0 > 0$, we have $\dim V/V_0 < n$, while the operator $f : V \to V$ induces the operator

$$\tilde{f} : V/V_0 \to V/V_0 : \tilde{f}(l + V_0) = f(l) + V_0.$$

(The correctness of the definition of f and its linearity are obvious.)

According to the induction hypothesis, $\tilde{f}$ has a Jordan basis. We can assume that it is non-empty. Otherwise $V = V_0$ and any basis of V_0 will be a Jordan basis for $\tilde{f}$. Let us construct the diagram $\tilde{D}$ for elements of the Jordan basis of $\tilde{f}$. We take the uppermost vector $\tilde{e}_i$, $i = 1, \dots, m$ in each column, and set $\tilde{e}_i = e_i + V_0$, $e_i \in V$. We shall now construct the diagram D of vectors of the space V as follows. For $i = 1, \dots, m$ the ith column in the diagram D will consist (top to bottom) of the vectors $e_i, f_{e_i}, \dots, f^{h_i - 1}(e_i), f^{h_i}(e_i)$, where h_i is the height of the ith column in the diagram $\tilde{D}$. Since $\tilde{f}^{h_i}(\tilde{e}_i) = 0$, $f^{h_i}(e_i) \in V_0$ and $f^{h_i+1}(e_i) = 0$. We select a basis of the linear span of the vectors $f^{h_1}(e_1), \dots, f^{h_m}(e_m)$ in V_0, extend it to a basis of V_0, and insert the additional vectors as additional columns (of unit height) in the bottom row of the diagram D; f transforms them into zero.

Thus the diagram D consisting of vectors of the space V together with the action of f on its elements has exactly the form required for a Jordan basis. We have only to check that the elements of D actually form a basis of V.

We shall first show that the linear span of D equals V. Let $l \in V$, $\tilde{l} = l + V_0$. By assumption, $\tilde{l} = \sum_{i=1}^{m} \sum_{j=0}^{h_i-1} a_{ij} \tilde{f}^j(\tilde{e}_i)$. Since V_0 is invariant under f, it follows that

$$l - \sum_{i=1}^{m} \sum_{j=0}^{h_i-1} a_{ij} f^j(e_i) \in V_0.$$

But all the vectors $f^j(e_i)$, $j \le h_i - 1$ lie in the rows of the diagram D, beginning with the second from the bottom, and the subspace V_0 is generated by the elements of the first row of D by construction. Therefore l can be represented as a linear combination of the elements of D.

It remains to verify that the elements of D are linearly independent. First of all, the elements in the bottom row of D are linearly independent. Indeed, if some non-trivial linear combination of them equals zero, then it must have the form $\sum_{i=1}^{m} a_i f^{h_i}(e_i) = 0$, because the remaining elements of the bottom row extend the basis of the linear span of $\{f^{h_1}(e_1), \ldots, f^{h_m}(e_m)\}$ up to a basis of V_0 . But all the $h_i \ge 1$, therefore

$$f\left(\sum_{i=1}^{m} a_i f^{h_i-1}(e_i)\right) = 0,$$

so that

$$\sum_{i=1}^{m} a_i f^{h_i-1}(e_i) \in V_0 \quad \text{and} \quad \sum_{i=1}^{m} a_i \tilde{f}^{h_i-1}(\tilde{e}_i) = 0.$$

It follows from the last relation that all the $a_i = 0$, because the vectors $\tilde{f}^{h_i-1}(\tilde{e}_i)$ comprise the bottom row of the diagram $\tilde{D}$ and are part of a basis of V/V_0.

Finally, we shall show that if there exists a non-trivial linear combination of the vectors of D equal to zero, then it is possible to obtain from it a non-trivial linear dependence between the vectors in the bottom row of D. Indeed, consider the top row of D, which contains the non-zero coefficients of this imagined linear combination. Let the number of this row (counting from the bottom) be h. We apply to this combination the operator f^{h-1}. Evidently, the part of this combination corresponding to the hth row will transform into a non-trivial linear combination of elements of the bottom row, while the remaining terms will vanish. This completes the proof of the proposition. □

Now we have only to verify the part of Theorem A.5 that refers to uniqueness.

Let an arbitrary Jordan basis of the operator f be fixed. Any diagonal element of the matrix f in this basis is obviously one of the eigenvalues λ of this operator. Examine the part of the basis corresponding to all of the blocks of matrices with this value of λ and denote by V_λ its linear span. Since $(J_r(\lambda) - \lambda)^r = 0$, we have $V_\lambda \subset V(\lambda)$, where $V(\lambda)$ is the root space of V_0. In addition, $V = \oplus V_{\lambda_i}$ by definition of the Jordan basis and $V = \oplus V(\lambda_i)$ by Proposition A.2, where in both cases λ_i runs through all eigenvalues of f once. Therefore, $\dim V_{\lambda_i} = \dim V(\lambda_i)$ and $V_{\lambda_i} = V(\lambda_i)$. Hence, the sum of the dimensions of the Jordan blocks, corresponding to each λ – id, independent of the choice of Jordan basis and, moreover, the linear spans of the corresponding subsets of the basis V_{λ_i} are basis-independent. It is thus sufficient to check the uniqueness theorem for the case $V = V(\lambda)$ or even for $V = V(0)$.

We construct the diagram D corresponding to a given Jordan basis of $V = V(0)$. The dimensions of the Jordan blocks are the heights of its columns; if the columns in the diagram are arranged in decreasing order, these heights are uniquely determined if the lengths of the rows in the diagram are known, beginning with the bottom row, in decreasing order. We shall show that the length of the bottom row equals the dimension of $V_0 = \ker f$. Indeed, we take any eigenvector l for f and represent it as a linear combination of the elements of D. All vectors lying above the bottom row will appear in this linear combination with zero coefficients. Indeed, if the highest vectors with non-zero coefficients were to lie in a row with number $h \geq 2$, then the vector $f^{h-1}(l) = 0$ would be a non-trivial linear combination of the elements of the bottom row of D (cf. the end of the proof of Proposition A.3), and this contradicts the linear independence of the elements of D. This means that the bottom row of D forms a basis of L_0, so that its length equals $\dim L_0$; hence, this length is the same for all Jordan bases. In exactly the same way, the length of the second row does not depend on the choice of basis, so that, in the notation used in this section, it equals the dimension of $\ker \tilde{f}$ in V/V_0. This completes the proof of uniqueness and of Theorem A.5.

Remarks. a) Let the operator f be represented by a matrix A in some basis. Then the problem of reducing A to Jordan form can be solved as follows.

Calculate the characteristic polynomial of A and its roots.

Calculate the dimensions of the Jordan blocks, corresponding to the roots λ. For this, it is sufficient to calculate the lengths of the rows of the corresponding diagrams, that is,

$$\dim\ker(A - \lambda I),\ \dim\ker(A - \lambda I)^2 - \dim\ker(A - \lambda I),$$
$$\dim\ker(A - \lambda I)^3 - \dim\ker(A - \lambda I)^2, \ldots .$$

Construct the Jordan form J of the matrix A and solve the matrix equation $AX - XJ = 0$. The space of solutions of this linear system of equations will, generally speaking, be multidimensional, and the solutions will also include singular matrices. But according to the existence theorem, non-singular solutions necessarily exist; any one can be chosen.

b) One of the most important applications of the Jordan form is for the calculation of functions of a matrix (thus far we have considered only polynomial functions). Assume, for example, that we must find a large power A^N of the matrix A. Since the degree of the Jordan matrix is easy to calculate, an efficient method is to use the formula $A^N = XJ^NX^{-1}$, where $A = XJX^{-1}$. The point is that the matrix X is calculated once and for all and does not depend on N. The same formula can be used to estimate the growth of the elements of the matrix A^N.

c) It is easy to calculate the minimal polynomial of a matrix A in terms of the Jordan form. Indeed, we shall restrict ourselves for simplicity to the case of a field with zero characteristic. Then the minimal polynomial of $J_r(\lambda)$ equals $(t - \lambda)^r$, the minimal polynomial of the block matrix $(J_{r_i}(\lambda))$ equals $(t - \lambda)^{\max(r_i)}$, and finally the minimal polynomial of the general Jordan matrix with diagonal elements $\lambda_1, \ldots, \lambda_s$, ($\lambda_i \neq \lambda_j$ for $i \neq j$) equals $\prod_{j=1}^{s}(t - \lambda_j)^{r_j}$, where r_j is the smallest dimension of the Jordan block corresponding to λ_j.

APPENDIX B

Elements of Representation Theory

Roughly speaking, representation theory studies symmetry in linear spaces. It is a beautiful mathematical subject which has many applications, ranging from number theory and combinatorics to geometry, probability theory, quantum mechanics and quantum field theory. Representation theory was born in 1896 in the work of the German mathematician F. G. Frobenius. This work was triggered by a letter to Frobenius by R. Dedekind. In this letter Dedekind made the following observation: take the multiplication table of a finite group G and turn it into a matrix XG by replacing every entry g of this table by a variable xg. Then the determinant of XG factors into a product of irreducible polynomials in $\{xg\}$, each of which occurs with multiplicity equal to its degree. Dedekind checked this surprising fact in a few special cases, but could not prove it in general. So he gave this problem to Frobenius. In order to find a solution of this problem, Frobenius created representation theory of finite groups. In this section we introduce the notion of representations of associative algebras.

Definition B.1. *A representation of an algebra $\mathfrak{A}$ (also called a left $\mathfrak{A}$-module) is a vector space V together with a homomorphism of algebras $\rho : \mathfrak{A} \longrightarrow \mathrm{End}(V)$.*

Similarly, a right $\mathfrak{A}$-module is a space V equipped with an antihomomorphism $\rho : \mathfrak{A} \longrightarrow \mathrm{End}(V)$; i.e., ρ satisfies $\rho(ab) = \rho(b)\rho(a)$ and $\rho(1) = 1$.

The usual abbreviated notation for $\rho(a)v$ is av for a left module and va for the right module. Then the property that ρ is an (anti)homomorphism can be written as a kind of associativity law:

$(ab)v = a(bv)$ for left modules, and $(va)b = v(ab)$ for right modules.

Here are some examples of representations.

Example B.1.

1. $V = \{0\}$.

2. $V = \mathfrak{A}$, *and* $\rho : \mathfrak{A} \longrightarrow \mathrm{End}(\mathfrak{A})$ *is defined as follows:* $\rho(a)$ *is the operator of left multiplication by a, so that* $\rho(a)b = ab$ *(the usual product). This representation is called the regular representation of* $\mathfrak{A}$. *Similarly, one can equip* $\mathfrak{A}$ *with a structure of a right* $\mathfrak{A}$*-module by setting* $\rho(a)b := ba$.

3. $\mathfrak{A} = \mathbb{F}$. *Then a representation of* $\mathfrak{A}$ *is simply a vector space over* $\mathbb{F}$.

4. $\mathfrak{A} = \mathbb{F}[x_1, \dots, x_n]$. *Then a representation of* $\mathfrak{A}$ *is just a vector space V over* $\mathbb{F}$ *with a collection of arbitrary linear operators* $\rho(x_1), \dots, \rho(x_n) : V \longrightarrow V$.

Definition B.2. *A subrepresentation of a representation V of an algebra* $\mathfrak{A}$ *is a subspace* $W \subset V$ *which is invariant under all the operators* $\rho(a) : V \longrightarrow V$, $a \in \mathfrak{A}$.

For instance, $\{0\}$ and V are always subrepresentations.

Definition B.3. *A representation* $V \neq \{0\}$ *of* $\mathfrak{A}$ *is irreducible (or simple) if the only subrepresentations of V are* $\{0\}$ *and V.*

Definition B.4. *Let* V_1, V_2 *be two representations of an algebra* $\mathfrak{A}$. *A homomorphism (or intertwining operator)* $\varphi : V_1 \longrightarrow V_2$ *is a linear operator which commutes with the action of* $\mathfrak{A}$, *i.e.,* $\varphi(av) = a\varphi(v)$ *for any* $v \in V_1$. *A homomorphism* φ *is said to be an isomorphism of representations if it is an isomorphism of vector spaces. The set (space) of all homomorphisms of representations* $V_1 \longrightarrow V_2$ *is denoted by* $\mathrm{Hom}_{\mathfrak{A}}(V_1, V_2)$.

Note that if a linear operator $\varphi : V_1 \longrightarrow V_2$ is an isomorphism of representations then so is the linear operator $\varphi^{-1} : V_2 \longrightarrow V_1$. Two representations between which there exists an isomorphism are said to be isomorphic. For practical purposes, two isomorphic representations may be regarded as "the same", although there could be subtleties related to the fact that an isomorphism between two representations, when it exists, is not unique.

Definition B.5. *Let V_1, V_2 be representations of an algebra $\mathfrak{A}$. Then the space $V_1 \oplus V_2$ has an obvious structure of a representation of $\mathfrak{A}$, given by $a(v_1 \oplus v_2) = av_1 \oplus av_2$.*

Definition B.6. *A nonzero representation V of an algebra $\mathfrak{A}$ is said to be indecomposable if it is not isomorphic to a direct sum of two nonzero representations.*

It is obvious that an irreducible representation is indecomposable. On the other hand, the converse statement is false in general. One of the main problems of representation theory is to classify irreducible and indecomposable representations of a given algebra up to isomorphism. This problem is usually difficult and often can be solved only partially (say, for finite dimensional representations). Below we will give a number of examples in which this problem is partially or fully solved for specific algebras. We will now give Schur's lemma. Although it is very easy to prove, it is fundamental in the whole subject of representation theory.

Proposition B.1. *(Schur's lemma) Let V_1, V_2 be representations of an algebra $\mathfrak{A}$ over any field $\mathbb{F}$ (which need not be algebraically closed). Let $\varphi : V_1 \longrightarrow V_2$ be a nonzero homomorphism of representations. Then:*

(i). *If V_1 is irreducible then φ is injective;*

(ii). *If V_2 is irreducible φ is surjective.*

Thus, if both V_1 and V_2 are irreducible, φ is an isomorphism.

Corollary B.1. *(Schur's lemma for algebraically closed fields) Let V be a finite dimensional irreducible representation of an algebra A over an algebraically closed field $\mathbb{F}$, and $\varphi : V \longrightarrow V$ be an intertwining operator. Then $\varphi = \lambda \cdot \mathrm{Id}$ for some $\lambda \in \mathbb{F}$ (a scalar operator).*

Remark. Note that this Corollary is false over the field of real numbers: it suffices to take $\mathfrak{A} = \mathbb{C}$ (regarded as an $\mathbb{R}$-algebra), and $V = \mathfrak{A}$.

Corollary B.2. *Let $\mathfrak{A}$ be a commutative algebra. Then every irreducible finite dimensional representation V of $\mathfrak{A}$ is one-dimensional.*

Remark. Note that a one-dimensional representation of any algebra is automatically irreducible.

Example B.2.

1. $\mathfrak{A} = \mathbb{F}$. *Since representations of A are simply vector spaces, $V = \mathfrak{A}$ is the only irreducible and the only indecomposable representation.*

2. $\mathfrak{A} = \mathbb{F}[x]$. *Since this algebra is commutative, the irreducible representations of $\mathfrak{A}$ are its one-dimensional representations. They are defined by a single operator $\rho(x)$. In the one-dimensional case, this is just a number from $\mathbb{F}$. So all irreducible representations of $\mathfrak{A}$ are $V_\lambda = \mathbb{F}$, $\lambda \in \mathbb{F}$, in which the action of $\mathfrak{A}$ defined by $\rho(x) = \lambda$. Clearly, these representations are pairwise non-isomorphic.*

APPENDIX C

Zariski Topology

Let $\mathbb{F}$ be a fixed algebraically closed field. We define affine space over $\mathbb{F}$ denoted $\mathbb{A}^n$, to be the set of all $n-$tuples of elements of $\mathbb{F}$. An element $P \in \mathbb{A}^n$ is called a point, and if $P = (a_1, \ldots, a_n)$ with $a_i \in \mathbb{F}$, then the a_i are called the coordinates of P.

Let $A = \mathbb{F}[x_1, \ldots, x_n]$ be the polynomial ring in n variables over $\mathbb{F}$. We will interpret the elements of A as functions from the affine $n-$space to $\mathbb{F}$, by defining $f(P) = f(a_1, \ldots, a_n)$, where $f \in A$ and $P \in \mathbb{A}^n$. Thus if $f \in A$ is a polynomial, we can talk about the set of zeros of f, namely

$$Z(f) = \{P \in \mathbb{A}^n \mid f(P) = 0\}.$$

More generally, if T is any subset of A, we define the zero set of T to be the common zeros of all the elements of T:

$$Z(T) = \{P \in \mathbb{A}^n \mid f(P) = 0 \text{ for all } f \in T\}.$$

Clearly, if $\mathfrak{I}$ is the ideal of A generated by T, then $Z(T) = Z(\mathfrak{I})$. Furthermore, since A is a noetherian ring, any ideal $\mathfrak{I}$ has a finite set of generators $f_1, \ldots, f_r$ we can regard $Z(T)$ as the common zeros of the finite set of polynomials $f_1, \ldots, f_r$.

Definition C.1. *A subset Y of $\mathbb{A}^n$ is an algebraic set if there exists a subset $T \subset A$ such that $Y = Z(T)$.*

Example C.1. *The space $\mathbb{A}^n$ is an algebraic set defined by $f(x_1, x_2, \ldots, x_n) \equiv 0$.*

Example C.2. *The set $\mathbb{A}^n \setminus \{0\}$ is not algebraic since any polynomial vanishing on all points $\mathbb{A}^n \setminus \{0\}$ is only zero polynomial.*

Example C.3. *If $X \subset \mathbb{A}^n$ and $Y \subset \mathbb{A}^m$ are algebraic subsets then $X \times Y \subset \mathbb{A}^{n+m}$ is also an algebraic subset. If both X, Y are irreducible then $X \times Y$ is also irreducible.*

Example C.4. *Let us describe all algebraic sets X in $\mathbb{A}^1$. Such a set is defined as a set of solutions of the system of equations*

$$f_1(x) = 0, f_2(x) = 0, \dots, f_m(x) = 0,$$

where $f_1(x), f_2(x), \dots, f_m(x)$ are polynomials of singe variable x.

- *If all of $f_1(x), f_2(x), \dots, f_m(x)$ are zero then $X = \mathbb{A}^1$.*
- *If $f_1(x), f_2(x), \dots, f_m(x)$ are relatively prime then there is no common root, hence $X = \emptyset$.*
- *If $gcd(f_1(x), f_2(x), \dots, f_m(x)) = d(x)$ then $d(x) = (x - \alpha_1)(x - \alpha_2)\dots(x - \alpha_m)$ and $X = \{\alpha_1, \alpha_2, \dots \alpha_m\}$, i.e., it consists of finite numbers of points.*

Proposition C.1. *The union of two algebraic sets is an algebraic set. The intersection of any family of algebraic sets is an algebraic set. The empty set and the whole space are algebraic sets.*

We define **the Zariski topology on** $\mathbb{A}^n$ by taking the open subsets to be the complements of the algebraic sets. This is a topology, because according to the proposition, the intersection of two open sets is open, and the union of any family of open sets is open. Furthermore, the empty set and the whole space are both open.

Example C.5. *The Zariski topology in $\mathbb{A}^n$ is the smallest topology in $\mathbb{A}^n$ for which any polynomial $f : \mathbb{A}^n \to \mathbb{A}$ function is continuous. Note that a continuous function $f : \mathbb{A}^n \to \mathbb{A}$ need not be polynomial. For example, any one-to-one function $f(x_1) : \mathbb{A} \to \mathbb{A}$ is continuous but it may be not polynomial. In general, a function $f : \mathbb{A} \to \mathbb{A}$ is continuous, if the coimage $f^{-1}(y) = \{x \in \mathbb{A} \mid f(x) = y\}$ is finite for any $y \in \mathbb{A}$.*

Example C.6. *Note that a polynomial map F defined by*

$$F = (f_1(\mathrm{x}), f_2(\mathrm{x}), \dots, f_m(\mathrm{x})) : \mathbb{A}^n \to \mathbb{A}^m$$

is continuous, where $f_i \in A = \mathbb{F}[\mathrm{x}]$ and $\mathrm{x} = (x_1, x_2, ..., x_n)$. Indeed, if

$$Y = Z(g_i(y_1, y_2, \dots, y_m), i = 1, 2, \dots, k)$$

is a closed subset in $\mathbb{A}^m$ *then*

$$\begin{aligned} F^{-1}(Y) &= \{\mathrm{x} = (x_1, x_2, \ldots, x_n) \in \mathbb{A}^n \mid F(\mathrm{x}) \in Y\} \\ &= \{\mathrm{x} = (x_1, x_2, \ldots, x_n) \in \mathbb{A}^n \mid g_i(f_1(\mathrm{x}), \ldots, f_m(\mathrm{x})) \\ &= 0,\ i = 1, 2, \ldots, k\}. \end{aligned}$$

Definition C.2. *A nonempty subset* Y *of a topological space* X *is irreducible if it cannot be expressed as a union* $Y = Y_1 \cup Y_2$ *of its two proper closed subsets* Y_1 *and* Y_2*. The empty set is not considered to be irreducible.*

Note that this definition is equivalent to each of the following:

(i) Every nonempty open subset of Y is dense.

(ii) Any pair of nonempty open subsets of Y intersect.

Example C.7. $\mathbb{A}^1$ *is irreducible, because its proper closed subsets are only finite sets, however* $\mathbb{A}^1$ *itself is infinite (because* $\mathbb{F}$ *is algebraically closed, hence infinite).*

Example C.8. *Any nonempty open subset of an irreducible space is irreducible and dense.*

Proposition C.2. *If* Y *is an irreducible subset of* X*, then its closure in* X *is also irreducible.*

Definition C.3. *An affine algebraic variety or simply affine variety is an irreducible closed subset of* $\mathbb{A}^n$ *(with the induced topology).*

Now we need to explore the relationship between subsets of $\mathbb{A}^n$ and ideals in A more deeply. So for any subset $Y \subset \mathbb{A}^n$, let us define the ideal of Y in A by

$$I(Y) = \{f \in A \mid f(P) = 0 \text{ for all } P \in Y\}.$$

Now we have a function Z that maps subsets A to algebraic sets, and a function I which maps subsets of $\mathbb{A}^n$ to ideals. Their properties are summarized in the following proposition.

Proposition C.3.

a) *If* $T_1 \subseteq T_2$ *are subsets of* A*, then* $Z(T_1) \supseteq Z(T_2)$.

b) *If* $Y_1 \subseteq Y_2$ *are subsets of* $\mathbb{A}^n$*, then* $I(Y_1) \supseteq I(Y_2)$.

c) *For any two subsets* Y_1, Y_2 *of* $\mathbb{A}^n$*, we have* $I(Y_1 \bigcup Y_2) = I(Y_1) \bigcap I(Y_2)$.

d) *For any ideal* $\mathfrak{J} \subseteq A$ *one has* $I(Z(\mathfrak{J})) = \sqrt{\mathfrak{J}}$*, where* $\sqrt{\mathfrak{J}}$ *is the radical of* $\mathfrak{J}$.

e) *For any subset* $Y \subseteq \mathbb{A}^n$ *we have* $Z(I(Y)) = \overline{Y}$*, where* $\overline{Y}$ *the closure of* Y.

Theorem C.1. *(Hilbert's Nullstellensatz). Let* $\mathbb{F}$ *be an algebraically closed field,* $\mathfrak{J}$ *be an ideal in* $A = \mathbb{F}[x^1, \ldots, x^n]$*, and* $f \in A$ *be a polynomial which vanishes at all points of* $Z(\mathfrak{J})$*. Then* $f^r \in \mathfrak{J}$ *for some integer* $r > 0$.

Proof. See Atiyah-Macdonald [22] or Zariski-Samuel [184]. □

Corollary C.1. *There is a one-to-one inclusion-reversing correspondence between algebraic sets in* $\mathbb{A}^n$ *and radical ideals (i.e., ideals which are equal to their own radical) in* A*, given by* $Y \longmapsto I(Y)$ *and* $\mathfrak{J} \longmapsto Z(\mathfrak{J})$*. Furthermore, an algebraic set is irreducible if and only if its ideal is prime.*

Example C.9. *Let* f *be an irreducible polynomial in* $A = \mathbb{F}[x, y]$*. Then* f *generates a prime ideal in* A*, since* A *is a unique factorization domain, so the zero set* $Y = Z(f)$ *is irreducible. It is called the affine curve defined by the equation* $f(x, y) = 0$*. If* f *has degree* d*, then* Y *is said to be a curve of degree* d.

Example C.10. *More generally, if* f *is an irreducible polynomial in* $A = \mathbb{F}[x_1, \ldots, x_n]$*, we obtain an affine variety* $Y = Z(f)$*, which is called a surface if* $n = 3$*, or a hypersurface if* $n > 3$.

Example C.11. *A maximal ideal* $\mathfrak{m}$ *of* $A = \mathbb{F}[x_1, \ldots, x_n]$ *corresponds to a minimal irreducible closed subset of* $\mathbb{A}^n$*, which must be a point, say* $P = (a_1, \ldots, a_n)$*. This shows that every maximal ideal of* A *is of the form* $\mathfrak{m} = (x_1 - a_1, \ldots, x_n - a_n)$*, for some* $a_1, \ldots, a_n \in \mathbb{F}$.

Example C.12. *If* $\mathbb{F}$ *is not algebraically closed, these results do not hold. For example, if* $\mathbb{F} = \mathbb{R}$*, the curve* $x^2 + y^2 + 1 = 0$ *in* $\mathbb{A}^2$ *has no points.*

Definition C.4. *If* $Y \subseteq \mathbb{A}^n$ *is an affine algebraic set, we define the affine coordinate ring* $K[Y]$ *(sometime* $A(Y)$*) of* Y *to be* $A/I(Y)$.

Remark. If Y is an affine variety, then $A(Y)$ is an integral domain. Furthermore, $A(Y)$ is a finitely generated $\mathbb{F}$-algebra. Conversely, any finitely generated $\mathbb{F}$-algebra B which is a domain is the affine coordinate ring of some affine variety. Indeed, write B as the quotient of a polynomial ring $A = \mathbb{F}[x_1, \ldots, x_n]$ by an ideal $\mathfrak{J}$, and let $Y = Z(\mathfrak{J})$.

Example C.13.

- *If X is a singleton then $A(X) = \mathbb{F}$.*
- *If $X = \mathbb{A}^n$ then $I(X) = 0$ and $A(X) = \mathbb{F}[x^1, \ldots, x^n]$.*
- *Let $X = \{(x_1, x_2) \in \mathbb{F} \times \mathbb{F} \mid x_1 x_2 = 1\}$. Then*

$$A(X) = \mathbb{F}\left(x_1, {x_1}^{-1}\right) = \left\{\frac{f(x_1)}{{x_1}^m} \mid m \geq 0 \text{ and } f(x_1) \in \mathbb{F}[x_1]\right\}.$$

Remark.

- The ring $A(X)$ for an algebraic set X being an image of a noetherian ring is noetherian.
- In $A(X)$ the following version of Hilbert's Nullstellensatz also holds, i.e., if a function $f(x) \in A(X)$ vanishes on those points, where functions $g_1(x), g_2(x), \ldots, g_m(x)$ vanish then

$$f^r \in (g_1, g_2, \ldots, g_m) \text{ for some } r > 0.$$

Definition C.5. *The function $\varphi : X \to Y$ from an affine variety X into an affine variety Y is said to be a morphism (or regular mapping) if $f \circ \varphi$ is a regular function on Y for any $f \in A(X)$.*

Example C.14. *Let $F_1, F_2, \ldots, F_m$ be polynomials in $\mathbb{F}[T_1, \ldots, T_n]$ then $\varphi : \mathbb{A}^n \to \mathbb{A}^m$ defined by*

$$\varphi(x) = \varphi(x_1, \ldots, x_n) = (F_1(x), F_2(x) \ldots, F_m(x))$$

is a morphism.

Example C.15. *Let* $X = Z(y^2 - x^3 + 1)$ *be affine algebraic variety in* $\mathbb{A}^2$ *with the coordinate system* (x, y) *and* $Y = Z((t^3 - s^2 + 1), (r - s^2))$ *be affine algebraic variety in* $\mathbb{A}^3$ *with the coordinate system* (s, t, r)*. Then the function* φ *defined by* $\varphi(x, y) = (x, y, x^2)$ *is a morphism from X into Y. Indeed, if* $f(s, t, r) = t^3 - s^2 + 1$ *then* $(f \circ \varphi)(x, y) = f(x, y, x^2) = y^3 - x^2 + 1$ *is identically zero on X. If* $f(s, t, r) = r - s^2$ *then* $(f \circ \varphi)(x, y) = f(x, y, x^2) = x^2 - x^2$ *is identically zero on X.*

Example C.16. *Let* $X = Z(y^2 - x^3 + 1)$ *be affine algebraic variety in* $\mathbb{A}^2$ *with the coordinate system* (x, y) *and* $Y = Z((t^3 - s^2 + 1), (r - s^2))$ *be affine algebraic variety in* $\mathbb{A}^3$ *with the coordinate system* (s, t, r)*. Then the function* φ *defined by* $\varphi(x, y) = (x, y, x^2)$ *is a morphism from X into Y.*

Definition C.6. *The morphism* $\varphi : X \to Y$ *is called isomorphism if there exists a morphism from Y into X such that if* $f \circ g = \mathrm{id}_Y$ *and* $g \circ f = \mathrm{id}_X$.

Example C.17. *The parabola* $X = \{(x, y) \in \mathbb{F} \times \mathbb{F} \mid y = x^k\}$*, where k is a fixed natural number, is isomorphic to the straight line* $Y = \mathbb{A}$*, corresponding morphisms are* $f(x, y) = x$ *and* $g(t) = (t, t^k)$*. Indeed,*

$$(f \circ g)(t) = f(g(t)) = f(t, t^k) = t \text{ for any } t \in Y$$

and

$$(g \circ f)(x, y) = g(f(x, y)) = g(x) = (x, x^k) = (x, y) \text{ for any } (x, y) \in X.$$

Example C.18. *The mapping* $f(t) = (t^2, t^3)$ *of the straight line* $Y = \mathbb{A}$ *to the curve* $X = \{(x, y) \in \mathbb{F} \times \mathbb{F} \mid x^3 = y^2\}$ *is one to one. However, it is not isomorphism since the inverse mapping* $g(x, y) = \frac{y}{x}$ *is not regular at the origin.*

Example C.19. *The set* $GL(n, \mathbb{F}) = \{g = (g_{ij})_{i,j=1,2,\dots,n} \mid \det(g) \neq 0\}$ *can be regarded as an affine variety as follows:* $GL(n, \mathbb{F})$ *is identified with*

$$X = \{(g, t) = ((g_{ij})_{i,j=1,2,\dots,n}, t) \in \mathbb{F}^{n^2+1} \mid \det(g)t - 1 = 0\}.$$

Note that in this case the product $(g_1, t_1)(g_2, t_2) = (g_1 g_2, t_1 t_2)$ *map from* $X \times X$ *to X and the inverse* $(g, t)^{-1} = (g^{-1}, t^{-1})$ *map from X to* $X \times X$ *are morphisms.*

C.1 ACTION OF A GROUP

Let G be a group and X be a nonempty set.

Definition C.7. *An action of the group G on X is a function*

$$\sigma : G \times X \to X$$

with:

i) $\sigma(e, x) = x$, *where e is the unit element of G and*

ii) $\sigma(g, \sigma(h, x)) = \sigma(gh, x)$, *for any $g, x \in G$ and $x \in X$.*

We briefly write gx for $\sigma(g, x)$, and call X a G-set.

Let $\mathbb{F}$ be a field and $\mathbb{F}[X] = \{f : X \to \mathbb{F}\}$ be the set of all functions on X. It is an algebra over $\mathbb{F}$ with respect to point wise addition, multiplication and multiplication by scalar operations.

Definition C.8. *A function $f : X \to \mathbb{F}$ is said to be invariant if $f(gx) = f(x)$ for any $g \in G$ and $x \in X$.*

The set of invariant functions on X, denoted by $\mathbb{F}[X]^G$, is a subalgebra of $\mathbb{F}[X]$. The set

$$\text{Orb}(x) = \{y \in X \mid \text{ there exists } g \in G \text{ such that } y = gx\}$$

is called the orbit of the element x under the action of G.

Theorem C.2. *Let X be a G-set. Define a relation $\sim$ on X for $x, y \in X$ as follows: $x \sim y$ if and only if $\sigma(g, x) = y$ for some $g \in G$. Then $\sim$ is an equivalence relation on X.*

It is evident that the equivalence classes with respect to equivalence relation $\sim$ are the orbits under the action of the group G. The invariant functions are functions on $X/\sim$.

C.2 ALGEBRAIC GROUPS

Definition C.9.

- *An algebraic group is an affine variety G equipped with morphisms of varieties $\mu : G \times G \longrightarrow G$, $\iota : G \longrightarrow G$ that give G the structure of a group.*

- *A morphism $f : G \longrightarrow H$ of algebraic groups is a morphism of varieties that is also a group homomorphism.*

Proposition C.4.

- *The kernel of a morphism $f : G \longrightarrow H$ of algebraic groups is a closed subgroup of G, so it is an algebraic group in its own right. The same will turn out to be true about the images.*
- *Translation by an element $g \in G$ is an isomorphism of varieties, so all geometric properties at one point can be transferred to any other point. For example, as G has simple points, G is smooth.*

Example C.20.

(i) *The additive group $\mathbb{G}_a$ is the group $(\mathbb{F}, +)$, i.e., the affine variety $\mathbb{A}^1$ under addition;*

(ii) *The multiplicative group $\mathbb{G}_m$ is the group $(\mathbb{F}^*, \times)$, i.e., the principal open subset $\mathbb{A}^1 \setminus \{0\}$ under multiplication;*

(iii) *The group $GL_n = GL_n(\mathbb{F})$ is the group of all invertible $n \times n$ matrices over $\mathbb{F}$. As a variety, this is a principal open set in $M_n(\mathbb{F}) = \mathbb{A}^{n^2}$ corresponding to the determinant. Since the formulas for matrix multiplication and inversion involve only polynomials in the matrix entries and $1/\det$, the group structure maps are morphisms of varieties.*

Example C.21.

(iv) *Let V be an n-dimensional vector space over $\mathbb{F}$. Then by fixing a basis we can define a structure of an algebraic group on $GL(V)$ which is independent of the choice of basis. Of course, $GL(V) \cong GL_n(\mathbb{F})$;*

(v) *The group $SL_n = SL_n(\mathbb{F})$ is the closed subgroup of GL_n defined by the zeros of $f(x_{11}, x_{12}, \ldots, x_{nn}) = \det(A) - 1$, where $A \in GL_n(\mathbb{F})$;*

(vi) *The group D_n of invertible diagonal matrices is a closed subgroup of GL_n. It is isomorphic to the direct product $\mathbb{G}_m \times \cdots \times \mathbb{G}_m$ (n copies).*

Here are more examples of algebraic groups.

Example C.22.

(vii) *The group* U_n *of upper unitriangular matrices is another closed subgroup of* GL_n*;*

(viii) *The orthogonal group* $\mathrm{O}_n(\mathbb{F}) = \{A \in \mathrm{GL}_n \mid AA^\top = I\}$*, with* $\mathrm{char}(\mathbb{F}) \neq 2$*;*

(ix) *The special orthogonal group* $\mathrm{SO}_n = \mathrm{O}_n \cap \mathrm{SL}_n$ *is a normal subgroup of* O_n *of index 2;*

(x) *The symplectic group* $\mathrm{Sp}_{2n} = \{A \in \mathrm{GL}_{2n} \mid A^\top JA = J\}$*, where* $\begin{pmatrix} 0 & I_n \\ I_n & 0 \end{pmatrix}$ *is another closed subgroup.*

Bibliography

[1] Abdukassymova A.S., Dzhumadil'daev A.S., *Simple Leibniz algebras of rank* 1, Abstract presented to the IX International Conference of the Representation Theory of Algebras, Beijing, China, 2000, 17–18.

[2] Adashev J.Q., Camacho L.M., Gómez-Vidal S., Karimdjanov I.A., Naturally graded Zinbiel algebras with nilindex $n-3$, *Linear Algebra and its Applications*, 443, 2014, 86–104.

[3] Adashev J.Q., Ladra M., Omirov B.A., Solvable Leibniz algebras with naturally graded non-Lie p-filiform nilradicals, *Communications in Algebra*, 45(10), 2017, 4329–4347.

[4] Adashev J.Q., Khudoyberdiev A.Kh., Omirov B.A., Classification of complex naturally graded quasi-filiform Zinbiel algebras, *Contemporary Math.*, AMS, 483, 2009.

[5] Adashev J.Q., Omirov B.A., Khudoyberdiev A.Kh., Classification of some classes of Zinbiel algebras, *J. Gen. Lie Theor. and Appl.*, 4, 2010.

[6] Agaoka Y., On the variety of 3-dimensional Lie algebras, *Lobachevski Journal of Math.*, 1999, 3, 5–17.

[7] Agaoka Y., An algorithm to determine the isomorphism classes of 4-dimensional complex Lie algebras, *Linear Algebra and its Applications*, 2002, 345, 85–118.

[8] Aguiar M., Pre-Poisson algebras, *Lett. Math. Phys.*, 2000, 54, 263–277.

[9] Albeverio S., Ayupov Sh.A., Omirov B.A., Turdibaev R.M., Cartan subalgebras of Leibniz n-algebras, *Communications in Algebra*, 2009, 37(6), 2080–2096.

[10] ALBEVERIO S., AYUPOV SH.A., OMIROV B.A., KHUDOYBERDIYEV A., n-dimensional filiform Leibniz algebras of length (n-1) and their derivations, *Journal of Algebra*, 2008, 319, 2471–2488.

[11] ALBEVERIO S., AYUPOV SH.A., OMIROV B.A., On Cartan subalgebras, weight spaces and criterion of solvability of finite dimensional Leibniz algebras, *Revista Matemática Complutense*, 2006, 19(1), 183–195.

[12] ALBEVERIO S., AYUPOV. SH.A., OMIROV B.A., On nilpotent and simple Leibniz algebras, *Communications in Algebra*, 2005, 33(1), 159–172.

[13] ALBEVERIO S., OMIROV B.A., RAKHIMOV I.S., Classification of 4-dimensional nilpotent complex Leibniz algebras, *Extracta Mathematicae*, 2006, 21(3), 197–210.

[14] ALBEVERIO S., OMIROV B.A., RAKHIMOV I.S., Varieties of nilpotent complex Leibniz algebras of dimension less than five, *Communications in Algebra*, 2005, 33, 1575–1585.

[15] ANCOCHEA BERMUDEZ J.M., CAMPOAMOR-STURSBERG R., Cohomologically rigid solvable Lie algebras with a nilradical of arbitrary characteristic sequence, *Linear Algebra and its Applications*, 2016, 488, 135–147.

[16] ANCOCHEA BERMUDEZ J.M., CAMPOAMOR-STURSBERG R., GARCIA VERGNOLLE L., Solvable Lie algebras with naturally graded nilradicals and their invariants, *J. Phys. A: Math. Gen.*, 2006, 39, 1339–1355.

[17] ANCOCHEA BERMUDEZ J.M., CAMPOAMOR-STURSBERG R., GARCÍA VERGNOLLE L. Classification of Lie algebras with naturally graded quasi-filiform nilradicals, *Journal of Geom. Phys.*, 2011, 61, 2168–2186.

[18] ANCOCHEA BERMUDEZ J.M., CAMPOAMOR-STURSBERG R. On a complete rigid Leibniz non-Lie algebra in arbitrary dimension, *Linear Algebra and its Applications,* 2013, 438(8), 3397–3407.

[19] ANCOCHEA BERMUDEZ J.M., CAMPOAMOR-STURSBERG R., GARCÍA L. Indecomposable Lie algebras with nontrivial Levi decom-

position cannot have filiform radical, *Int. Math. Forum*, 2006, 1(5–8), 309–316.

[20] Ancochea Bermudez J.M., Goze M., Classification des algèbres de Lie filiformes de dimension 8, *Journal of Arch. Math.*, 1988, 50, 511–525.

[21] Ancochea Bermudez J.M., Goze M., Classification des algèbres de Lie nilpotentes complexes de dimension 7, *Journal of Arch. Math.*, 1989, 52(2), 157–185.

[22] Atiyah M.F., Macdonald I.G., *Introduction to Commutative Algebras*, Addison-Wesley, 1969.

[23] Ayupov Sh.A., Omirov B.A., On Leibniz algebras, *Algebra and Operators Theory, Proceeding of the Colloquium in Tashkent*, 1997, Kluwer Academic Publishers, 1998, 1–13.

[24] Ayupov Sh.A., Omirov B.A., On some classes of nilpotent Leibniz algebras, *Sibirsk. Mat. Zh.* (in Russian), 2001, 42(1), 18–29 (English translation in *Siberian Math. J.*, 2001, 42(1), 15–24).

[25] Ayupov Sh.A., Omirov B.A., The nilpotency properties of the Leibniz algebra $M_n(\mathbb{C})_D$, *Sib. Math. J.*, 2004, 45(3), 399–409.

[26] Ayupov Sh.A., Omirov B.A., Khudoyberdiyev A. Kh., The classification of filiform Leibniz superalgebras of nilindex $n + m$, *Acta Mathematica Sinica* (English Series), 2009, 25(2), 171–190.

[27] Ayupov Sh.A., Omirov B.A., On 3-dimensional Leibniz algebras, *Uzbek Math. Journal*, 1999, 9–14.

[28] Balavoine D., Déformations des algèbres de Leibniz, *C.R. Acad. Sci. Paris*, 1994, 319, Serie I, 783–788.

[29] Balavoine D., Déformations et rigidité géométrique des algèbres de Leibniz, *Communications in Algebra*, 1996, 24(3), 1017–1034

[30] Barnes D.W., Some theorems on Leibniz algebras, *Communications in Algebra*, 2011, 39(7), 2463–2472.

[31] Barnes D.W., Faithful representations of Leibniz algebras, *Proc. Amer. Math. Soc.*, 2013, 141(9), 2991–2995.

[32] Barnes D.W., On Levi's theorem for Leibniz algebras, *Bull. Aust. Math. Soc.*, 86(2), 2012, 184–185.

[33] Basri W., Rikhsiboev I.M., On low dimensional diassociative algebras, *Proceedings of Third Conference on Research and Education in Mathematics* (ICREM-3), UPM, Malaysia, 2007, 164–170.

[34] Beck R.E., Kolman B., Constructions of nilpotent Lie algebras over arbitrary fields, In Paul S.Wang, editor, *Proceedings of* 1981 *ACM Symposium on Symbolic and Algebraic Computation*, ACM New York, 169–174.

[35] Blokh A.M., On a generalization of the concept of Lie algebra, *Dokl. Akad. Nauk SSSR*, 165, 1965, 471–473 (in Russian).

[36] Blokh A.M., Cartan-Eilenberg homology theory for a generalized class of Lie algebras, *Dokl. Akad. Nauk SSSR*, 175, 1967, 824–826 (in Russian).

[37] Boyko V., Patera J., Popovych R., Invariants of solvable Lie algebras with triangular nilradicals and diagonals nilindependent elements, *Linear Algebra and its Applications*, 428(4), 2008, 834–854.

[38] Boza L., Fedriani E. M., Núñez J. A new method for classifying complex filiform Lie algebras, *Appl. Math. and Computation*, 121(2–3), 2001, 169–175.

[39] Burde D., Degeneration of 7-dimensional nilpotent Lie algebras, *Communications in Algebra*, 2005, 33(4), 1259–1277.

[40] Burde D., Beneš T., Degenerations of pre-Lie algebras, *Journal of Mathematical Physics*, 50(11), 2009, 112102-1–112102-9.

[41] Burde D., Steinhoff C., Classification of orbit closures of 4-dimensional complex Lie algebras, *Journal of Algebra*, 214, 1999, 729–739.

[42] Cabezas J. M., Gómez J.R., and Jimenez-Merchan A., Family of p-filiform Lie algebras, *in Algebra and Operator Theory, Proceedings of the Colloquium in Tashkent*, 1997, Kluwer Academic Publishers, 1998, 93–102.

[43] Camacho L.M., Cañete E.M., Gómez-Vidal S., Omirov B.A., p-Filiform Zinbiel algebras, *Linear Algebra and its Applications*, 438(7), 2013, 2958–2972.

[44] Camacho L.M., Gómez-Vidal S., Omirov B.A., Leibniz algebras associated to extensions of sl_2. *Communications in Algebra*, 2015, 43(10), 4403–4414.

[45] Camacho L.M., Gómez-Vidal S., Omirov B.A., Karimjanov I.A., Leibniz algebras whose semisimple part is related to $\mathfrak{sl}_2$, *Bull. Malays. Math. Sci. Soc.*, 2017, 40(2), 599–615.

[46] Camacho L.M., Gómez J.R., González A.J., Omirov B.A., Naturally graded quasi-filiform Leibniz algebras, *Journal Symb. Comput.*, 2009, 527–539.

[47] Camacho L.M., Gómez J.R., González A.J., Omirov B.A., Naturally graded 2-filiform Leibniz algebras, *Communications in Algebra*, 2010, 38, 3671–3685.

[48] Camacho L.M., Gómez J.R., González A.J., Omirov B.A., The classification of naturally graded p-filiform Leibniz algebras, *Communications in Algebra*, 2011, 39, 153–168.

[49] Campoamor-Stursberg R., Solvable Lie algebras with an $\mathbb{N}$-graded nilradical of maximum nilpotency degree and their invariants, *J. Phys.* A., 43(14), 2010, 145–202.

[50] Cañete E.M., Khudoyberdiyev A.Kh., The classification of 4-dimensional Leibniz algebras, *Linear Algebra and its Applications*, 439(1), 2013, 273–288.

[51] Carles R., Sur la structure des algèbres de Lie rigides, *Ann. Inst. Fourier Grenoble*, 1984, 34(3), 65–82.

[52] Carles R., Weight systems for nilpotent Lie algebras of dimension 7 over complex field, Technical report, Département de Mathématiques, Université de Poitiers, 1989.

[53] Carles R., Diakité Y., Sur les variétés d'algèbres de Lie de dimension ≤ 7, *Journal of Algebra*, 91, 1984, 53–63.

[54] Casas J.M, Ladra M., Omirov B.A., Karimjanov I.A., Classification of solvable Leibniz algebras with null-filiform nilradical, *Linear and Multilinear Algebra*, 2013, 61(6), 758–774.

[55] Casas J.M, Ladra M., Omirov B.A., Karimjanov I.A., Classification of solvable Leibniz algebras with naturally graded filiform nilradical, *Linear Algebra and its Applications*, 2013, 438(7), 2973–3000.

[56] Casas J. M., Ladra M., Non-abelian tensor product of Leibniz algebras and an exact sequence in Leibniz homology, *Communications in Algebra*, 31(9), 2003, 4639–4646.

[57] Casas J.M., Insua M.A., Ladra M., Ladra S., An algorithm for the classification of 3-dimensional complex Leibniz algebras, *Linear Algebra and its Applications*, 2012, 436(9), 3747–3756.

[58] Chevally C., Eilenberg S., Cohomology theory of Lie groups and Lie algebras, *Trans. Amer. Math. Soc.*, 1948, 63, 84–124.

[59] Cuvier C., Algèbres de Leibnitz: définitions, propriétés, *Ann. Scient. Éc. Norm. Sup.*, 4^a série, 1994, 27, 1–45.

[60] Dixmier J. Cohomologie des algèbres de Lie nilpotentes, *Acta Sci. Math. Szeged*, 1955, 16, 246–250.

[61] Dixmier J., Sur les représentations unitaires des groupes de Lie nilpotents III, *Canadian Journal of Mathematics*, 1958, 10, 321–348.

[62] Druhl K., A theory of classical limit for quantum theories which are defined by real Lie algebras, *J. Math. Phys.*, 1978, 19(7), 1600–1606.

[63] Dzhumadil'daev A., Abdykassymova S., Leibniz algebras in characteristic p, *C.R. Acad. Sci.*, Serie I, Paris, 2001, 332, 1047–1052.

[64] Dzhumadildaev A.S., Tulenbaev K.M., Nilpotency of Zinbiel Algebras, *Journal of Dynamical and Control Systems*, 2005, 11(2), 195–213.

[65] Ebrahimi-Fard K., Loday-type algebras and the Rota-Baxter relation, *Lett. in Math. Phys.*, 2002, 61(2), 139–147.

[66] Erdmann K., Wildom M.J., *Introduction to Lie algebras*, Springer Undergraduate Mathematics Series, 2006, Springer-Verlag London Limited.

[67] Favre G., Système de poids sur une algèbre de Lie nilpotente, *Manuscripta Math.*, 1973, 9, 53–90.

[68] Feldvoss J., Leibniz algebras as non-associative algebras, arXiv:1802.07219v3 [math.RA] 17 Oct 2018.

[69] Feldvoss J., Wagemann, F., On Leibniz cohomology, arXiv:1902.06128v2 [math.AT] 23 Apr 2019.

[70] Fialowski A., O'Halloran J., A comparison of deformations and orbit closure, *Communications in Algebra*, 1990, 18, 4121–4140.

[71] Fialowski A., Magnin L., Mandal A., About Leibniz cohomology and deformations of Lie algebras, *Journal of Algebra*, 2013, 383(1), 63–77.

[72] Fialowski A., Mandal A., Mukherjee G., Versal deformations of Leibniz algebras, *J. K-Theory*, 2009, 3(2), 327–358.

[73] Fialowski A., The module space and versal deformations of three dimensional Lie algebras, *Algebras, rings and their Representations*, *Proc. Intern. Algebra Conference*, 2006, 79–92.

[74] Fialowski A., de Montigny M., On deformations and contractions of Lie algebras, SIGMA, 2006, 2, paper 048, 10 p.

[75] Fialowski A., Penkava M., The moduli space of 3-dimensional associative algebras, *Communications in Algebra*, 2009, 7(10), 3666–3685.

[76] Fialowski A., Penkava M., Versal deformations of four dimensional Lie algebras, *Communications in Contemporary Mathematics*, 2007, 9(1), 41–79.

[77] Filippov A.F., A short proof of the reduction to Jordan form, *Moscow Univ. Math. Bull.*, 1971, 26, 70–71.

[78] Flanigan F.J., Algebraic geography: varieties of structure constants, *Pacific Journal Math.*, 1968, 27, 71–79.

[79] Fox T., An introduction to algebraic deformation theory, *Journal of Pure and Appl. Algebra*, 1993, 84, 17–41.

[80] Frabetti A., Leibniz homology of dialgebras of matrices, *Journal of Pure and Appl. Algebra*, 1998, 129, 123–141.

[81] Gabriel P., Finite representation type is open, *Proceedings of* ICRAI, *Ottawa* 1974, Lecture Notes in Mathematics, Springer, 1975, 488.

[82] Galitski L.Yu., Timashev D.A., On classification of matabelian Lie algebras, *Journal of Lie Theory*, 1999, 9, 125–156.

[83] Gauger M.A., On the classification of metabelian Lie algebras, *Trans. Amer. Math. Soc.*, 1973, 87, 293–329.

[84] Gerstenhaber M., On the deformations of rings and algebras, *Ann. of Math.*, 1964, 79, 59–104.

[85] Gerstenhaber M., Schack S., Algebraic Cohomology and deformation theory, *Deformation Theory of Algebras and Structures and Applications*, Kluwer Academic Publishers, 1988, 11–264.

[86] Ginzburg V., Kapranov M.M., Koszul duality for operads, *Duke Math. J.*, 1994, 76, 203–272.

[87] Gómez J.R., Jimenez-Merchan A., Khakimdjanov Yu.B., Low-dimensional filiform Lie Algebras, *Journal of Pure and Applied Algebra*, 1998, 130, 133–158.

[88] Gómez J.R., Omirov B.A., On classification of complex filiform Leibniz algebras, *Algebra Colloquium*, 2015, 22 (spec01), 757–774.

[89] Gong M.-P., Classification of Nilpotent Lie algebras of Dimension 7 (Over Algebraically Closed Fields and $\mathbb{R}$), PhD thesis, University, Waterloo, Canada, 1998.

[90] Gorbatsevich V.V., Contractions and degenerations of finite-dimensional algebras, *Izvestiia Vyssih Uchebnyh Zavedenii, Matematika*, 1991, 10, 19–27 (Translated in Soviet Math. (Iz. VUZ), 35(10), 17–24.)

[91] Gorbatsevich V.V., Anticommutative finite-dimensional algebras of the first three levels of complexity, *Algebra i Analiz*, 1993, 5(3), 100–118 (Translated in 1994 *St. Petersburg Math. Journal*, 5, 505–521).

[92] Gorbatsevich V.V., On the level of some solvable Lie algebras, *Siberian Math. Journal*, 1998, 39, 872–883.

[93] Gorbatsevich V.V., On liezation of the Leibniz algebras and its applications, *Russian Mathematics* (*Izv. VUZ*), 2016, 60(4), 10–16 (Original Russian text published in *Izvestiya Visshikh Uchebnykh Zavedenii. Matematika*, 2016, 4, 14–22).

[94] Goze M., Ancochea Bermudez J.M., On the varieties of nilpotent Lie algebras of dimension 7 and 8, *Journal of Pure and Appl. Algebra*, 1992, 77, 131–140.

[95] Goze M., Ancochea Bermudez J.M., On the classification of rigid Lie algebras, *Journal of Algebra*, 2001, 245(1), 68–91.

[96] Goze M., Khakimdjanov Yu.B., Sur les algèbres de Lie nilpotentes admettant un tore de dérivations, *Manuscripta Math.*, 1994, 84, 115–124.

[97] Goze M., Khakimjanov Yu. *Nilpotent Lie algebras.* Kluwer Academic Publishers, 361, 1996, 336 p.

[98] De Graaf W.A. Classification of nilpotent associative algebras of small dimension, *International Journal of Algebra and Computation*, 2018, 28(1), 133–161.

[99] Grunewald F., O'Halloran J., Varieties of nilpotent Lie algebras of dimension less than six, *Journal of Algebra*, 1988, 112(2), 315–326.

[100] Grunewald F., O'Halloran J., Deformations of Lie algebras, *Journal of Algebra*, 1993, 162, 210–224.

[101] Guo L., Ebrahimi-Fard K., Rota-Baxter algebras and dendriform algebras, *Journal of Pure and Appl. Algebra*, 2008, 212, 320–339.

[102] Hall M., *Combinatorial Theory*, John Wiley & Sons Inc, 1986.

[103] Hall B.C., *Lie groups, Lie algebras, and representations: An elementary introduction*, Graduate Texts in Mathematics, Springer, 2015, 222, 2nd ed.

[104] Hochschild G., Serre J.P., Cohomology of Lie algebras, *Ann. of Math.*, 1953, 57(2), 591–603.

[105] Humphreys J., *Introduction to Lie Algebras and Representation Theory*, Springer, 1972.

[106] Jacobson N. *Lie Algebras*, Wiley Interscience, 1962.

[107] Kashuba I., Shestakov I., Jordan algebras of dimension three: geometric classification and representation type, *Revista Matem. Iberoamer.*, 2007, 295–315.

[108] Kashuba I., Variety of Jordan algebras in small dimensions, *Algebra Discrete Math.*, 2006, 2, 62–76.

[109] Kaygorodov I, Popov Y., Pozhidaev A., Volkov Y., Degenerations of Zinbiel and nilpotent Leibniz algebras, *Linear and Multilinear Algebra*, 2018, 66(4), 704–716.

[110] Khudoyberdiyev A.Kh., Ladra M., Omirov B.A., On solvable Leibniz algebras whose nilradical is a direct sum null-filiform algebras, *Linear and Multilinear Algebra*, 2014, 62(9), 1220–1239.

[111] Khudoyberdiyev A.Kh., Rakhimov I.S., Said Husain Sh.K., On classification of 5-dimensional solvable Leibniz algebras, *Linear Algebra and its Applications*, 2014, 457, 428–454.

[112] Kinyon M.K., Weinstein A., Leibniz algebras, Courant algebroids, and multiplications on reductive homogeneous spaces, *Amer. Journal Math.*, 2001, 123(3), 525–550.

[113] Kosman-Schwarzbach Y., From Poisson algebras to Gerstenhaber algebras, *Ann. Inst. Fourier (Grenoble)*, 1996, 46, 1243–1274.

[114] Kostrikin A.I., Manin Yu.I., *Linear Algebra and Geometry*, Gordon & Breach Science Publishers, 1997, 308 pp.

[115] Kudoybergenov K.K., Ladra M., Omirov B.A., On Levi-Malcev theorem for Leibniz algebras, *Linear and Multilinear Algebra*, 2019, 67(7), 1471–1482.

[116] Leger G., Luks E., Cohomology theorems for Borel-like solvable Lie algebras in arbitrary characteristic, *Canadian J. Math.*, 1972, 24(6), 1019–1026.

[117] Loday J.-L., Une version non commutative des algèbres de Lie: les algèbres de Leibniz, *L'Ens. Math.*, 1993, 39, 269–293.

[118] Loday J.-L., Frabetti A., Chapoton F., and Goichot F., *Dialgebras and Related Operads*, Lecture Notes in Mathematics, IV, Springer, 2001, 1763.

[119] Loday J.-L., Pirashvili T., Universal enveloping algebras of Leibniz algebras and (co)homology, *Math. Ann.*, 1993, 296, 139–158.

[120] Loday J.-L., Cup product for Leibniz cohomology and dual Leibniz algebras, *Math. Scand.*, 1995, 77, 189–196.

[121] Makhlouf A., Goze M., *Classification of Rigid Associative Algebras in Low Dimensions* (Preprint), Universite Louis Pasteur, Strasbourg, 1996.

[122] Makhlouf A., The irreducible components of the nilpotent associative algebras, *Revista Mat. de la Univ. Compl. de Madrid*, 1993, 6(1), 27–40.

[123] Malcev A.I., On the representation of an algebra as a direct sum of the radical and a semi-simple subalgebra, *Dokl. Akad. Nauk SSSR*, 1942, 36(2), 42–45 (in Russian).

[124] Malcev A.I., On semisimple subgroups of Lie groups, *Izvestiya AN SSSR, Ser. Matem.*, 1944, 8(4), 143–174 (in Russian).

[125] Malcev A. I., Solvable Lie algebras, *Amer. Math. Soc. Trans.*, 1950, 27, 36 pp.

[126] Mason G., Yamskulma G., Leibniz algebras and Lie algebras, *SIGMA*, 2013, 9, p. 063 (10 pages).

[127] Mazzola G., The algebraic and geometric classification of associative algebras of dimension five, *Manuscripta Math.*, 1979, 27, 1–21.

[128] Moody R.V., Patera J., Discrete and continuous graded contractions of representations of Lie algebras, *J. Phys. A: Math. Gen.*, 1991, 24, 2227–2257.

[129] Morozov V.V., Classification of nilpotent Lie algebras of 6-th order, *Izvestiia Vyssih Uchebnyh Zavedenii, Matematika*, 1958, 161–171 (in Russian).

[130] Mubarakzjanov G.M., The classification of the real structure of five-dimensional Lie algebras, *Izvestiya Vysshikh Uchebnykh Zavedenii. Matematika*, 1963, 3(34), 99–106 (in Russian).

[131] Mubarakzjanov G.M., On solvable Lie algebras, *Izvestiya Vysshikh Uchebnykh Zavedenii. Matematika*, 1963, 1(32), 114–123 (in Russian).

[132] Ndogmo J.C., Winternitz P., Solvable Lie algebras with abelian nilradicals, *J. Phys.* A, 1994, 27, 405–423.

[133] Nijenhuis A., Richardson R., Cohomology and deformations in graded Lie algebras, *Bull. Amer. Math. Soc.*, 1966, 72, 1–29.

[134] Nijenhuis A., Richardson R., Cohomology and deformations of Lie algebra structures, *J. Math. and Mech.*, 1967, 17, 89–106.

[135] Ntolo P., Homologie de Leibniz d'algèbres de Lie semi-simples, *C.R. Acad. Sci. Paris*, Série I, 1994, 318, 707–710.

[136] Ombolo R., Deformations of Leibniz Algebras and Lie Bialgebras, PhD thesis, Howard University, Washington, DC, 1997.

[137] Omirov B.A., Classification of two-dimensional Zinbiel algebras, *Uzbek Math. Journal*, 2002, 2, 55–59 (in Russian).

[138] OMIROV B.A., Conjugacy of Cartan subalgebras of complex finite-dimensional Leibniz algebras, *Journal of Algebra*, 2006, 302(2), 887–896.

[139] OMIROV B.A., RAKHIMOV I.S., TURDIBAEV R.M., On description of Leibniz algebras corresponding to sl_2, *Algebras and Representation Theory*, 2013, 16(5), 1507–1519.

[140] OMIROV B.A., RAKHIMOV I.S., On Lie-like complex filiform Leibniz algebras, *Bull. Aust. Math. Soc.*, 2009, 79(3), 391–404.

[141] OMIROV B.A., WAGEMANN F., A rigid Leibniz algebra with non-trivial HL^2, arXiv:1508.06877v6 [math.KT] 20 Feb 2019.

[142] PAGE S, RICHARDSON R.W., Stable subalgebras of Lie algebras and associative algebras, *Trans. Amer. Math. Soc.*, 1967, 127, 302–312.

[143] PATSOURAKOS A., On nilpotent properties of Leibniz algebras, *Communications in Algebra*, 2007, 35, 3828–3834.

[144] PATSOURAKOS A., On solvable Leibniz algebras, *Bulletin of the Greek Mathematical Society*, 2008, 55, 59–63.

[145] PIERCE B., Linear associative algebras, *Amer. Math. Journal*, 1881, 4(97).

[146] PIRASHVILI T., On Leibniz homology, *Ann. Inst. Fourier*, 1994, 44, 401–411.

[147] POJIDAEV A.P., Classification of simple finite-dimensional commutative n-ary Leibniz algebras of characteristic 0, *Actas do* IX ENAL. USP.*San Paulo, Brazil,* 2001, 1–6.

[148] RAKHIMOV I.S., BEKBAEV U.D., On isomorphism classes and invariants of finite dimensional complex filiform Leibniz algebras, *Communications in Algebra*, 2010, 38(12), 4705–4738.

[149] RAKHIMOV I.S., HASSAN M.A., On low-dimensional Lie-like filiform Leibniz algebras and their invariants, *Bulletin of the Malaysian Mathematical Science Society*, 2011, 34(3), 475–485.

[150] Rakhimov I.S., Hassan M.A., On one-dimensional central extension of a filiform Lie algebra, *Bulletin of the Australian Mathematical Society*, 2011, 83(3), 205–224,

[151] Rakhimov I.S., Hassan M.A., On isomorphism criteria for Leibniz central extensions of a linear deformation of μ_n, *International Journal of Algebra and Computations*, 2011, 21(5), 715–729.

[152] Rakhimov I.S., Rikhsiboev I.M., Basri W., Classification of 3-dimensional complex diassociative algebras, *International Advanced of Technology Congress (ATCi)*, PWTC, Malaysia, November 3–5, 2009.

[153] Rakhimov I.S., Rikhsiboev I.M., Basri W., Classification of 4-dimensional complex nilpotent diassociative algebras, *Malaysian Journal of Math. Sciences*, 2010, 4(2), 241–254.

[154] Rakhimov I.S., Rikhsiboev I.M., Mohammed M.A., An algorithm for classifications of three-dimensional Leibniz algebras over arbitrary fields, *JP Journal of Algebra, Number Theory and Applications*, 2018, 40(2), 181–198.

[155] Rakhimov I.S., Said Husain S.K., On isomorphism classes and invariants of low dimensional complex filiform Leibniz algebras, *Linear and Multilinear Algebra*, 2011, 59(2), 205–220.

[156] Rakhimov I.S., Said Husain S.K., Classification of a subclass of nilpotent Leibniz algebras, *Linear and Multilinear algebra*, 2011, 59(3), 339–354.

[157] Rakhimov I.S., Rikhsiboev I.M., Basri W., Complete lists of low-dimensional complex associative algebras, arXiv:0910.0932v2 [math.RA] 7 Oct 2009.

[158] Revoy R., Algèbres de Lie métabéliennes, *Annales de la Faculté des Sciences de Toulouse*, 1980, 11, 93–100.

[159] Richardson R., The regidity of semi-direct product of Lie algebras, *Pacific J. Math.*, 1967, 22, 339–344.

[160] RIKHSIBOEV I.M., RAKHIMOV I.R., Classification of three dimensional complex Leibniz algebras, *International Journal of Modern Physics: Conference Proceedings*, 2012.

[161] RIKHSIBOEV I.M., RAKHIMOV I.S., BASRI W., The description of dendriform algebra structures on two-dimensional complex space, *Journal of Algebra, Number Theory: Advances and Applications*, 2010, 4(1), 1–18.

[162] ROMDHANI M., Classification of real and complex nilpotent Lie algebras of dimension 7, *Linear and Multilinear Algebra*, 1989, 24, 167–189.

[163] SAFIULLINA E.N., Classification of nilpotent Lie algebras of dimension 7, *Math. Mech. Phys.*, 1964, 66–69. (in Russian).

[164] SCHEUNEMAN J., Two-step nilpotent Lie algebras, *Journal of Algebra*, 1967, 7, 152–159.

[165] SCORZA B.G., Le algebre del 3 ordine, *Atti. Nap.*, 1938, 20(13), 20(14) (in Italian).

[166] SEELEY G., Degenerations of 6-dimentional nilpotent Lie algebras over $\mathbb{C}$, *Communications in Algebra*, 1990, 8(10), 3493–3505.

[167] SEELEY G., Seven-dimensional nilpotent Lie algebras over the complex numbers, PhD thesis, University of Illinois at Chicago, 1988.

[168] SEGAL I.E., A class of operator algebras which are determined by groups, *Duke Mathematical Journal*, 1951, 18, 221–256.

[169] SHABANSKAYA A., Solvable extensions of naturally graded quasi-filiform Leibniz algebras of second type L^1 and L^3, *Communications in Algebra*, 2017, 45(10), 4492–4520.

[170] SHAFAREVICH I. P., *Fundamentals of Algebraic Geometry*, Vol. 1, Nauka, Moscow, 1988 (in Russian).

[171] SHEDLER G.S., On the classification of nilpotent Lie algebras of dimension six, Master's thesis, Tufts University, Boston, 1964.

[172] Skjelbred T., Sund T., Sur la classification des algébres de Lie nilpotentes, *C.R. Acad. Sci. Paris*, 1978, 286, 241–242.

[173] Šnobl L. Winternitz P., A class of solvable Lie algebras and their Casimir invariants, *J. Phys.* A, 2005, 38, 2687–2700.

[174] Sozan J.Obaiys, Isamiddin S.Rakhimov, Kamel A.M.Atan, *Classification of First Class of Complex Filiform Leibniz Algebras*, Lambert Academic Publishing, 2010, 184 pp.

[175] Tolpygo A.K., On cohomology of parabolic Lie algebras, *Math. Notes*, 1973, 12, 585–587.

[176] Tremblay P., Winternitz P., Solvable Lie algebras with triangular nilradicals, *J. Phys.* A., 1998, 31(2), 789–806.

[177] Umlauf K.A., Uber die Zusammensetzung der endlichen contiuierlichen Transformationgruppen insbesondere der Gruppen vom Range Null, PhD thesis, University of Leipzig, Germany, 1891 (in German).

[178] Vergne M., Cohomologie des algébres de Lie nilpotentes. Application *à* l'étude de la variété des algébres de Lie nilpotentes. *Bull. Soc. Math. France,* 1970, 98, 81–116.

[179] Vergne M., *Variete des algebres de Lie nilpotentes*, PhD thesis, University of Paris, 1966.

[180] Wan Z., *Lie Algebras*, Pergamon Press Ltd, 1975.

[181] Wang Y., Lin J., Deng S., Solvable Lie algebras with quasifiliform nilradicals, *Communications in Algebra*, 2008, 36(11), 4052–4067.

[182] Weimar-Woods E., Contractions of invariants of Lie algebras with applications to classical inhomogeneous Lie algebras, *J. Math. Phys.*, 2008, 49, 033507; http://dx.doi.org/10.1063/1.2839911.

[183] Yan Wang, Jie Lin, Shaoqiang Deng., Solvable Lie Algebras with Quasifiliform Nilradicals, *Communications in Algebra*, 2008, 36(11), 4052–4067.

[184] Zariski O., Samuel P., *Commutative Algebra*, I, II, Van Nostrand, Princeton, 1958, 1960.

[185] Zhelobenko D.P., *Compact Lie Groups and Their Representations*, AMS, Providence, 1973.

[186] Zinbiel G.W., Encyclopedia of types of algebras 2010, Bai Cheng Ming; Guo Li; Loday Jean-Louis, Operads and universal algebra, *Nankai Series in Pure, Applied Mathematics and Theoretical Physics*, 2012, 9, 217–298.

Index

9781032337722